Multiple Solutions of Boundary Value Problems
A Variational Approach

TRENDS IN ABSTRACT AND APPLIED ANALYSIS

Series Editor: John R. Graef
The University of Tennessee at Chattanooga, USA

Published

Vol. 1   Multiple Solutions of Boundary Value Problems:
A Variational Approach
by John R. Graef & Lingju Kong

Trends in Abstract
and Applied Analysis
Volume **1**

Multiple Solutions of Boundary Value Problems

A Variational Approach

John R Graef
Lingju Kong

The University of Tennessee at Chattanooga, USA

 **World Scientific**

NEW JERSEY · LONDON · SINGAPORE · BEIJING · SHANGHAI · HONG KONG · TAIPEI · CHENNAI · TOKYO

Published by

World Scientific Publishing Co. Pte. Ltd.
5 Toh Tuck Link, Singapore 596224
USA office: 27 Warren Street, Suite 401-402, Hackensack, NJ 07601
UK office: 57 Shelton Street, Covent Garden, London WC2H 9HE

Library of Congress Cataloging-in-Publication Data
Graef, John R., 1942–
   Multiple solutions of boundary value problems : a variational approach / by John R. Graef &
Lingju Kong (The University of Tennessee at Chattanooga, USA).
     pages cm. -- (Trends in abstract and applied analysis ; v. 1)
   Includes bibliographical references and index.
   ISBN 978-9814696548 (alk. paper)
   1. Boundary value problems. 2. Variational method. 3. Differential equations. I. Kong, Lingju.
II. Title.
   QA379.G72 2015
   518'.63--dc23

                        2015025797

British Library Cataloguing-in-Publication Data
A catalogue record for this book is available from the British Library.

In-house Editors: V. Vishnu Mohan/Kwong Lai Fun

Typeset by Stallion Press
Email: enquiries@stallionpress.com

Printed in Singapore

To Frances, without whose help and support this work would never have been finished, and to our sons John, Chris, and Kevin.

Dedicated to my loving wife Zhen and son Michael.

Preface

Variational methods and their generalizations have proved to be a useful tool in proving the existence of solutions to a variety of boundary value problems for ordinary, impulsive, and partial differential equations as well as for difference equations. In this monograph, we look at how variational methods can be used in all these settings. In our first chapter, we gather the basic notions and fundamental theorems that will be applied in the remainder of this monograph. While many of these items are easily available in the literature, we gather them here both for the convenience of the reader and for the purpose of making this volume somewhat self contained. Subsequent chapters deal with Sturm-Liouville problems, multi-point boundary value problems, problems with impulses, partial differential equations, and difference equations. The chapters are presented in such a way that, except for references to the basic results in Chapter 1, each is essentially a stand-alone entity that can be read with little reference to other chapters. Similarly, many of the sections in each chapter can be treated in that same independent way.

Chapter 2 starts by presenting the basic setting for Sturm-Liouville boundary value problems. In the second section, sufficient conditions a for the existence of a nontrivial solution to nonlinear Sturm-Liouville systems are presented. Section 3 is concerned with the existence of multiple solutions, while Section 4 presents sufficient conditions for the existence of infinitely many solutions.

In Chapter 3, multipoint problems are discussed. After a brief introduction, the next three sections parallel those in Chapter 2 by examining sufficient conditions for the existence of a nontrivial solution, the existence of multiple solutions, and the existence of infinitely many solutions, respectively. Section 5 deals with two parameter systems and the final section in the chapter contains results on the existence of solutions proved by using the dual action principle.

Chapter 4 is concerned with impulsive boundary values problems, and in addition to discussing the existence of one and infinitely many solutions, the last section in the chapter is devoted to anti-periodic problems.

Partial differential equations is the focus of Chapter 5. Topics include Kirchhoff problems with two parameters, biharmonic systems, elliptic problems with nonstandard growth conditions, and elliptic systems of Kirchhoff type.

Chapter 6 is devoted to difference equations. Periodic problems with one parameter are discussed in Section 2, periodic problems with two parameters in Section 3, multipoint problems with several parameters in Section 4, and homoclinic solutions in Section 5. The final section in the chapter is concerned with anti-periodic solutions of higher order difference equations.

Some brief notes are given in Chapter 7. More than 300 references to the current literature are included.

The authors especially want to thank their colleagues Johnny Henderson of Baylor University, Shapour Heidarkhani of Razi University (Iran), Min Wang formerly of The University of Tennessee at Chattanooga and now with Equifax Inc., and Bo Yang of Kennesaw State University, whose collaboration on the application of variational methods to boundary value problems have inspired this monograph.

The authors also wish to thank Lai Fun Kwong, Vishnu Mohan, and the staff at World Scientific for all their help in bringing this project to fruition.

John R. Graef
Lingju Kong

Contents

Chapter 1

Mathematical Preliminaries

1.1 Mathematical Preliminaries

In this chapter, we gather together some of the primary mathematical tools that will be used throughout this monograph in proving the existence of solutions to boundary value problems of different types. For the reader's convenience, in the second section in this chapter we present definitions of some of the concepts used in these theorems.

We are interested in results that guarantee the existence of at least one solution, the existence of multiple solutions, and the existence of infinitely many solutions to various types of boundary value problems. Theorems 1.1.1–1.1.16 below will be extensively used in this process. The first two are consequences of an existence result for a local minimum of a functional ([39, Theorem 3.1]) that was inspired by Ricceri's variational principle (see [244]).

To begin, we let X be a nonempty set and let Φ, $\Psi : X \to \mathbb{R}$ be functionals. Define

$$\eta(r_1, r_2) = \inf_{v \in \Phi^{-1}(r_1, r_2)} \frac{\sup_{u \in \Phi^{-1}(r_1, r_2)} \Psi(u) - \Psi(v)}{r_2 - \Phi(v)} \tag{1.1.1}$$

and

$$\rho(r_1, r_2) = \sup_{v \in \Phi^{-1}(r_1, r_2)} \frac{\Psi(v) - \sup_{u \in \Phi^{-1}(-\infty, r_1)} \Psi(u)}{\Phi(v) - r_1} \tag{1.1.2}$$

for all $r_1, r_2 \in \mathbb{R}$ with $r_1 < r_2$, and

$$\rho_2(r) = \sup_{v \in \Phi^{-1}(r, \infty)} \frac{\Psi(v) - \sup_{u \in \Phi^{-1}(-\infty, r)} \Psi(u)}{\Phi(v) - r}$$

for all $r \in \mathbb{R}$. In what follows, X^* will always denote the dual space of X.

1

Theorem 1.1.1. ([39, Theorem 5.1]) *Let X be a reflexive real Banach space, $\Phi : X \to \mathbb{R}$ be a sequentially weakly lower semicontinuous, coercive, and continuously Gâteaux differentiable functional whose Gâteaux derivative admits a continuous inverse on X^*, and let $\Psi : X \to \mathbb{R}$ be a continuously Gâteaux differentiable functional whose Gâteaux derivative is compact. Let $I_\lambda = \Phi - \lambda\Psi$ and assume that there are $r_1, r_2 \in \mathbb{R}$ with $r_1 < r_2$ such that*

$$\eta(r_1, r_2) < \rho(r_1, r_2).$$

Then, for each $\lambda \in (1/\rho(r_1, r_2), 1/\eta(r_1, r_2))$, there exists $u_{0,\lambda} \in \Phi^{-1}(r_1, r_2)$ such that $I_\lambda(u_{0,\lambda}) \leq I_\lambda(u)$ for all $u \in \Phi^{-1}(r_1, r_2)$ and $I_\lambda'(u_{0,\lambda}) = 0$.

Another result in this same spirit is the following.

Theorem 1.1.2. ([39, Theorem 5.3]) *Let X be a real Banach space, $\Phi : X \to \mathbb{R}$ be a continuously Gâteaux differentiable functional whose Gâteaux derivative admits a continuous inverse on X^*, and let $\Psi : X \to \mathbb{R}$ be a continuously Gâteaux differentiable functional whose Gâteaux derivative is compact. Choose r so that $\inf_X \Phi < r < \sup_X \Phi$, $\rho_2(r) > 0$, and for each $\lambda > \frac{1}{\rho_2(r)}$, the functional $I_\lambda := \Phi - \lambda\Psi$ is coercive. Then, for each $\lambda \in (\frac{1}{\rho_2(r)}, \infty)$, there exists $u_{0,\lambda} \in \Phi^{-1}(r, \infty)$ such that $I_\lambda(u_{0,\lambda}) \leq I_\lambda(u)$ for all $u \in \Phi^{-1}(r, \infty)$ and $I_\lambda'(u_{0,\lambda}) = 0$.*

These results and their many variations have often been used to obtain multiplicity results for nonlinear problems of a variational nature. See, for example, [23, 24, 39, 40, 244] and the references therein.

To prove the existence of multiple solutions to Sturm-Liouville problems, we will make use of the two following three critical point theorems. The first one was obtained in [22] and is a more precise version of Theorem 3.2 in [24]. It requires the coercivity of the functional $I_\lambda = \Phi - \lambda\Psi$. The second result appeared in [40] and needs a certain sign condition on the functionals.

Theorem 1.1.3. ([40, Theorem 3.2], [42, Theorem 2.2]) *Let X be a reflexive real Banach space, $\Phi : X \longrightarrow \mathbb{R}$ be a coercive and continuously Gâteaux differentiable functional whose derivative admits a continuous inverse on X^*, and $\Psi : X \longrightarrow \mathbb{R}$ be a continuously Gâteaux differentiable functional whose derivative is compact with*

$$\inf_X \Phi = \Phi(0) = \Psi(0) = 0. \qquad (1.1.3)$$

Assume that there is a positive constant r and an element $\overline{v} \in X$, with $2r < \Phi(\overline{v})$, such that:

(a_1) $\dfrac{\sup_{u \in \Phi^{-1}(-\infty, r)} \Psi(u)}{r} < \dfrac{2}{3} \dfrac{\Psi(\overline{v})}{\Phi(\overline{v})}$;

(a_2) *for all* $\lambda \in \left(\dfrac{3}{2} \dfrac{\Phi(\overline{v})}{\Psi(\overline{v})}, \dfrac{r}{\sup_{u \in \Phi^{-1}(-\infty, r)} \Psi(u)} \right)$, *the functional* $\Phi - \lambda\Psi$ *is coercive.*

Then, for each $\lambda \in \left(\dfrac{3}{2} \dfrac{\Phi(\overline{v})}{\Psi(\overline{v})}, \dfrac{r}{\sup_{u \in \Phi^{-1}(-\infty, r)} \Psi(u)} \right)$, *the functional* $\Phi - \lambda\Psi$ *has at least three distinct critical points.*

Theorem 1.1.4. ([40, Theorem 3.3], [42, Theorem 2.3]) *Let* X *be a reflexive real Banach space,* $\Phi : X \longrightarrow \mathbb{R}$ *be a convex, coercive and continuously Gâteaux differentiable functional whose derivative admits a continuous inverse on* X^*, *and* $\Psi : X \longrightarrow \mathbb{R}$ *be a continuously Gâteaux differentiable functional whose derivative is compact with:*

(i) $\inf_X \Phi = \Phi(0) = \Psi(0) = 0$;

(ii) *for each* $\lambda > 0$ *and for every pair of local minima* u_1, u_2 *of the functional* $\Phi - \lambda\Psi$ *such that* $\Psi(u_1) \geq 0$ *and* $\Psi(u_2) \geq 0$, *we have*

$$\inf_{s \in [0,1]} \Psi(su_1 + (1-s)u_2) \geq 0.$$

Assume that there are two positive constants r_1, r_2 *and* $\overline{v} \in X$, *with* $2r_1 < \Phi(\overline{v}) < \frac{r_2}{2}$, *such that:*

(b_1) $\dfrac{\sup_{u \in \Phi^{-1}(-\infty, r_1)} \Psi(u)}{r_1} < \dfrac{2}{3} \dfrac{\Psi(\overline{v})}{\Phi(\overline{v})}$;

(b_2) $\dfrac{\sup_{u \in \Phi^{-1}(-\infty, r_2)} \Psi(u)}{r_2} < \dfrac{1}{3} \dfrac{\Psi(\overline{v})}{\Phi(\overline{v})}$.

Then, for each

$$\lambda \in \left(\dfrac{3}{2} \dfrac{\Phi(\overline{v})}{\Psi(\overline{v})}, \ \min\left\{ \dfrac{r_1}{\sup_{u \in \Phi^{-1}(-\infty, r_1)} \Psi(u)}, \ \dfrac{\frac{r_2}{2}}{\sup_{u \in \Phi^{-1}(-\infty, r_2)} \Psi(u)} \right\} \right),$$

the functional $\Phi - \lambda\Psi$ *has at least three distinct critical points that lie in* $\Phi^{-1}(-\infty, r_2)$.

Since we are also interested in obtaining sufficient conditions for the existence of infinitely many solutions to Sturm-Liouville problems, we will need the following critical points theorem; see [54, Theorem 2.1] and [244, Theorem 2.5].

Theorem 1.1.5. *Let* X *be a reflexive real Banach space and let* Φ, Ψ : $X \to \mathbb{R}$ *be two Gâteaux differentiable functionals such that* Φ *is sequentially* *weakly lower semicontinuous, strongly continuous and coercive, and Ψ is sequentially weakly upper semicontinuous. For every $r > \inf_X \Phi$, let*

$$\varphi(r) := \inf_{u \in \Phi^{-1}(-\infty, r)} \frac{\sup_{v \in \Phi^{-1}(-\infty, r)} \Psi(v) - \Psi(u)}{r - \Phi(u)}, \qquad (1.1.4)$$

and

$$\gamma := \liminf_{r \to \infty} \varphi(r), \quad \delta := \liminf_{r \to (\inf_X \Phi)^+} \varphi(r).$$

Then:

(a) *For every $r > \inf_X \Phi$ and every $\lambda \in (0, 1/\varphi(r))$, the restriction of the functional $I_\lambda := \Phi - \lambda\Psi$ to $\Phi^{-1}(-\infty, r)$ admits a global minimum, which is a critical point (local minimum) of I_λ in X.*

(b) *If $\gamma < \infty$, then for each $\lambda \in (0, 1/\gamma)$, the following alternative holds: either*

(b$_1$) *I_λ possesses a global minimum, or*

(b$_2$) *there is a sequence $\{u_n\}$ of critical points (local minima) of I_λ such that*

$$\lim_{n \to \infty} \Phi(u_n) = \infty.$$

(c) *If $\delta < \infty$, then for each $\lambda \in (0, 1/\delta)$, the following alternative holds: either*

(c$_1$) *there is a global minimum of Φ which is a local minimum of I_λ, or*

(c$_2$) *there is a sequence $\{u_n\}$ of pairwise distinct critical points (local minima) of I_λ that converges weakly to a global minimum of Φ.*

The following three critical points theorem due to Ricceri [248] is useful for showing the existence of three solutions to systems of multi-point boundary value problems.

Theorem 1.1.6. ([248, Theorem 1]) *Let X be a real reflexive Banach space, $I \subseteq R$ an interval, $\Phi : X \longrightarrow R$ a sequentially weakly lower semicontinuous C^1 functional that is bounded on each bounded subset of X and whose derivative admits a continuous inverse on X^*, and let $J : X \longrightarrow R$ be a C^1 functional with a compact derivative. Assume that*

$$\lim_{\|x\| \to \infty} (\Phi(x) + \lambda J(x)) = \infty \qquad (1.1.5)$$

for all $\lambda \in I$, and that there exists $\rho \in R$ such that

$$\sup_{\lambda \in I} \inf_{x \in X} (\Phi(x) + \lambda(J(x) + \rho)) < \inf_{x \in X} \sup_{\lambda \in I} (\Phi(x) + \lambda(J(x) + \rho)). \qquad (1.1.6)$$

Then, there exist a non-empty open interval $A \subseteq I$ and a positive real number q with the following property: for every $\lambda \in A$ and every C^1 functional $\Psi : X \longrightarrow R$ with compact derivative, there exists $\delta > 0$ such that, for each $\mu \in [0, \delta]$, the equation

$$\Phi'(x) + \lambda J'(x) + \mu \Psi'(x) = 0$$

has at least three solutions in X whose norms are less than q.

In order to apply Theorem 1.1.6 above, the following result, which is a restatement of Proposition 1.3 in [39] (J is replaced by $-J$), is useful in showing that inequality (1.1.6) holds.

Theorem 1.1.7. ([39, Proposition 1.3]) *Let X be a non-empty set and Φ and J be two real valued functionals defined on X. Assume that $\Phi(u) \geq 0$ for every $u \in X$ and there exists $u_0 \in X$ such that $\Phi(u_0) = J(u_0) = 0$. Furthermore, assume that there exist $w \in X$ and $r > 0$ such that $\Phi(w) > r$ and*

$$\sup_{\Phi(u) < r} (-J(u)) < r \frac{-J(w)}{\Phi(w)}.$$

Then, for every $h > 1$ and for every $\rho \in R$ satisfying

$$\sup_{\Phi(u) < r} (-J(u)) + \frac{r \frac{-J(w)}{\Phi(w)} - \sup_{\Phi(u) < r}(-J(u))}{h} < \rho < r \frac{-J(w)}{\Phi(w)},$$

we have

$$\sup_{\lambda \in R} \inf_{u \in X} (\Phi(u) + \lambda(J(u) + \rho)) < \inf_{u \in X} \sup_{\lambda \in [0,v]} (\Phi(u) + \lambda(J(u) + \rho)),$$

where

$$v = \frac{hr}{r \frac{-J(u_1)}{\Phi(u_1)} - \sup_{\Phi(u) < r}(-J(u))}.$$

We next define what is meant by the Palais–Smale condition.

Definition 1.1.1. *Let X be a real Banach space. The functional $J \in C^1(X, \mathbb{R})$ is said to satisfy the Palais–Smale (PS) condition if every sequence $\{u_n\} \subset X$, such that $J(u_n)$ is bounded and $J'(u_n) \to 0$ as $n \to \infty$, has a convergent subsequence. Here, the sequence $\{u_n\}$ is called a (PS) sequence.*

The following result is the well-known Mountain Pass Theorem of Ambrosetti and Rabinowitz [241] (also see [185, Theorem 7.1]). Here, $B_r(u)$ is the open ball centered at $u \in X$ with radius $r > 0$, $\overline{B}_r(u)$ is its closure, and $\partial B_r(u)$ is its boundary.

Theorem 1.1.8. *Let $(X, \|\cdot\|)$ be a real Banach space and $I \in C^1(X, \mathbb{R})$. Assume that I satisfies the (PS) condition and there exist $u_0, u_1 \in X$ and $\rho > 0$ such that:*

(A1) $u_1 \notin \overline{B}_\rho(u_0)$;
(A2) $\max\{I(u_0), I(u_1)\} < \inf_{u \in \partial B_\rho(u_0)} I(u)$.

Then, I possesses a critical value which can be characterized as

$$c = \inf_{\gamma \in \Gamma} \max_{s \in [0,1]} I(\gamma(s)) \geq \inf_{u \in \partial B_\rho(u_0)} I(u),$$

where

$$\Gamma = \{\gamma \in C([0,1], X) \, : \, \gamma(0) = u_0, \ \gamma(1) = u_1\}.$$

In the following definition, we present an important notion known as the Ambrosetti–Rabinowitz condition.

Definition 1.1.2. *A function $f : \mathbb{Z} \times \mathbb{R} \to \mathbb{R}$ is said to satisfy an Ambrosetti–Rabinowitz type condition if*

(AR) *there exists $\mu > p$ and $R > 0$ such that*

$$0 \leq \mu F(k,t) \leq t f(k,t) \quad \text{for } (k,t) \in \mathbb{Z} \times \mathbb{R} \text{ with } |t| \geq R, \quad (1.1.7)$$

where $F(k,t) = \int_0^t f(k,s)ds$.

Condition (AR) is often used to show that any (PS) sequence of the corresponding energy functional is bounded. This plays an important role in the application of critical point theory.

The following three theorems will be useful in proving existence results for boundary value problems for difference equations. Theorem 1.1.9 below can be found in [227, 289], Theorem 1.1.10 in [66], and Theorem 1.1.11 in [89, 241].

Theorem 1.1.9. *Let X be a real reflexive Banach space, and let J be a weakly upper (lower, respectively) semicontinuous functional such that*

$$\lim_{\|u\| \to \infty} J(u) = -\infty \quad \left(\lim_{\|u\| \to \infty} J(u) = \infty, \text{ respectively}\right).$$

Then, there exists $u_0 \in X$ such that

$$J(u_0) = \sup_{u \in X} J(u) \quad \left(J(u_0) = \inf_{u \in X} J(u), \text{ respectively} \right).$$

Furthermore, if $J \in C^1(X, R)$, then $J'(u_0) = 0$.

Theorem 1.1.10. *Let X be a real Banach space with a direct sum decomposition $X = X_1 \oplus X_2$ with $\dim X_2 < \infty$. Suppose that $J \in C^1(X, \mathbb{R})$ satisfies the (PS) condition and is bounded from below, $J(0) = 0$, and $\inf_{u \in X} J(u) < 0$. In addition, assume that J has a local linking at 0, that is, for some $R > 0$,*

$$J(u) \geq 0 \quad \text{for } u \in X_1 \text{ with } \|u\| \leq R,$$

$$J(u) \leq 0 \quad \text{for } u \in X_2 \text{ with } \|u\| \leq R.$$

Then, J has at least two nontrivial critical points.

Theorem 1.1.11. *Let X be a real Banach space and $J \in C^1(X, R)$ be even, bounded from below, and satisfy the (PS) condition. Suppose that $J(0) = 0$ and there is a set $K \subset X$ such that K is homeomorphic to S^{n-1} by an odd map and $\sup_{u \in K} J(u) < 0$, where S^{n-1} is the $n-1$ dimensional unit sphere. Then, J has at least n disjoint pairs of nontrivial critical points.*

We wish to present another three critical points theorem due to Ricceri [244]. To do so, we need to introduce the following class of functionals.

Definition 1.1.3. *Let E be a real Banach space with the norm $\| \cdot \|_E$ and let $\mathcal{W}_E$ be the class of all functionals $\Phi : E \to \mathbb{R}$ with the property: if $\{u_n\}$ is a sequence in E converging weakly to u and $\liminf_{n \to \infty} \Phi(u_n) \leq \Phi(u)$, then $\{u_n\}$ has a subsequence converging strongly to u.*

Remark 1.1.1. It is easy to see that if E is uniformly convex and $h : \mathbb{R}^+ \to \mathbb{R}$ is a continuous and strictly increasing function, then the functional $u \to h(\|u\|_E)$ belongs to the class $\mathcal{W}_E$.

Theorem 1.1.12. ([244, Theorem 2]) *Let E be a separable and reflexive real Banach space with the norm $\| \cdot \|_E$. Let $\Phi : E \to \mathbb{R}$ be a coercive, sequentially weakly lower semicontinuous C^1 functional belonging to $\mathcal{W}_E$ that is bounded on each bounded subset of E and whose derivative admits a continuous inverse on E^*. Let $J : E \to \mathbb{R}$ be a C^1 functional with compact derivative. Assume that Φ has a strict local minimum u_0 with $\Phi(u_0) = J(u_0) = 0$. Finally, let*

$$\delta = \max \left\{ 0, \limsup_{\|u\|_E \to \infty} \frac{J(u)}{\Phi(u)}, \limsup_{\|u\|_E \to u_0} \frac{J(u)}{\Phi(u)} \right\},$$

$$\eta = \sup_{u \in \Phi^{-1}(0,\infty)} \frac{J(u)}{\Phi(u)},$$

and assume that $\delta < \eta$. Then, for each compact interval $[a,b] \subset (1/\eta, 1/\delta)$ (with the conventions $1/0 = \infty$ and $1/\infty = 0$), there exists $K > 0$ with the following property: for every $x \in [a,b]$ and every C^1 functional $\Psi : E \to \mathbb{R}$ with compact derivative, there exists $\zeta > 0$ such that for each $y \in [0, \zeta]$, the equation

$$\Phi'(u) - y\Psi'(u) - xJ'(u) = 0$$

has at least three solutions in E whose norms are less than K.

We next define the notion of genus.

Definition 1.1.4. *Let X be a Banach space and A be a subset of X. We say that A is symmetric if $u \in A$ implies $-u \in A$. For a closed symmetric set A with $0 \notin A$, the genus $\gamma(A)$ of A is defined as the smallest integer k such that there exists an odd continuous mapping from A to $\mathbb{R}^k \setminus \{0\}$. If there does not exist such k, we define $\gamma(A) = \infty$. Moreover, we set $\gamma(\emptyset) = 0$. Let Γ_k denote the family of closed symmetric subsets of X such that if $A \in \Gamma_k$, then $0 \notin A$ and $\gamma(A) \geq k$.*

The next result follows from [187, Theorem 1].

Theorem 1.1.13. *Let X be an infinite dimensional Banach space and $I \in C^1(X, \mathbb{R})$ satisfy the following two conditions:*

(A3) *$I(u)$ is even, bounded from below, $I(0) = 0$, and $I(u)$ satisfies the (PS) condition;*

(A4) *For each $n \in \mathbb{N}$, there exists $A_n \in \Gamma_n$ such that $\sup_{u \in A_n} I(u) < 0$.*

Then, $I(u)$ has a sequence of critical points $\{u_n\}$ such that

$$I(u_n) \leq 0, \quad u_n \neq 0, \quad \text{and} \quad \lim_{n \to \infty} u_n = 0.$$

The next concept is known as Cerami's condition [75, 77, 213].

Definition 1.1.5. *Let X be a Banach space and $I \in C^1(X, \mathbb{R})$. We say that I satisfies Cerami's condition if for any $c \in \mathbb{R}$, we have:*

(i) *any bounded sequence $\{u_n\} \subset X$ such that $I(u_n) \to c$ and $I'(u_n) \to 0$ has a convergent subsequence;*

(ii) *there exist positive constants δ, η, and R such that*

$$\|I'(u)\|\|u\| \geq \eta \quad \text{for all } u \in I^{-1}([c - \delta, c + \delta]) \text{ with } \|u\| \geq R.$$

Let X be a reflexive Banach space. It is well known (see, for example, [299, page 233]) that there exist $e_i \in X$ and $e_i^* \in X^*$ such that

$$X = \overline{\text{span}\{e_i \ : \ i = 1, 2, \ldots\}}, \quad X^* = \overline{\text{span}\{e_i^* \ : \ i = 1, 2, \ldots\}}^{w^*},$$

and

$$\langle e_i^*, e_j \rangle = \begin{cases} 1 & \text{if } i = j, \\ 0 & \text{if } i \neq j. \end{cases}$$

Now, we let

$$X_i = \text{span}\{e_i\}, \quad Y_n = \bigoplus_{i=1}^{n} X_i, \quad \text{and} \quad Z_n = \bigoplus_{i=n}^{\infty} X_i. \tag{1.1.8}$$

The following result can be found in [213, Proposition 2.4] or [287, Proposition 2.8].

Theorem 1.1.14. *Assume that $I \in C^1(X, \mathbb{R})$ satisfies Cerami's condition and $I(u) = I(-u)$ for $u \in X$. In addition, assume that, for any $n \in \mathbb{N}$, there exist $\rho_n > r_n > 0$ such that:*

(a) $a_n := \inf_{u \in Z_n, \, \|u\| = r_n} I(u) \to \infty$ *as $n \to \infty$;*
(b) $b_n := \max_{u \in Y_n, \, \|u\| = \rho_n} I(u) \leq 0.$

Then, I has a sequence of critical values tending to ∞.

The next three critical points theorem is based on [251, Theorem 1].

Theorem 1.1.15. *Let X be a real reflexive Banach space, X^* be the dual space of X, $\Phi : X \to \mathbb{R}$ be a continuously Gâteaux differentiable and sequentially weakly lower semicontinuous functional that is bounded on bounded subsets of X and whose Gâteaux derivative admits a continuous inverse on X^*. Let $\Psi : X \to \mathbb{R}$ be a continuously Gâteaux differentiable functional whose Gâteaux derivative is compact and such that $\Phi(0) = \Psi(0) = 0$. Assume that there exist $r > 0$ and $\overline{x} \in X$, with $r < \Phi(\overline{x})$, such that*

(i) $\dfrac{\sup_{\Phi(x) \leq r} \Psi(x)}{r} < \dfrac{\Psi(\overline{x})}{\Phi(\overline{x})},$

(ii) *for each $\lambda \in \Lambda_r := \left(\dfrac{\Phi(\overline{x})}{\Psi(\overline{x})}, \dfrac{r}{\sup_{\Phi(x) \leq r} \Psi(x)} \right)$, the functional $\Phi - \lambda\Psi$ is coercive.*

Then, for each compact interval $[a, b] \subset \Lambda_r$, there exists $\rho > 0$ with the following property: for every $\lambda \in [a, b]$ and every C^1 functional $\Gamma : X \to \mathbb{R}$ with compact derivative, there exists $\delta > 0$ such that, for each $\mu \in [0, \delta]$, the functional $\Phi - \lambda\Psi - \mu\Gamma$ has at least three distinct critical points in X whose norms are less than ρ.

Our final three critical points theorem to be presented here is useful in showing the existence of multiple solutions to boundary value problems for partial differential equations.

Theorem 1.1.16. ([22, Theorem B]) *Let X be a real reflexive Banach space, X^* be the dual space of X, $\Phi : X \to \mathbb{R}$ a continuously Gâteaux differentiable and sequentially weakly lower semicontinuous functional whose Gâteaux derivative admits a continuous inverse on X^*, and $\Psi : X \to \mathbb{R}$ be a continuously Gâteaux differentiable functional whose Gâteaux derivative is compact. Assume that*

(a) $\lim_{\|u\| \to \infty}(\Phi(u) + \lambda\Psi(u)) = \infty$ *for all* $\lambda \in [0, \infty)$,
(b) there is $r \in \mathbb{R}$ *such that*

$$\inf_X \Phi < r$$

and

$$\varphi_1(r) < \varphi_2(r),$$

where

$$\varphi_1(r) := \inf_{u \in \Phi^{-1}((-\infty, r))} \frac{\Psi(u) - \inf_{\overline{\Phi^{-1}((-\infty, r))}^w} \Psi}{r - \Phi(u)} \qquad (1.1.9)$$

and

$$\varphi_2(r) := \inf_{u \in \Phi^{-1}((-\infty, r))} \sup_{v \in \Phi^{-1}([r, \infty))} \frac{\Psi(u) - \Psi(v)}{\Phi(v) - \Phi(u)}, \qquad (1.1.10)$$

with $\overline{\Phi^{-1}((-\infty, r))}^w$ *being the closure of* $\Phi^{-1}((-\infty, r))$ *in the weak topology.*

Then, for each $\lambda \in (1/\varphi_2(r), 1/\varphi_1(r))$ the functional $\Phi(u) + \lambda\Psi(u)$ has at least three critical points in X.

Another tool that has proved to be quite useful in showing the existence of solutions to partial differential equations is Ekeland's variation principle.

Theorem 1.1.17. ([227, Theorem 4.1]) *Let M be a complete metric space and let $J : M \to (-\infty, \infty]$ be a lower semicontinuous functional, bounded from below, and not identically equal to ∞. Let $\epsilon > 0$ be given and $z \in M$ be such that*

$$J(z) \leq \inf_M J + \epsilon.$$

Then, there exists $v \in M$ such that

$$J(v) \leq J(z) \leq \inf_M J + \epsilon,$$

$$d(z, v) \leq 1,$$

and for any $u \neq v$ in M,

$$J(v) < J(u) + \epsilon d(v, u),$$

where $d(\cdot, \cdot)$ denotes the distance between two elements in M.

These theorems and variations on them have frequently been used to obtain existence and multiplicity results for boundary value problems. For a small sampling of papers using this approach we refer the reader to [23, 24, 39–43, 48, 53, 54, 74, 97, 131, 171, 244]. An excellent bibliographical account can be found in [250].

1.2 Background Concepts

In this section, we present some of the background notions that are used in the theorems in the previous section of this chapter. For additional discussion, we refer the reader, for example, to the monographs of Mawhin and Willem [227] and Zeidler [289].

Let X be a normed space and $L : X \to \mathbb{R}$ be a functional. The functional $L : X \to \mathbb{R}$ is *coercive* if

$$L(u) \to \infty \quad \text{as} \quad \|u\| \to \infty.$$

The functional $L : X \to \mathbb{R}$ is *convex* if

$$L((1 - \lambda)u + \lambda v) \leq (1 - \lambda)L(u) + \lambda L(v),$$

and it is *concave* if

$$L((1 - \lambda)u + \lambda v) \geq (1 - \lambda)L(u) + \lambda L(v).$$

The functional $L : X \to \mathbb{R}$ is *sequentially lower semicontinuous* if for any sequence $u_n \to u$, we have

$$\liminf_{n \to \infty} L(u_n) \geq L(u),$$

and it is *weakly sequentially lower semicontinuous* if for any sequence $u_n \rightharpoonup u$,

$$\liminf_{n \to \infty} L(u_n) \geq L(u).$$

The functional $L : X \to \mathbb{R}$ is *sequentially upper semicontinuous* if for any sequence $u_n \to u$, we have

$$\limsup_{n \to \infty} L(u_n) \leq L(u),$$

and it is *weakly sequentially upper semicontinuous* if for any sequence $u_n \rightharpoonup u$,

$$\limsup_{n \to \infty} L(u_n) \leq L(u).$$

The set of all bounded linear operators $T : X \to \mathbb{R}$ on the Banach space X is called the *dual space* of X and is denoted by X^*. In addition, we say that a Banach space X is *reflexive* if $(X^*)^* = X$.

The functional $L : X \to \mathbb{R}$ is *Gâteaux differentiable* at $x \in X$ if there exists a mapping $T : X \to \mathbb{R}$ such that

$$L(x + ku) - L(x) = kTu + o(k) \quad \text{as} \quad k \to 0$$

for all $u \in X$ with $\|u\| = 1$.

The map T is called the *Gâteaux derivative* of L at x and we write $L'(x) = T$.

We say that the Gâteaux derivative Φ' of the functional Φ admits a continuous inverse on X^* if there exists a continuous operator $T : X^* \to X$ such that $T(\Phi'(u)) = u$ for all $u \in X$.

Let Ω be an open subset of $\mathbb{R}^n$. For any $m \in \mathbb{R}$ and $p \in [1, \infty]$, the Sobolev space

$$W^{m,p}(\Omega) = \{u \in L^p(\Omega) : D^\alpha u \in L^p(\Omega) \quad \text{for all} \quad |\alpha| \leq m\},$$

where $\alpha = (\alpha_1, \ldots, \alpha_N)$ is a multi-index, $|\alpha| = \sum_{i=1}^{N} \alpha_i$, and

$$D^\alpha u = \frac{\partial^{|\alpha|} u}{\partial^{\alpha_1} u_1 \cdots \partial^{\alpha_N} u_N}.$$

The space $W_0^{m,p}(\Omega)$ is the closure of $C_0^\infty(\Omega)$ in $W^{m,p}(\Omega)$, that is,

$$W_0^{m,p}(\Omega) = \{u \in W^{m,p}(\Omega) : D^\alpha u = 0 \text{ on } \partial\Omega \text{ for all } |\alpha| \leq m - 1\}.$$

Chapter 2

Sturm-Liouville Problems

2.1 Introduction

Sturm-Liouville problems have long been an object of study because of the many applications in which they arise. Recent results on proving the existence of solutions of such problems can be found in $[23, 35, 42, 99, 131, 137, 138, 158, 181, 258, 265, 268, 269, 300]$ and the references contained therein. In this chapter, we establish sufficient conditions for the system

$$\begin{cases} -(\phi_{p_i}(u_i'(x)))' = \lambda F_{u_i}(x, u_1, \ldots, u_n)h_i(u_i'(x)), & \text{in } (a, b), \\ \alpha_i u_i(a) - \beta_i u_i'(a) = 0, \quad \gamma_i u_i(b) + \sigma_i u_i'(b) = 0, \end{cases} \quad i = 1, \ldots, n.$$

$$(2.1.1)$$

to have at least one nontrivial classical solution, to have multiple solutions, and to have infinitely many solutions. Here, we assume that λ is a positive parameter, $p_i > 1$, $\phi_{p_i}(t) = |t|^{p_i - 2}t$, $\alpha_i, \gamma_i \geq 0$, $\beta_i, \sigma_i > 0$, $h_i : \mathbb{R} \to [0, \infty)$ is a bounded and continuous function with $\inf_{t \in \mathbb{R}} h_i(t) > 0$ for $i = 1, \ldots, n$. In addition, $F : [a, b] \times \mathbb{R}^n \to \mathbb{R}$ is a function with the property that the mapping $(t_1, t_2, \ldots, t_n) \to F(x, t_1, t_2, \ldots, t_n)$ is in C^1 in $\mathbb{R}^n$ for all $x \in [a, b]$, F_{t_i} is continuous in $[a, b] \times \mathbb{R}^n$ for $i = 1, \ldots, n$ (F_{t_i} denotes the partial derivative of F with respect to t_i), and $F(x, 0, \ldots, 0) = 0$ for all $x \in [a, b]$. In keeping with the theme of this volume, our approach is to use variational methods and critical point theory. In particular, we will invoke Theorem 1.1.1, Theorems 1.1.3–1.1.4, and Theorem 1.1.5, respectively, for each of these scenarios.

2.2 Nontrivial Solutions of Sturm-Liouville Systems

By a *classical solution* of system (2.1.1), we mean a function $u = (u_1, \ldots, u_n)$ such that, for $i = 1, \ldots, n$, $u_i(x) \in C^1[a, b]$, $\phi_{p_i}(u_i'(x)) \in C^1[a, b]$, and $u_i(x)$ satisfies (2.1.1).

We will take X to be the Cartesian product of n Sobolev spaces $W^{1,p_i}([a,b])$, $i = 1,\ldots,n$, i.e., $X = W^{1,p_1}([a,b]) \times \cdots \times W^{1,p_n}([a,b])$, endowed with the norm

$$||u|| = ||(u_1,\ldots,u_n)|| = \sum_{i=1}^{n} ||u_i||_1, \quad u = (u_1,\ldots,u_n) \in X,$$

where

$$||u_i||_1 = \left(\int_a^b (|u_i'(x)|^{p_i} + |u_i(x)|^{p_i}) dx \right)^{1/p_i}, \quad i = 1,\ldots,n.$$

Then, X is a reflexive real Banach space.

We will make use of the following notation. For $i = 1,\ldots,n$ and $v \in L^{p_i}([a,b])$, let

$$||v||_{L^{p_i}} = \left(\int_a^b |v(t)|^{p_i} dt \right)^{1/p_i},$$

$$m_i = \inf_{t \in \mathbb{R}} h_i(t), \quad M_i = \sup_{t \in \mathbb{R}} h_i(t),$$

$$\underline{m} = \min\{m_i : i = 1,\ldots,n\}, \quad \overline{M} = \max\{m_i : i = 1,\ldots,n\},$$

$$\underline{p} = \min\{p_i : i = 1,\ldots,n\}, \quad \overline{p} = \max\{p_i : i = 1,\ldots,n\}$$

and notice that $\overline{M} \geq \underline{m} > 0$. We also let q_i be the conjugates of p_i, i.e., $1/p_i + 1/q_i = 1$.

For $s \in \mathbb{R}$, we set

$$J_i(s) = \int_0^s \frac{(p_i - 1)|\delta|^{p_i - 2}}{h_i(\delta)} d\delta, \quad i = 1,\ldots,n.$$

For $u = (u_1,\ldots,u_n) \in X$, we define the functionals $\Phi, \Psi : X \to \mathbb{R}$ by

$$\Phi(u) = \sum_{i=1}^{n} \left[\int_a^b \left(\int_0^{u'(x)} J_i(s) ds \right) dx \right.$$

$$\left. + \frac{\beta_i}{\alpha_i} \int_0^{\frac{\alpha_i u_i(a)}{\beta_i}} J_i(s) ds + \frac{\sigma_i}{\gamma_i} \int_0^{-\frac{\gamma_i u_i(b)}{\sigma_i}} J_i(s) ds \right] \quad (2.2.1)$$

and

$$\Psi(u) = \int_a^b F(x, u_1(x),\ldots,u_n(x)) dx. \quad (2.2.2)$$

A straightforward calculation yields

$$\frac{1}{\overline{M}\,\overline{p}} \sum_{i=1}^{n} \left(\|u_i'\|_{L^{p_i}}^{p_i} + \frac{\alpha_i^{p_i-1}}{\beta_i^{p_i-1}} |u_i(a)|^{p_i} + \frac{\gamma_i^{p_i-1}}{\sigma_i^{p_i-1}} |u_i(b)|^{p_i} \right)$$

$$\leq \Phi(u) \leq \frac{1}{\underline{m}\underline{p}} \sum_{i=1}^{n} \left(\|u_i'\|_{L^{p_i}}^{p_i} + \frac{\alpha_i^{p_i-1}}{\beta_i^{p_i-1}} |u_i(a)|^{p_i} + \frac{\gamma_i^{p_i-1}}{\sigma_i^{p_i-1}} |u_i(b)|^{p_i} \right). \quad (2.2.3)$$

Definition 2.2.1. *We say that a function* $u = (u_1, \ldots, u_n) \in X$ *is a weak solution of system* (2.1.1) *if*

$$\sum_{i=1}^{n} \left[\int_a^b J_i(u_i'(x)) v_i'(x) dx + J_i \left(\frac{\alpha_i u_i(a)}{\beta_i} \right) v_i(a) - J_i \left(\frac{-\gamma_i u_i(b)}{\sigma_i} \right) v_i(b) \right.$$

$$\left. - \lambda \int_a^b F_{u_i}(x, u_1(x), \ldots, u_n(x)) v_i'(x) dx \right] = 0$$

for any $v = (v_1, \ldots, v_n) \in X$.

The following lemma can be found in [131] (also see [23, Lemma 2.1]).

Lemma 2.2.1. ([131, Lemma 2.1]) *A weak solution to* (2.1.1) *coincides with a classical solution to* (2.1.1).

Lemma 2.2.2. ([137, Lemma 2.2]) *Let the functionals* Φ, $\Psi : X \to \mathbb{R}$ *be defined by* (2.2.1) *and* (2.2.2). *Then,*

(a) Φ *is sequentially weakly lower semicontinuous, continuous, satisfies* $\lim_{\|u\| \to \infty} \Phi(u) = \infty$, *and its derivative at the point* $u = (u_1, \ldots, u_n) \in X$ *is the functional* $\Phi'(u)$ *given by*

$$\Phi'(u)(v) = \sum_{i=1}^{n} \left[\int_a^b J_i(u_i'(x)) v_i'(x) dx \right.$$

$$\left. + J_i \left(\frac{\alpha_i u_i(a)}{\beta_i} \right) v_i(a) - J_i \left(\frac{-\gamma_i u_i(b)}{\sigma_i} \right) v_i(b) \right]$$

for every $v = (v_1, \ldots, v_n) \in X$.

(b) Ψ *is sequentially weakly upper semicontinuous and its derivative at the point* $u = (u_1, \ldots, u_n) \in X$ *is the functional* $\Psi'(u)$ *given by*

$$\Psi'(u)(v) = \int_a^b \sum_{i=1}^{n} F_{u_i}(x, u_1(x), \ldots, u_n(x)) v_i(x) dx$$

for every $v = (v_1, \ldots, v_n) \in X$.

Remark 2.2.1. By Definition 2.2.1 and Lemmas 2.2.1 and 2.2.2, we see that $u \in X$ is a critical point of $\Phi - \lambda\Psi$ if and only u is a classical solution of system (2.1.1).

Lemma 2.2.3. ([137, Lemma 2.3]) *Assume that, for $u = (u_1, \ldots, u_n) \in X$, there exists $r > 0$ such that $\Phi(u) \leq r$. Then, we have*

$$\max_{x \in [a,b]} \sum_{i=1}^{n} |u_i(x)| \leq \sum_{i=1}^{n} \left(\sqrt[p_i]{\frac{\beta_i^{p_i-1}}{\alpha_i^{p_i-1}} \overline{M}\overline{p}r} + \sqrt[p_i]{\overline{M}\overline{p}r} \, (b-a)^{\frac{1}{q_i}} \right). \quad (2.2.4)$$

Proof. Note that, for $i = 1, \ldots, n$, $u_i \in W^{1,p_i}([a,b])$. Then, by Hölder's inequality,

$$|u_i(x)| = \left| u_i(a) + \int_a^x u_i'(t)dt \right| \leq |u_i(a)| + ||u_i'||_{L^{p_i}} (b-a)^{\frac{1}{q_i}}. \quad (2.2.5)$$

Since $\Phi(u) \leq r$, from (2.2.3), we have

$$\frac{1}{\overline{M}\overline{p}} \left(\sum_{i=1}^{n} ||u_i'||_{L^{p_i}}^{p_i} + \frac{\alpha_i^{p_i-1}}{\beta_i^{p_i-1}} |u_i(a)|^{p_i} + \frac{\gamma_i^{p_i-1}}{\sigma_i^{p_i-1}} |u_i(b)|^{p_i} \right) \leq r.$$

Thus,

$$||u_i'||_{L^{p_i}} \leq \sqrt[p_i]{\overline{M}\overline{p}r} \quad \text{and} \quad |u_i(a)| \leq \sqrt[p_i]{\frac{\beta_i^{p_i-1}}{\alpha_i^{p_i-1}} \overline{M}\overline{p}r} \quad \text{for } i = 1, \ldots, n.$$

Together with (2.2.5), this implies

$$|u_i(x)| \leq \sqrt[p_i]{\frac{\beta_i^{p_i-1}}{\alpha_i^{p_i-1}} \overline{M}\overline{p}r} + \sqrt[p_i]{\overline{M}\overline{p}r} \, (b-a)^{\frac{1}{q_i}}.$$

Hence,

$$\max_{x \in [a,b]} \sum_{i=1}^{n} |u_i(x)| \leq \sum_{i=1}^{n} \left(\sqrt[p_i]{\frac{\beta_i^{p_i-1}}{\alpha_i^{p_i-1}} \overline{M}\overline{p}r} + \sqrt[p_i]{\overline{M}\overline{p}r} \, (b-a)^{\frac{1}{q_i}} \right),$$

and this completes the proof of the lemma. $\qquad \square$

Lemma 2.2.4. *Assume that*

(H) *Either $\underline{p} \geq 2$ or $\overline{p} < 2$.*

Then, $\Phi' : X \to X^*$ admits a continuous inverse on X^*.

The scalar version of Lemma 2.2.4 with Dirichlet boundary conditions was proved in [23, Corollary 2.5]. Lemma 2.2.4 for systems with Dirichlet boundary conditions was proved in [137, Lemma 2.1]. For general systems with Sturm-Liouville boundary conditions, the proof is essentially the same, and the details are left to the reader.

In the remainder of this section, we assume condition (H) holds without further mention.

In order to prove the main result in this section, for any $\vartheta > 0$, we let

$$Q(\vartheta) = \left\{ (t_1, \ldots, t_n) \in \mathbb{R}^n \ : \ \sum_{i=1}^n |t_i| \leq \vartheta \sum_{i=1}^n \left(\sqrt[p_i]{\frac{\beta_i^{p_i-1}}{\alpha_i^{p_i-1}}} + (b-a)^{\frac{1}{q_i}} \right) \right\},$$

and for any $u = (u_1, \ldots, u_n) \in X$, we let

$$\Theta_u = \sum_{i=1}^n \left(||u_i'||_{L^{p_i}}^{p_i} + \frac{\alpha_i^{p_i-1}}{\beta_i^{p_i-1}} |u_i(a)|^{p_i} + \frac{\gamma_i^{p_i-1}}{\sigma_i^{p_i-1}} |u_i(b)|^{p_i} \right). \qquad (2.2.6)$$

For any constant $\nu \geq 0$ and function $u = (u_1, \ldots, u_n) \in X$ with

$$\frac{\nu^{\underline{p}}}{\overline{M}\overline{p}} \neq \frac{\Theta_u}{m\underline{p}} \quad \text{and} \quad \frac{\nu^{\overline{p}}}{\overline{M}\overline{p}} \neq \frac{\Theta_u}{m\underline{p}},$$

define

$$a_u(\nu) = \frac{\int_a^b \sup_{(t_1,\ldots,t_n) \in Q(\nu)} F(x, t_1, \ldots, t_n) dx - \int_a^b F(x, u_1(x), \ldots, u_n(x)) dx}{\frac{\nu^{\underline{p}}}{\overline{M}\overline{p}} - \frac{1}{m\underline{p}} \Theta_u}$$

$$(2.2.7)$$

if $\nu \geq 1$, and

$$a_u(\nu) = \frac{\int_a^b \sup_{(t_1,\ldots,t_n) \in Q(\nu)} F(x, t_1, \ldots, t_n) dx - \int_a^b F(x, u_1(x), \ldots, u_n(x)) dx}{\frac{\nu^{\overline{p}}}{\overline{M}\overline{p}} - \frac{1}{m\underline{p}} \Theta_u}$$

$$(2.2.8)$$

if $\nu < 1$.

Theorem 2.2.1. *Assume that there exist two constants $\nu_1 \geq 0$ and $\nu_2 > 0$, and a function $w = (w_1, \ldots, w_n) \in X$ such that*

(B1) $\nu_1^{\underline{p}} < \Theta_w$ *and* $\Theta_w/(m\underline{p}) < \nu_2^{\underline{p}}/(\overline{M}\overline{p})$ *if $\nu_2 \geq 1$; and $\nu_1^{\overline{p}} < \Theta_w$ *and* $\Theta_w/(m\underline{p}) < \nu_2^{\overline{p}}/(\overline{M}\overline{p})$ *if $\nu_2 < 1$;*

(B2) $a_w(\nu_2) < a_w(\nu_1)$.

Then, for each $\lambda \in (1/a_w(\nu_1), \; 1/a_w(\nu_2))$, *system* (2.1.1) *has at least one nontrivial classical solution* $u_0 = (u_{01}, \ldots, u_{0n}) \in X$ *such that* $r_1 < \Phi(u_0) < r_2$, *where* Φ *is defined by* (2.2.1) *and*

$$r_1 = \begin{cases} \dfrac{\nu_1^{\underline{p}}}{\overline{M}\,\overline{p}}, & \nu_1 \geq 1, \\[2ex] \dfrac{\nu_1^{\overline{p}}}{\overline{M}\,\overline{p}}, & \nu_1 < 1, \end{cases} \qquad r_2 = \begin{cases} \dfrac{\nu_2^{\underline{p}}}{\overline{M}\,\overline{p}}, & \nu_2 \geq 1, \\[2ex] \dfrac{\nu_2^{\overline{p}}}{\overline{M}\,\overline{p}}, & \nu_2 < 1. \end{cases} \qquad (2.2.9)$$

Proof. Define the functionals $\Phi, \Psi : X \to \mathbb{R}$ by (2.2.1) and (2.2.2), respectively. Then, in view of Lemmas 2.2.2 and 2.2.2, it is easy to see that Φ and Ψ satisfy all the regularity assumptions given in Theorem 1.1.1.

From (2.2.3) and (2.2.6), it follows that $\Theta_w/(\overline{M}\,\overline{p}) \leq \Phi(w) \leq \Theta_w/(\underline{m}\,\underline{p})$. Then (B1) and (2.2.9) imply

$$r_1 < \Phi(w) < r_2. \qquad (2.2.10)$$

In view of (2.2.4), we see that

$$\Phi^{-1}(-\infty, r_2) \subseteq \left\{ (u_1, \ldots, u_n) \in X : \sum_{i=1}^{n} |u_i(x)| \right.$$
$$\left. \leq \nu_2 \sum_{i=1}^{n} \left(\sqrt[p_i]{\frac{\beta_i^{p_i-1}}{\alpha_i^{p_i-1}}} + (b-a)^{\frac{1}{q_i}} \right) \text{ for } x \in [a,b] \right\},$$

so

$$\sup_{(u_1,\ldots,u_n) \in \Phi^{-1}(-\infty, r_2)} \Psi(u) = \sup_{(u_1,\ldots,u_n) \in \Phi^{-1}(-\infty, r_2)} \int_a^b F(x, u_1(x), \ldots, u_n(x)) dx$$
$$\leq \int_a^b \sup_{(t_1,\ldots,t_n) \in Q(\nu_2)} F(x, t_1, \ldots, t_n) dx.$$

Hence, from (1.1.1) and (2.2.7)–(2.2.10), we obtain that

$$\eta(r_1, r_2) \leq \frac{\sup_{u \in \Phi^{-1}(-\infty, r_2)} \Psi(u) - \Psi(w)}{r_2 - \Phi(w)}$$
$$\leq \frac{\int_a^b \sup_{(t_1,\ldots,t_n) \in Q(\nu_2)} F(x, t_1, \ldots, t_n) dx - \Psi(w)}{r_2 - \Phi(w)}$$
$$\leq a_w(\nu_2).$$

Note that (B2) implies that $a_w(\nu_1) > 0$, and so

$$\Psi(w) \geq \int_a^b \sup_{(t_1,\ldots,t_n) \in Q(\nu_1)} F(x, t_1, \ldots, t_n) dx.$$

Then, by similar reasoning, we also have

$$\rho(r_1, r_2) \geq \frac{\Psi(w) - \sup_{u \in \Phi^{-1}(-\infty, r_1)} \Psi(u)}{\Phi(w) - r_1}$$

$$\geq \frac{\Psi(w) - \int_a^b \sup_{(t_1, \ldots, t_n) \in Q(\nu_1)} F(x, t_1, \ldots, t_n) dx}{\Phi(w) - r_1}$$

$$\geq a_w(\nu_1).$$

Thus, from (B2),

$$\eta(r_1, r_2) \leq a_w(\nu_2) < a_w(\nu_1) \leq \rho(r_1, r_2).$$

Hence, Theorem 1.1.1 implies that, for each $\lambda \in (1/a_w(\nu_1), \ 1/a_w(\nu_2))$, $\Phi(u) - \lambda \Psi(u)$ has at least one nontrivial critical point $u_0 = (u_{01}, \ldots, u_{0n})$ in X satisfying $r_1 < \Phi(u_0) < r_2$. Invoking Remark 2.2.1 completes the proof of the theorem. $\qquad \square$

Conditions (B1)–(B2) in Theorem 2.2.1 are related to the function $w \in X$. A different choice of $w \in X$ would lead to different conditions. For example, for some $\tau > 0$, taking $w(t) = (\tau(x - a), \ldots, \tau(x - a)) \in X$ or $w(t) = (\tau, \ldots, \tau) \in X$ in Theorem 2.2.1 leads, respectively, to the following two results.

Corollary 2.2.1. *Assume there exist constants* $\nu_1 \geq 0$, $\nu_2 \geq 1$, *and* $\tau > 0$ *such that*

$$\nu_1^{\underline{p}} < \sum_{i=1}^n \left(\tau^{p_i} + \frac{\gamma_i^{p_i - 1}}{\sigma_i^{p_i - 1}} (\tau(b - a))^{p_i} \right),$$

$$\frac{1}{\underline{mp}} \sum_{i=1}^n \left(\tau^{p_i} + \frac{\gamma_i^{p_i - 1}}{\sigma_i^{p_i - 1}} (\tau(b - a))^{p_i} \right) < \frac{\nu_2^{\underline{p}}}{\overline{M} \overline{p}}$$

and

$$a_{2,\tau} := \left[\int_a^b \sup_{(t_1,\dots,t_n)\in Q(\nu_2)} F(x,t_1,\dots,t_n)dx \right.$$

$$\left. - \int_a^b F(x,\tau(x-a),\dots,\tau(x-a))dx \right]$$

$$\times \left[\frac{\nu_2^{\frac{p}{2}}}{\overline{M}\overline{p}} - \frac{1}{\underline{m}\underline{p}} \sum_{i=1}^n \left(\tau^{p_i} + \frac{\gamma_i^{p_i-1}}{\sigma_i^{p_i-1}}(\tau(b-a))^{p_i} \right) \right]^{-1}$$

$$< a_{1,\tau} := \left[\int_a^b \sup_{(t_1,\dots,t_n)\in Q(\nu_1)} F(x,t_1,\dots,t_n)dx \right.$$

$$\left. - \int_a^b F(x,\tau(x-a),\dots,\tau(x-a))dx \right]$$

$$\times \left[\frac{\nu_1^{\frac{p}{2}}}{\overline{M}\overline{p}} - \frac{1}{\underline{m}\underline{p}} \sum_{i=1}^n \left(\tau^{p_i} + \frac{\gamma_i^{p_i-1}}{\sigma_i^{p_i-1}}(\tau(b-a))^{p_i} \right) \right]^{-1}.$$

Then, for each $\lambda \in (1/a_{1,\tau},\ 1/a_{2,\tau})$, system (2.1.1) has at least one nontrivial classical solution $u_0 = (u_{01},\dots,u_{0n}) \in X$ such that $r_1 < \Phi(u_0) < r_2$, where Φ is defined by (2.2.1), and r_1 and r_2 are defined by (2.2.9).

Corollary 2.2.2. *Assume that there exist three constants $\nu_1 \geq 0$, $\nu_2 \geq 1$, and $\tau > 0$ such that*

$$\nu_1^{\frac{p}{2}} < \sum_{i=1}^n \left(\frac{\alpha_i^{p_i-1}}{\beta_i^{p_i-1}} + \frac{\gamma_i^{p_i-1}}{\sigma_i^{p_i-1}} \right) \tau^{p_i}, \qquad \frac{1}{\underline{m}\underline{p}} \sum_{i=1}^n \left(\frac{\alpha_i^{p_i-1}}{\beta_i^{p_i-1}} + \frac{\gamma_i^{p_i-1}}{\sigma_i^{p_i-1}} \right) \tau^{p_i} < \frac{\nu_2^{\frac{p}{2}}}{\overline{M}\overline{p}},$$

and

$$b_{2,\tau} := \frac{\int_a^b \sup_{(t_1,\dots,t_n)\in Q(\nu_2)} F(x,t_1,\dots,t_n)dx - \int_a^b F(x,\tau,\dots,\tau)dx}{\frac{\nu_2^{\frac{p}{2}}}{\overline{M}\overline{p}} - \frac{1}{\underline{m}\underline{p}} \sum_{i=1}^n \left(\frac{\alpha_i^{p_i-1}}{\beta_i^{p_i-1}} + \frac{\gamma_i^{p_i-1}}{\sigma_i^{p_i-1}} \right) \tau^{p_i}}$$

$$< b_{1,\tau} := \frac{\int_a^b \sup_{(t_1,\dots,t_n)\in Q(\nu_1)} F(x,t_1,\dots,t_n)dx - \int_a^b F(x,\tau,\dots,\tau)dx}{\frac{\nu_1^{\frac{p}{2}}}{\overline{M}\overline{p}} - \frac{1}{\underline{m}\underline{p}} \sum_{i=1}^n \left(\frac{\alpha_i^{p_i-1}}{\beta_i^{p_i-1}} + \frac{\gamma_i^{p_i-1}}{\sigma_i^{p_i-1}} \right) \tau^{p_i}}.$$

Then, for each $\lambda \in (1/b_{1,\tau},\ 1/b_{2,\tau})$, system (2.1.1) has at least one nontrivial classical solution $u_0 = (u_{01},\dots,u_{0n}) \in X$ such that $r_1 < \Phi(u_0) < r_2$, where Φ is defined by (2.2.1), and r_1 and r_2 are defined by (2.2.9).

By choosing $\nu_1 = 0$ and $\nu_2 = \nu$, the following two results follow directly from Corollaries 2.2.1 and 2.2.2.

Corollary 2.2.3. *Assume that there exist two constants $\nu \geq 1$ and $\tau > 0$ such that*

$$\frac{1}{\underline{mp}} \sum_{i=1}^{n} \left(\tau^{p_i} + \frac{\gamma_i^{p_i-1}}{\sigma_i^{p_i-1}} (\tau(b-a))^{p_i} \right) < \frac{\nu^{\underline{p}}}{\overline{M\overline{p}}}$$

and

$$c_{2,\tau} := \left[\int_a^b \sup_{(t_1,\ldots,t_n) \in Q(\nu)} F(x,t_1,\ldots,t_n)dx \right.$$

$$\left. - \int_a^b F(x, \tau(x-a), \ldots, \tau(x-a))dx \right]$$

$$\times \left[\frac{\nu^{\underline{p}}}{\overline{M\overline{p}}} - \frac{1}{\underline{mp}} \sum_{i=1}^{n} \left(\tau^{p_i} + \frac{\gamma_i^{p_i-1}}{\sigma_i^{p_i-1}} (\tau(b-a))^{p_i} \right) \right]^{-1}$$

$$< c_{1,\tau} := \frac{\int_a^b F(x, \tau(x-a), \ldots, \tau(x-a))dx}{\frac{1}{\underline{mp}} \sum_{i=1}^{n} \left(\tau^{p_i} + \frac{\gamma_i^{p_i-1}}{\sigma_i^{p_i-1}} (\tau(b-a))^{p_i} \right)}. \tag{2.2.11}$$

Then, for each $\lambda \in (1/c_{1,\tau}, \ 1/c_{2,\tau})$, system (2.1.1) has at least one nontrivial classical solution $u_0 = (u_{01}, \ldots, u_{0n}) \in X$ such that $0 < \Phi(u_0) < \nu^{\underline{p}}/(\overline{M\overline{p}})$, where Φ is defined by (2.2.1).

Corollary 2.2.4. *Assume there exist constants $\nu \geq 1$ and $\tau > 0$ such that*

$$\frac{1}{\underline{mp}} \sum_{i=1}^{n} \left(\frac{\alpha_i^{p_i-1}}{\beta_i^{p_i-1}} + \frac{\gamma_i^{p_i-1}}{\sigma_i^{p_i-1}} \right) \tau^{p_i} < \frac{\nu^{\underline{p}}}{\overline{M\overline{p}}}$$

and

$$d_{2,\tau} := \frac{\int_a^b \sup_{(t_1,\ldots,t_n) \in Q(\nu)} F(x,t_1,\ldots,t_n)dx - \int_a^b F(x,\tau,\ldots,\tau)dx}{\frac{\nu^{\underline{p}}}{\overline{M\overline{p}}} - \frac{1}{\underline{mp}} \sum_{i=1}^{n} \left(\frac{\alpha_i^{p_i-1}}{\beta_i^{p_i-1}} + \frac{\gamma_i^{p_i-1}}{\sigma_i^{p_i-1}} \right) \tau^{p_i}}$$

$$< d_{1,\tau} := \frac{\int_a^b F(x,\tau,\ldots,\tau)dx}{\frac{1}{\underline{mp}} \sum_{i=1}^{n} \left(\frac{\alpha_i^{p_i-1}}{\beta_i^{p_i-1}} + \frac{\gamma_i^{p_i-1}}{\sigma_i^{p_i-1}} \right) \tau^{p_i}}. \tag{2.2.12}$$

Then, for each $\lambda \in (1/d_{1,\tau}, \ 1/d_{2,\tau})$, system (2.1.1) has at least one nontrivial classical solution $u_0 = (u_{01}, \ldots, u_{0n}) \in X$ such that $0 < \Phi(u_0) < \nu^{\underline{p}}/(\overline{M\overline{p}})$, where Φ is defined by (2.2.1).

Remark 2.2.2. The inequality in (2.2.11) is equivalent to

$$\int_a^b \sup_{(t_1,\ldots,t_n)\in Q(\nu)} F(x,t_1,\ldots,t_n)dx < \frac{\underline{m}p\nu^{\underline{p}} \int_a^b F(x,\tau(x-a),\ldots,\tau(x-a))dx}{\overline{M}\overline{p} \sum_{i=1}^n \left(\tau^{p_i} + \frac{\gamma_i^{p_i-1}}{\sigma_i^{p_i-1}}(\tau(b-a))^{p_i} \right)},$$

and (2.2.12) is equivalent to

$$\int_a^b \sup_{(t_1,\ldots,t_n)\in Q(\nu)} F(x,t_1,\ldots,t_n)dx < \frac{\underline{m}p\nu^{\underline{p}} \int_a^b F(x,\tau,\ldots,\tau)dx}{\overline{M}\overline{p} \sum_{i=1}^n \left(\frac{\alpha_i^{p_i-1}}{\beta_i^{p_i-1}} + \frac{\gamma_i^{p_i-1}}{\sigma_i^{p_i-1}} \right) \tau^{p_i}}.$$

These forms are somewhat easier to verify than the original ones.

Next we give a simple version of Corollary 2.2.4 with $n = 1$. Let α and γ be nonnegative constants, β and σ be positive constants, and $p > 1$. Let $f : [a,b] \times \mathbb{R} \to \mathbb{R}$ and $h : \mathbb{R} \to [0,\infty)$ be continuous with

$$0 < m := \inf_{t \in \mathbb{R}} h(t) \leq M := \inf_{t \in \mathbb{R}} h(t) < \infty.$$

Define F by

$$F(x,t) = \int_0^t f(x,s)ds \quad \text{for each } (x,t) \in [a,b] \times \mathbb{R}$$

and for any $\vartheta > 0$, let

$$W(\vartheta) = \left\{ t \in \mathbb{R} \; : \; |t| \leq \vartheta \left(\sqrt[p]{\frac{\beta^{p-1}}{\alpha^{p-1}}} + (b-a)^{\frac{1}{q}} \right) \right\},$$

where $1/p + 1/q = 1$.

The following result is an obvious consequence of Corollary 2.2.4.

Corollary 2.2.5. *Assume that there exist $\nu \geq 1$ and $\tau > 0$ such that*

$$\left(\frac{\alpha^{p-1}}{\beta^{p-1}} + \frac{\gamma^{p-1}}{\sigma^{p-1}} \right) \tau^p < \frac{m\nu^p}{M} \tag{2.2.13}$$

and

$$k_{2,\tau} := \frac{\int_a^b \sup_{t\in W(\nu)} F(x,t)dx - \int_a^b F(x,\tau)dx}{\frac{\nu^p}{Mp} - \frac{1}{mp}\left(\frac{\alpha^{p-1}}{\beta^{p-1}} + \frac{\gamma^{p-1}}{\sigma^{p-1}} \right)\tau^p}$$

$$< k_{1,\tau} := \frac{\int_a^b F(x,\tau)dx}{\frac{1}{mp}\left(\frac{\alpha^{p-1}}{\beta^{p-1}} + \frac{\gamma^{p-1}}{\sigma^{p-1}} \right)\tau^p}. \tag{2.2.14}$$

Then, for each $\lambda \in (1/k_{1,\tau},\ 1/k_{2,\tau})$, *the problem*

$$\begin{cases} -(\phi_p(u'))' = \lambda f(x,u)h(u') & in\ (a,b), \\ \alpha u(a) - \beta u'(a) = 0, \quad \gamma u(b) + \sigma u'(b) = 0, \end{cases} \qquad (2.2.15)$$

has at least one nontrivial classical solution $u \in W^{1,p}([a,b])$ *such that* $0 < \Phi_1(u) < \nu^{\underline{p}}/(\overline{M}\overline{p})$, *where*

$$\begin{aligned} \Phi_1(u) = \Bigg[&\int_a^b \int_0^{u'(x)} \int_0^s \frac{(p-1)|\delta|^{p-2}}{h(\delta)} d\delta ds dx \\ &+ \frac{\beta}{\alpha} \int_0^{\frac{\alpha u(a)}{\beta}} \int_0^s \frac{(p-1)|\delta|^{p-2}}{h(\delta)} d\delta ds \\ &+ \frac{\sigma}{\gamma} \int_0^{-\frac{\gamma u(b)}{\sigma}} \int_0^s \frac{(p-1)|\delta|^{p-2}}{h(\delta)} d\delta ds \Bigg]. \end{aligned}$$

Remark 2.2.3. It is not hard to see that (2.2.14) is equivalent to

$$\int_a^b \sup_{t \in W(\nu)} F(x,t)dx < \frac{m\nu^p \int_a^b F(x,\tau)dx}{M \left(\frac{\alpha^{p-1}}{\beta^{p-1}} + \frac{\gamma^{p-1}}{\sigma^{p-1}} \right) \tau^p}, \qquad (2.2.16)$$

which would be easier to verify in applications.

Additional results corresponding to Corollaries 2.2.1–2.2.3 can be formulated in a similar fashion. We leave this to the interested reader.

To conclude this section, we give a special case of Corollary 2.2.5.

Corollary 2.2.6. *Let* $f_1 : [a,b] \to (0,\infty)$ *and* $f_2 : \mathbb{R} \to [0,\infty)$ *be continuous functions such that* $\lim_{t \to 0^+} f_2(t)/t^{p-1} = \infty$. *Let* $h : \mathbb{R} \to [0,\infty)$ *be a continuous function such that*

$$0 < m := \inf_{t \in \mathbb{R}} h(t) \le M := \inf_{t \in \mathbb{R}} h(t) < \infty.$$

Then, for each

$$\lambda \in \left(0,\ \frac{1}{Mp \int_a^b f_1(x)dx} \sup_{\nu \ge 1} \frac{\nu^p}{\int_0^{\kappa\nu} f_2(\xi)d\xi} \right),$$

where

$$\kappa = \sqrt[p]{\frac{\beta^{p-1}}{\alpha^{p-1}}} + (b-a)^{\frac{1}{q}},$$

the problem

$$\begin{cases} -(\phi_p(u'))' = \lambda f_1(x)f_2(u)h(u') & in\ (a,b), \\ \alpha u(a) - \beta u'(a) = 0, \quad \gamma u(b) + \sigma u'(b) = 0, \end{cases} \qquad (2.2.17)$$

has at least one nontrivial classical solution $u \in W^{1,p}([a,b])$.

Proof. Let $f(x,t) = f_1(x)f_2(t)$. We see that (2.2.17) is a special case of (2.2.15). For fixed λ as in the conclusion, there exists a constant $\nu \geq 1$ such that

$$\lambda < \frac{1}{Mp \int_a^b f_1(x)dx \int_0^{\kappa\nu} f_2(\xi)d\xi} \frac{\nu^p}{} = \frac{\frac{\nu^p}{Mp}}{\int_a^b f_1(x)dx \sup_{t\in W(\nu)} \int_0^t f_2(\xi)d\xi}.$$

Since $\lim_{t\to 0+} \frac{f_2(t)}{t^{p-1}} = \infty$, we have $\lim_{t\to 0+} \frac{\int_0^t f_2(\xi)d\xi}{t^p} = \infty$. Then, for the above ν, in view of the fact that $F(x,t) = f_1(x)\int_0^t f_2(s)ds$, we see that there exists $\tau^* > 0$ such that (2.2.13) and (2.2.16) hold, and

$$\lambda < \frac{\frac{\nu^p}{Mp} - \frac{1}{mp}\left(\frac{\alpha^{p-1}}{\beta^{p-1}} + \frac{\gamma^{p-1}}{\sigma^{p-1}}\right)\tau^p}{\int_a^b f_1(x)dx \sup_{t\in W(\nu)} \int_0^t f_2(\xi)d\xi - \int_a^b f_1(x)dx \int_0^\tau f_2(s)ds} = \frac{1}{k_{1,\tau}}$$

for all $\tau \in (0, \tau^*)$. By Remark 2.2.3, (2.2.14) holds if $\tau \in (0, \tau^*)$. Note that $\lim_{\tau\to 0+} k_{2,\tau} = \infty$. The conclusion then follows from Corollary 2.2.5. □

2.3   Multiple Solutions of Sturm-Liouville Systems

Our first multiplicity result for the problem (2.1.1) in this section is contained in the following theorem. As in the previous section, we will assume that condition (H) in Lemma 2.2.4 holds here as well.

Theorem 2.3.1. *Assume there exist $\nu > 0$ and $w = (w_1, \ldots, w_n) \in X$ such that:*

(B3) $2\nu^{\underline{p}} < \Theta_w$ *if $\nu \geq 1$ and $2\nu^{\overline{p}} < \Theta_w$ if $\nu < 1$, where Θ_w is given in (2.2.6);*

(B4) $\dfrac{\int_a^b \sup_{(t_1,\ldots,t_n)\in Q(\nu)} F(x,t_1,\ldots,t_n)dx}{r} < \dfrac{2}{3}\dfrac{\int_a^b F(x,w_1(x),\ldots,w_n(x))dx}{\frac{1}{m\overline{p}}\Theta_w}$, *where*

$$r = \begin{cases} \dfrac{\nu^{\underline{p}}}{M\overline{p}}, & \nu \geq 1, \\[3mm] \dfrac{\nu^{\overline{p}}}{M\underline{p}}, & \nu < 1; \end{cases} \qquad (2.3.1)$$

(B5) $\limsup_{|t_1|\to\infty,\ldots,|t_n|\to\infty} \dfrac{\sup_{x\in[a,b]} F(x,t_1,\ldots,t_n)}{\sum_{i=1}^n |t_i|^{p_i}} \leq 0.$

Then, for each

$$\lambda \in \Lambda := \left(\frac{3}{2}\frac{\frac{1}{m\underline{p}}\Theta_w}{\int_a^b F(x,w_1(x),\ldots,w_n(x))dx},\right.$$

$$\left.\frac{r}{\int_a^b \sup_{(t_1,\ldots,t_n)\in Q(\nu)} F(x,t_1,\ldots,t_n)dx}\right),$$

the system (2.1.1) admits at least three classical solutions.

Proof. Let the functionals Φ, $\Psi : X \to \mathbb{R}$ be defined by (2.2.1) and (2.2.2), respectively. From Lemmas 2.2.2 and 2.2.4, it is easy to see that Φ and Ψ satisfy all the covering assumptions given in Theorem 1.1.3. Also, (2.2.2) and (2.2.6) implies that (1.1.3) holds. From (2.2.3) and (2.2.6), it follows that

$$\frac{1}{\overline{Mp}}\Theta_w \leq \Phi(w) \leq \frac{1}{\underline{mp}}\Theta_w. \tag{2.3.2}$$

Then, from (B3), (2.3.1), and (2.3.2), we have $\Phi(w) > 2r$. From (2.2.4), we see that

$$\Phi^{-1}(-\infty, r) \subseteq \left\{ (u_1, \dots, u_n) \in X \ : \ \sum_{i=1}^{n} |u_i(x)| \right.$$

$$\left. \leq \nu \sum_{i=1}^{n} \left(\sqrt[p_i]{\frac{\beta_i^{p_i-1}}{\alpha_i^{p_i-1}}} + (b-a)^{\frac{1}{q_i}} \right) \text{ for all } x \in [a,b] \right\}. \tag{2.3.3}$$

Then,

$$\sup_{(u_1,\dots,u_n)\in\Phi^{-1}(-\infty,r)} \Psi(u) = \sup_{(u_1,\dots,u_n)\in\Phi^{-1}(-\infty,r)} \int_a^b F(x, u_1(x), \dots, u_n(x))dx$$

$$\leq \int_a^b \sup_{(t_1,\dots,t_n)\in Q(\nu)} F(x, t_1, \dots, t_n)dx.$$

Therefore, in view of (B4) and (2.3.2), we see that

$$\frac{\sup_{u\in\Phi^{-1}(-\infty,r)} \Psi(u)}{r} = \frac{\sup_{(u_1,\dots,u_n)\in\Phi^{-1}(-\infty,r)} \int_a^b F(x, u_1(x), \dots, u_n(x))dx}{r}$$

$$\leq \frac{\int_a^b \sup_{(t_1,\dots,t_n)\in Q(\nu)} F(x, t_1, \dots, t_n)dx}{r}$$

$$< \frac{2}{3} \frac{\int_a^b F(x, w_1(x), \dots, w_n(x))dx}{\frac{1}{\underline{mp}}\Theta_w}$$

$$\leq \frac{2}{3} \frac{\Psi(w)}{\Phi(w)}.$$

Now for $\lambda > 0$, fix $\eta > 0$ such that

$$\frac{1}{\overline{M}\,\overline{p}\lambda(b-a)} \min\left\{ \frac{\alpha_i^{p_i-1}}{2^{p_i-1}\beta_i^{p_i-1}}, \ \frac{1}{2^{p_i-1}(b-a)^{\frac{p_i}{q_i}}} \ : \ i=1,\dots,n \right\} > \eta. \tag{2.3.4}$$

From (B5), there exists a function $g \in L^1([a,b])$ such that

$$F(x, t_1, \dots, t_n) \leq \eta \sum_{i=1}^{n} |t_i|^{p_i} + g(x) \text{ for all } x \in [a,b] \text{ and all } (t_1, \dots, t_n) \in \mathbb{R}^n.$$

Fix $(u_1, \ldots, u_n) \in X$. Then, for all $x \in [a, b]$,

$$F(x, u_1(x), \ldots, u_n(x)) \leq \eta \sum_{i=1}^{n} |u_i(x)|^{p_i} + g(x)$$

$$\leq \eta \sum_{i=1}^{n} \left(|u_i(a)| + ||u_i'||_{L^{p_i}} (b - a)^{\frac{1}{q_i}} \right)^{p_i} + g(x).$$

From this and (2.2.3), we have

$$\Phi(u) - \lambda\Psi(u)$$

$$\geq \frac{1}{\overline{M}\, \overline{p}} \sum_{i=1}^{n} \left(||u_i'||_{L^{p_i}}^{p_i} + \frac{\alpha_i^{p_i-1}}{\beta_i^{p_i-1}} |u_i(a)|^{p_i} + \frac{\gamma_i^{p_i-1}}{\sigma_i^{p_i-1}} |u_i(b)|^{p_i} \right)$$

$$- \lambda(b - a)\eta \sum_{i=1}^{n} \left(|u_i(a)| + ||u_i'||_{L^{p_i}} (b - a)^{\frac{1}{q_i}} \right)^{p_i} - \lambda ||g||_{L^1}$$

$$\geq \frac{1}{\overline{M}\, \overline{p}} \sum_{i=1}^{n} \left(||u_i'||_{L^{p_i}}^{p_i} + \frac{\alpha_i^{p_i-1}}{\beta_i^{p_i-1}} |u_i(a)|^{p_i} + \frac{\gamma_i^{p_i-1}}{\sigma_i^{p_i-1}} |u_i(b)|^{p_i} \right)$$

$$- \lambda(b - a)\eta \sum_{i=1}^{n} 2^{p_i-1} \left(|u_i(a)|^{p_i} + ||u_i'||_{L^{p_i}}^{p_i} (b - a)^{\frac{p_i}{q_i}} \right) - \lambda ||g||_{L^1}$$

$$= \sum_{i=1}^{n} \left(\frac{1}{\overline{M}\, \overline{p}} - \lambda\eta 2^{p_i-1} (b - a)^{\frac{p_i}{q_i}+1} \right) ||u_i'||_{L^{p_i}}^{p_i}$$

$$+ \sum_{i=1}^{n} \left(\frac{1}{\overline{M}\, \overline{p}} \frac{\alpha_i^{p_i-1}}{\beta_i^{p_i-1}} - \lambda(b - a)\eta 2^{p_i-1} \right) |u_i(a)|^{p_i}$$

$$+ \frac{1}{\overline{M}\, \overline{p}} \sum_{i=1}^{n} \left(\frac{\gamma_i^{p_i-1}}{\sigma_i^{p_i-1}} |u_i(b)|^{p_i} \right) - \lambda ||g||_{L^1}.$$

Condition (2.3.4) then implies

$$\lim_{||u|| \to \infty} (\Phi(u) - \lambda\Psi(u)) = \infty,$$

that is, the functional $\Phi - \lambda\Psi$ is coercive for every $\lambda > 0$. Therefore, conditions $(A1)$ and $(A2)$ in Theorem 1.1.3 are satisfied, so by Remark 2.2.1, the system (2.1.1) admits at least three distinct classical solutions in X. $\qquad\square$

Our second result on the existence of multiple solutions is as follows.

Theorem 2.3.2. *Let* $F : [a, b] \times \mathbb{R}^n \to \mathbb{R}$ *satisfy the condition* $F(x, t_1, \ldots, t_n) \geq 0$ *for all* $(x, t_1, \ldots, t_n) \in [a, b] \times [0, \infty)^n$. *Assume that there exist a function* $w = (w_1, \ldots, w_n) \in X$ *and two positive constants* ν_1 *and* ν_2 *such that:*

(B6) $2\nu_1^{\overline{p}} < \Theta_w$ *if* $\nu_1 < 1$, $2\nu_1^{\underline{p}} < \Theta_w$ *if* $\nu_1 \geq 1$, $\frac{1}{m\underline{p}}\Theta_w < \frac{\nu_2^{\underline{p}}}{2M\overline{p}}$ *if* $\nu_2 < 1$,

and $\frac{1}{m\underline{p}}\Theta_w < \frac{\nu_2^{\underline{p}}}{2M\overline{p}}$ *if* $\nu_2 \geq 1$;

(B7)

$$\max\left\{\frac{\int_a^b \sup_{(t_1,\ldots,t_n)\in Q(\nu_1)} F(x,t_1,\ldots,t_n)dx}{r_1},\right.$$

$$\left.\frac{2\int_a^b \sup_{(t_1,\ldots,t_n)\in Q(\nu_2)} F(x,t_1,\ldots,t_n)dx}{r_2}\right\}$$

$$< \frac{2}{3}\frac{\int_a^b F(x,w_1(x),\ldots,w_n(x))dx}{\frac{1}{m\underline{p}}\Theta_w},$$

where

$$r_1 = \begin{cases} \frac{\nu_1^{\underline{p}}}{M\overline{p}}, & \nu_1 \geq 1, \\ \frac{\nu_1^{\overline{p}}}{M\overline{p}}, & \nu_1 < 1, \end{cases} \qquad r_2 = \begin{cases} \frac{\nu_2^{\underline{p}}}{M\overline{p}}, & \nu_2 \geq 1, \\ \frac{\nu_2^{\overline{p}}}{M\overline{p}}, & \nu_2 < 1. \end{cases} \qquad (2.3.5)$$

Then, for each

$$\lambda \in \left(\frac{3}{2}\frac{\frac{1}{m\underline{p}}\Theta_w}{\int_a^b F(x,w_1(x),\ldots,w_n(x))dx},\right.$$

$$\min\left\{\frac{r_1}{\int_a^b \sup_{(t_1,\ldots,t_n)\in Q(\nu_1)} F(x,t_1,\ldots,t_n)dx},\right.$$

$$\left.\left.\frac{\frac{r_2}{2}}{\int_a^b \sup_{(t_1,\ldots,t_n)\in Q(\nu_2)} F(x,t_1,\ldots,t_n)dx}\right\}\right)$$

the system (2.1.1) *admits at least three non-negative classical solutions* $v^j = (v_1^j, v_2^j, \ldots, v_n^j)$, $j = 1,2,3$, *such that*

$$\sum_{i=1}^n |v_i^j(x)| \leq \nu_2 \sum_{i=1}^n \left(\sqrt[p_i]{\frac{\beta_i^{p_i-1}}{\alpha_i^{p_i-1}}} + (b-a)^{\frac{1}{q_i}}\right) \quad \text{for each } x \in [a,b], \ j = 1,2,3.$$

Proof. We will use Theorem 1.1.4 to prove this theorem, so let Φ and Ψ be defined by (2.2.1) and (2.2.2), respectively. Clearly, Φ and Ψ satisfy condition (i) of Theorem 1.1.4. To show that the functional $\Phi - \lambda\Psi$ satisfies (ii), let $u^* = (u_1^*,\ldots,u_n^*)$ and $u^{**} = (u_1^{**},\ldots,u_n^{**})$ be two local minima for $\Phi - \lambda\Psi$. Then u^* and u^{**} are critical points for $\Phi - \lambda\Psi$, and so, they are classical solutions of the system (2.1.1). Since $F(x,t_1,\ldots,t_n) \geq 0$ for all $(x, t_1, \ldots, t_n) \in [a, b] \times [0, \infty)^n$, from the Maximum Principle (see, for instance, [65] or [128]), we have $u_i^*(x) \geq 0$ and $u_i^{**}(x) \geq 0$ for every $x \in \mathbb{R}$ for $1 \leq i \leq n$. It then follows that $su_i^* + (1 - s)u_i^{**} \geq 0$ for all $s \in [0, 1]$ for $1 \leq i \leq n$, and that $F(x, su^* + (1 - s)u^{**}) \geq 0$ for $1 \leq i \leq n$. Consequently, $\Psi(su^* + (1 - s)u^{**}) \geq 0$ for all $s \in [0, 1]$. Then, by condition (B6), (2.3.2), and (2.3.5), we have $2r_1 < \Phi(w) < \frac{r_2}{2}$. In view of (2.3.3), we see that

$$\sup_{(u_1, \ldots, u_n) \in \Phi^{-1}(-\infty, r_1)} \Psi(u) = \sup_{(u_1, \ldots, u_n) \in \Phi^{-1}(-\infty, r_1)} \int_a^b F(x, u_1(x), \ldots, u_n(x)) dx$$

$$\leq \int_a^b \sup_{(t_1, \ldots, t_n) \in Q(\nu_1)} F(x, t_1, \ldots, t_n) dx.$$

From (B7) and (2.3.2), we have

$$\frac{\sup_{u \in \Phi^{-1}(-\infty, r_1)} \Psi(u)}{r_1} = \frac{\sup_{(u_1, \ldots, u_n) \in \Phi^{-1}(-\infty, r_1)} \int_a^b F(x, u_1(x), \ldots, u_n(x)) dx}{r_1}$$

$$\leq \frac{\int_a^b \sup_{(t_1, \ldots, t_n) \in Q(\nu_1)} F(x, t_1, \ldots, t_n) dx}{r_1}$$

$$< \frac{2}{3} \frac{\int_a^b F(x, w_1(x), \ldots, w_n(x)) dx}{\frac{1}{mp} \Theta_w}$$

$$\leq \frac{2}{3} \frac{\Psi(w)}{\Phi(w)}.$$

Again from (B7) and (2.3.2), we obtain

$$2 \frac{\sup_{u \in \Phi^{-1}(-\infty, r_2)} \Psi(u)}{r_2}$$

$$= \frac{2 \sup_{(u_1, \ldots, u_n) \in \Phi^{-1}(-\infty, r_2)} \int_a^b F(x, u_1(x), \ldots, u_n(x)) dx}{r_2}$$

$$\leq \frac{2 \int_a^b \sup_{(t_1, \ldots, t_n) \in Q(\nu_2)} F(x, t_1, \ldots, t_n) dx}{r_1}$$

$$< \frac{2}{3} \frac{\int_a^b F(x, w_1(x), \ldots, w_n(x)) dx}{\frac{1}{mp} \Theta_w}$$

$$\leq \frac{2}{3} \frac{\Psi(w)}{\Phi(w)}.$$

Thus, conditions $(A3)$ and $(A4)$ in Theorem 1.1.4 are satisfied, so by Remark 2.2.1, the conclusion of the theorem follows. $\qquad\square$

One limitation of Theorems 2.3.1 and 2.3.2 is that they require knowledge of a test function w satisfying (B3)–(B5) or (B6)–(B7). The following two corollaries give sufficient conditions for applying Theorems 2.3.1 and 2.3.2 without this restriction. We accomplish this by $w(t) = (\tau(x-a), \ldots, \tau(x-a)) \in X$ for some $\tau > 0$.

Corollary 2.3.1. *In addition to condition (B5), assume that there exist two positive constants ν and τ such that*

$$2\nu^{\underline{p}} < \sum_{i=1}^{n} \left(\tau^{p_i} + \frac{\gamma_i^{p_i-1}}{\sigma_i^{p_i-1}} (\tau(b-a))^{p_i} \right) \quad \text{if } \nu \geq 1,$$

$$2\nu^{\overline{p}} < \sum_{i=1}^{n} \left(\tau^{p_i} + \frac{\gamma_i^{p_i-1}}{\sigma_i^{p_i-1}} (\tau(b-a))^{p_i} \right) \quad \text{if } \nu < 1,$$

and

$$\frac{\int_a^b \sup_{(t_1,\ldots,t_n) \in Q(\nu)} F(x,t_1,\ldots,t_n) dx}{r} < \frac{2 \int_a^b F(x, \tau(x-a), \ldots, \tau(x-a)) dx}{\frac{3}{m\underline{p}} \sum_{i=1}^{n} \left(\tau^{p_i} + \frac{\gamma_i^{p_i-1}}{\sigma_i^{p_i-1}} (\tau(b-a))^{p_i} \right)},$$

where r is given as in (2.3.1). Then, for each

$$\lambda \in \Lambda := \left(\frac{3}{2} \frac{\frac{1}{m\underline{p}} \sum_{i=1}^{n} \left(\tau^{p_i} + \frac{\gamma_i^{p_i-1}}{\sigma_i^{p_i-1}} (\tau(b-a))^{p_i} \right)}{\int_a^b F(x, \tau(x-a), \ldots, \tau(x-a)) dx}, \right.$$

$$\left. \frac{r}{\int_a^b \sup_{(t_1,\ldots,t_n) \in Q(\nu)} F(x,t_1,\ldots,t_n) dx} \right),$$

the system (2.1.1) admits at least three classical solutions.

Corollary 2.3.2. *Let $F : [a,b] \times \mathbb{R}^n \to \mathbb{R}$ satisfy the condition $F(x,t_1,\ldots,t_n) \geq 0$ for all $(x,t_1,\ldots,t_n) \in [a,b] \times [0,\infty)^n$. Assume that there exist three positive constants ν_1, ν_2, and τ such that*

$$2\nu_1^{\overline{p}} < \sum_{i=1}^{n} \left(\tau^{p_i} + \frac{\gamma_i^{p_i-1}}{\sigma_i^{p_i-1}} (\tau(b-a))^{p_i} \right) \quad \text{if } \nu_1 < 1,$$

$$2\nu_1^{\underline{p}} < \sum_{i=1}^{n} \left(\tau^{p_i} + \frac{\gamma_i^{p_i-1}}{\sigma_i^{p_i-1}} (\tau(b-a))^{p_i} \right) \quad \text{if } \nu_1 \geq 1,$$

$$\frac{1}{m\underline{p}} \sum_{i=1}^{n} \left(\tau^{p_i} + \frac{\gamma_i^{p_i-1}}{\sigma_i^{p_i-1}} (\tau(b-a))^{p_i} \right) < \frac{\nu_2^{\overline{p}}}{2\overline{M}\overline{p}} \quad \text{if } \nu_2 < 1,$$

$$2 \frac{1}{mp} \sum_{i=1}^{n} \left(\tau^{p_i} + \frac{\gamma_i^{p_i-1}}{\sigma_i^{p_i-1}} (\tau(b-a))^{p_i} \right) < \frac{\nu_2^{\frac{p}{2}}}{2M\overline{p}} \quad if \ \nu_2 \geq 1,$$

and

$$\max \left\{ \frac{\int_a^b \sup_{(t_1,\ldots,t_n) \in Q(\nu_1)} F(x, t_1, \ldots, t_n) dx}{r_1}, \right.$$

$$\left. \frac{2 \int_a^b \sup_{(t_1,\ldots,t_n) \in Q(\nu_2)} F(x, t_1, \ldots, t_n) dx}{r_2} \right\}$$

$$< \frac{2}{3} \frac{\int_a^b F(x, \tau(x-a), \ldots, \tau(x-a)) dx}{\frac{1}{mp} \sum_{i=1}^n \left(\tau^{p_i} + \frac{\gamma_i^{p_i-1}}{\sigma_i^{p_i-1}} (\tau(b-a))^{p_i} \right)},$$

where r_1 and r_2 are given in (2.3.5). Then, for each

$$\lambda \in \left(\frac{3}{2} \frac{\frac{1}{mp} \sum_{i=1}^n \left(\tau^{p_i} + \frac{\gamma_i^{p_i-1}}{\sigma_i^{p_i-1}} (\tau(b-a))^{p_i} \right)}{\int_a^b F(x, \tau(x-a), \ldots, \tau(x-a)) dx}, \right.$$

$$\left. \min \left\{ \frac{r_1}{\int_a^b \sup_{(t_1,\ldots,t_n) \in Q(\nu_1)} F(x, t_1, \ldots, t_n) dx}, \right. \right.$$

$$\left. \left. \frac{\frac{r_2}{2}}{\int_a^b \sup_{(t_1,\ldots,t_n) \in Q(\nu_2)} F(x, t_1, \ldots, t_n) dx} \right\} \right),$$

the system (2.1.1) admits at least three non-negative classical solutions $v^j = (v_1^j, v_2^j, \ldots, v_n^j)$, $j = 1, 2, 3$, such that

$$\sum_{i=1}^n |v_i^j(x)| \leq \nu_2 \sum_{i=1}^n \left(\sqrt[p_i]{\frac{\beta_i^{p_i-1}}{\alpha_i^{p_i-1}}} + (b-a)^{\frac{1}{q_i}} \right) \quad for \ each \ x \in [a, b], \ j = 1, 2, 3.$$

If $n = 1$, the above two corollaries yield the following special cases. Let α, γ be two non-negative constants, β, σ be two positive constants, and $p > 1$. Let $f : [0, 1] \times \mathbb{R} \to \mathbb{R}$ be a continuous function and $h : \mathbb{R} \to [0, \infty)$ be a bounded continuous function with

$$0 < m := \inf_{t \in \mathbb{R}} h(t) \leq M := \inf_{t \in \mathbb{R}} h(t) < \infty.$$

Let the function F be defined by

$$F(x, t) = \int_0^t f(x, s) ds \quad for \ each \ (x, t) \in [a, b] \times \mathbb{R}.$$

For any $\vartheta > 0$, we set

$$W(\vartheta) = \left\{ t \in \mathbb{R} \ : \ |t| \leq \vartheta \left(\sqrt[p]{\frac{\beta^{p-1}}{\alpha^{p-1}}} + (b-a)^{\frac{1}{q}} \right) \right\},$$

where $1/p + 1/q = 1$.

Corollary 2.3.3. *Assume there exist two positive constants ν and τ such that*

$$2\nu^p < \left(\tau^p + \frac{\gamma^{p-1}}{\sigma^{p-1}} (\tau(b-a))^p \right),$$

$$\frac{\int_a^b \sup_{t \in W(\nu)} F(x,t) dx}{\frac{\nu^p}{Mp}} < \frac{2}{3} \frac{\int_a^b F(x, \tau(x-a)) dx}{\frac{1}{mp} \left(\tau^p + \frac{\gamma^{p-1}}{\sigma^{p-1}} (\tau(b-a))^p \right)},$$

and

$$\limsup_{|t| \to \infty} \frac{\max_{x \in [a,b]} f(x,t)}{|t|^{p-1}} \leq 0.$$

Then, for each

$$\lambda \in \Lambda := \left(\frac{3}{2} \frac{\frac{1}{mp} \left(\tau^p + \frac{\gamma^{p-1}}{\sigma^{p-1}} (\tau(b-a))^p \right)}{\int_a^b F(x, \tau(x-a)) dx}, \ \frac{\frac{\nu^p}{Mp}}{\int_a^b \sup_{t \in W(\nu)} F(x,t) dx} \right),$$

the problem

$$\begin{cases} -(\phi_p(u'))' = \lambda f(x,u) h(u') & in \ (a,b), \\ \alpha u(a) - \beta u'(a) = 0, \quad \gamma u(b) + \sigma u'(b) = 0, \end{cases} \qquad (2.3.6)$$

admits at least three classical solutions.

Corollary 2.3.4. *Let f satisfy the condition $f(x,t) \geq 0$ for all $(x,t) \in [a,b] \times [0,\infty)$ and assume that there exist three positive constants ν_1, ν_2, and τ such that*

$$2\nu_1^p < \left(\tau^p + \frac{\gamma^{p-1}}{\sigma^{p-1}} (\tau(b-a))^p \right), \quad \frac{1}{mp} \left(\tau^p + \frac{\gamma^{p-1}}{\sigma^{p-1}} (\tau(b-a))^p \right) < \frac{\nu_2^p}{2Mp},$$

and

$$\max \left\{ \frac{\int_a^b \sup_{t \in W(\nu_1)} F(x,t) dx}{\frac{\nu_1^p}{Mp}}, \ \frac{2 \int_a^b \sup_{t \in W(\nu_2)} F(x,t) dx}{\frac{\nu_2^p}{Mp}} \right\}$$

$$< \frac{2}{3} \frac{\int_a^b F(x, \tau(x-a)) dx}{\frac{1}{mp} \left(\tau^p + \frac{\gamma^{p-1}}{\sigma^{p-1}} (\tau(b-a))^p \right)}.$$

Then, for each

$$\lambda \in \left(\frac{3}{2} \frac{\frac{1}{mp}\left(\tau^p + \frac{\gamma^{p-1}}{\sigma^{p-1}}(\tau(b-a))^p\right)}{\int_a^b F(x, \tau(x-a))dx}, \right.$$

$$\left. \min\left\{ \frac{\frac{\nu_1^p}{Mp}}{\int_a^b \sup_{t \in W(\nu_1)} F(x,t)dx}, \frac{\frac{\nu_2^p}{2Mp}}{\int_a^b \sup_{t \in W(\nu_2)} F(x,t)dx} \right\} \right),$$

the problem (2.3.6) admits at least three non-negative classical solutions $v^j = (v_1^j, v_2^j, \ldots, v_n^j)$, $j = 1, 2, 3$, such that

$$|v^j(x)| \le \nu_2 \left(\sqrt[p]{\frac{\beta^{p-1}}{\alpha^{p-1}}} + (b-a)^{\frac{1}{q}} \right) \text{ for each } x \in [a,b], \ j = 1, 2, 3.$$

Using the above corollaries, we can also prove the following results.

Corollary 2.3.5. *Let $g_i : \mathbb{R}^n \to \mathbb{R}$ for $1 \le i \le n$ be n continuous functions such that the differential 1-form $w := \sum_{i=1}^n g_i(\xi_1, \ldots, \xi_n)d\xi_i$ is integrable and let F be a primitive of w such that $F(0, \ldots, 0) = 0$. Assume that*

$$\liminf_{(\xi_1, \ldots, \xi_n) \to (0, \ldots, 0)} \frac{F(\xi_1, \ldots, \xi_n)}{\sum_{i=1}^n |\xi_i|^{p_i}} = \limsup_{|\xi_1| \to \infty, \ldots, |\xi_n| \to \infty} \frac{F(\xi_1, \ldots, \xi_n)}{\sum_{i=1}^n |\xi_i|^{p_i}} = 0.$$

Then, there is $\lambda^* > 0$ such that for each $\lambda > \lambda^*$, the problem

$$\begin{cases} -(\phi_{p_i}(u_i'))' = \lambda g_i(u_1, \ldots, u_n)h_i(u_i'), & \text{in } (a,b), \\ \alpha_i u_i(a) - \beta_i u_i'(a) = 0, \quad \gamma_i u_i(b) + \sigma_i u_i'(b) = 0, \end{cases} \quad i = 1, \ldots, n,$$

admits at least three classical solutions.

Corollary 2.3.6. *Let $1 < p \le q$ and $F : \mathbb{R}^2 \to \mathbb{R}$ be a C^1 function satisfying the condition $F(t_1, t_2) \ge 0$ for all $(t_1, t_2) \in [0, \infty)^2$. Let $h_1, h_2 : \mathbb{R} \to [0, \infty)$ be bounded continuous functions with $\inf_{t \in \mathbb{R}} h_i(t) > 0$ for $i = 1, 2$, and set $m = \min\{\inf_{t \in \mathbb{R}} h_i(t) : i = 1, 2\}$ and $M = \max\{\sup_{t \in \mathbb{R}} h_i(t) : i = 1, 2\}$. Assume that there exist two positive constants $\nu > 1$ and τ such that $\tau^p + \tau^q < \frac{mp}{8Mq}\nu^p$,*

$$\lim_{x \to 0^+} \frac{\max_{|t_1| + |t_2| \le 4x} F(t_1, t_2)}{x^q} = 0,$$

and

$$\frac{\max_{|t_1| + |t_2| \le 4\nu} F(t_1, t_2)}{\nu^p} < \frac{1}{6} \frac{mp}{Mq(\tau^p + \tau^q)} \int_0^1 F(\tau x, \tau x)dx.$$

Then, for every $\lambda \in \left(\dfrac{\frac{3}{mp}(\tau^p + \tau^q)}{\int_0^1 F(\tau x, \tau x)dx}, \dfrac{\frac{\nu^p}{2Mq}}{\max_{|t_1|+|t_2| \le 4\nu} F(t_1, t_2)} \right)$, *the problem*

$$\begin{cases} -(\phi_p(u_1'))' = \lambda F_{u_1}(u_1, u_2)h_1(u_1'), & in \ (0,1), \\ -(\phi_q(u_2'))' = \lambda F_{u_2}(u_1, u_2)h_2(u_2'), & in \ (0,1), \\ u_i(0) - u_i'(0) = 0, \quad u_i(1) + u_i'(1) = 0, & i = 1,2, \end{cases}$$

admits at least three non-negative classical solutions $v^j = (v_1^j, v_2^j)$, $j = 1, 2, 3$, *such that*

$$|v_1^j(x)| + |v_2^j(x)| \le 4\nu \ \text{for each } x \in [0,1], \ j = 1,2,3.$$

2.4 Infinitely Many Solutions of Sturm-Liouville Systems

Our first result in this section is based on part (b) of Theorem 1.1.5.

Theorem 2.4.1. *Assume that*

$$\liminf_{\xi \to \infty} \frac{\int_a^b \sup_{(t_1, \dots, t_n) \in Q(\xi)} F(x, t_1, \dots, t_n)dx}{\xi^{\underline{p}}}$$

$$< \frac{m \ p}{\overline{Mp}} \limsup_{t \to \infty} \frac{\int_a^b F(x, t, \dots, t)dx}{\sum_{i=1}^n \left(\frac{\alpha_i^{p_i - 1}}{\beta_i^{p_i - 1}} + \frac{\gamma_i^{p_i - 1}}{\sigma_i^{p_i - 1}} \right) t^{p_i}}. \tag{2.4.1}$$

Then, for each $\lambda \in \Lambda$, *system* (2.1.1) *has an unbounded sequence of classical solutions in* X, *where*

$$\Lambda = \left(\frac{1}{\underline{m} \ \underline{p} \limsup_{t \to \infty} \frac{\int_a^b F(x, t, \dots, t)dx}{\sum_{i=1}^n \left(\frac{\alpha_i^{p_i - 1}}{\beta_i^{p_i - 1}} + \frac{\gamma_i^{p_i - 1}}{\sigma_i^{p_i - 1}} \right) t^{p_i}}}, \right.$$

$$\left. \frac{1}{\overline{Mp} \liminf_{\xi \to \infty} \frac{\int_a^b \sup_{(t_1, \dots, t_n) \in Q(\xi)} F(x, t_1, \dots, t_n)dx}{\xi^{\underline{p}}}} \right). \tag{2.4.2}$$

Proof. As before, we define the functionals Φ, $\Psi : X \to \mathbb{R}$ by (2.2.1) and (2.2.2) and is easy to see from Lemma 2.2.2 that the regularity conditions in Theorem 1.1.5 are satisfied.

Let $\{\xi_k\}$ be a sequence of positive numbers such that $\xi_k > 1$ for $k \in \mathbb{N}$, $\xi_k \to \infty$ as $k \to \infty$, and

$$\lim_{k \to \infty} \frac{\int_a^b \sup_{(t_1, \dots, t_n) \in Q(\xi_k)} F(x, t_1, \dots, t_n)dx}{\xi_k^{\underline{p}}}$$

$$= \liminf_{\xi \to \infty} \frac{\int_a^b \sup_{(t_1, \dots, t_n) \in Q(\xi)} F(x, t_1, \dots, t_n)dx}{\xi^{\underline{p}}}. \tag{2.4.3}$$

Let $r_k = \frac{\xi_k^p}{M\overline{p}}$ for $k \in \mathbb{N}$; then, $\overline{M}\overline{p}r_k = \xi_k^p > 1$. Lemma 2.2.3 implies

$$\max_{x \in [a,b]} \sum_{i=1}^n |u_i(x)| \leq (\overline{M}\overline{p}r_k)^{\frac{1}{p}} \sum_{i=1}^n \left(\sqrt[p_i]{\frac{\beta_i^{p_i-1}}{\alpha_i^{p_i-1}}} + (b-a)^{\frac{1}{q_i}} \right)$$

$$= \xi_k \sum_{i=1}^n \left(\sqrt[p_i]{\frac{\beta_i^{p_i-1}}{\alpha_i^{p_i-1}}} + (b-a)^{\frac{1}{q_i}} \right) \qquad (2.4.4)$$

for $u = (u_1, \ldots, u_n) \in X$ with $\Phi(u) \leq r_k$. Now $\Phi(0,\ldots,0) = \Psi(0,\ldots,0) = 0$, so

$$0 \leq \varphi(r_k) = \inf_{u \in \Phi^{-1}(-\infty, r_k)} \frac{\sup_{v \in \Phi^{-1}(-\infty, r_k)} \Psi(v) - \Psi(u)}{r_k - \Phi(u)}$$

$$\leq \frac{\sup_{v \in \Phi^{-1}(-\infty, r_k)} \Psi(v)}{r_k}$$

$$\leq \overline{M}\overline{p} \frac{\int_a^b \sup_{(t_1,\ldots,t_n) \in Q(\xi_k)} F(x, t_1, \ldots, t_n) dx}{\xi_k^p}.$$

Then, in view of (2.4.1) and (2.4.2), we see that

$$\gamma \leq \liminf_{k \to \infty} \varphi(r_k) \leq \overline{M}\overline{p} \lim_{k \to \infty} \frac{\int_a^b \sup_{(t_1,\ldots,t_n) \in Q(\xi_k)} F(x, t_1, \ldots, t_n) dx}{\xi_k^p} < \infty. \qquad (2.4.5)$$

From (2.4.2) and (2.4.5), we have $\Lambda \subseteq (0, 1/\gamma)$. Let $\lambda \in \Lambda$ be fixed. By Theorem 1.1.5 (b), one of the following alternatives holds:

(b₁) either $I_\lambda := \Phi - \lambda\Psi$ has a global minimum, or
(b₂) there exists a sequence $\{u_k\} = \{(u_{1k}, \ldots, u_{nk})\}$ of critical points of I_λ such that

$$\lim_{k \to \infty} \Phi(u_k) = \infty. \qquad (2.4.6)$$

To show that (b₁) does not hold, note that $\lambda \in \Lambda$, so (2.4.2) implies

$$\frac{1}{\lambda} < \underline{m}\,\underline{p} \limsup_{t \to \infty} \frac{\int_a^b F(x, t, \ldots, t) dx}{\sum_{i=1}^n \left(\frac{\alpha_i^{p_i-1}}{\beta_i^{p_i-1}} + \frac{\gamma_i^{p_i-1}}{\sigma_i^{p_i-1}} \right) t^{p_i}}.$$

Then, there exist a real sequence $\{d_k\}$ of positive numbers, $\nu > 0$, and $N \in \mathbb{N}$ such that $d_k \to \infty$ as $k \to \infty$ and

$$\frac{1}{\lambda} < \nu < \frac{\int_a^b F(x, d_k, \ldots, d_k) dx}{\frac{1}{\underline{m}\,\underline{p}} \sum_{i=1}^n \left(\frac{\alpha_i^{p_i-1}}{\beta_i^{p_i-1}} + \frac{\gamma_i^{p_i-1}}{\sigma_i^{p_i-1}} \right) d_k^{p_i}} \qquad (2.4.7)$$

for each $k \geq N$. For $k \in \mathbb{N}$, let $w_k = (d_k, \ldots, d_k) \in X$. Then, from (2.2.2) and (2.2.3),

$$\Phi(w_k) \leq \frac{1}{\underline{m}\,\underline{p}} \sum_{i=1}^{n} \left(\frac{\alpha_i^{p_i-1}}{\beta_i^{p_i-1}} + \frac{\gamma_i^{p_i-1}}{\sigma_i^{p_i-1}} \right) d_k^{p_i} \tag{2.4.8}$$

and

$$\Psi(w_k) = \int_a^b F(x, d_k, \ldots, d_k)dx. \tag{2.4.9}$$

From (2.4.7)–(2.4.9), it follows that

$$I_\lambda(w_k) \leq \frac{1}{\underline{m}\,\underline{p}} \sum_{i=1}^{n} \left(\frac{\alpha_i^{p_i-1}}{\beta_i^{p_i-1}} + \frac{\gamma_i^{p_i-1}}{\sigma_i^{p_i-1}} \right) t^{p_i} - \lambda \int_a^b F(x, d_k, \ldots, d_k)dx$$

$$< (1 - \lambda\nu)\frac{1}{\underline{m}\,\underline{p}} \sum_{i=1}^{n} \left(\frac{\alpha_i^{p_i-1}}{\beta_i^{p_i-1}} + \frac{\gamma_i^{p_i-1}}{\sigma_i^{p_i-1}} \right) d_k^{p_i}$$

for every $k \geq N$. From the fact that $\lambda\nu > 1$ (by (2.4.7)) and $d_k \to \infty$, we see that $\lim_{k\to\infty} I_\lambda(w_k) = -\infty$, i.e., the functional I_λ is unbounded from below. This shows that alternative (b$_1$) in Theorem 1.1.5 (b) does not hold. Hence, there exists a sequence $\{u_k\} = \{(u_{1k}, \ldots, u_{nk})\}$ of critical points of I_λ such that (2.4.6) holds. By (2.2.3), (2.4.6), and the definition of the norm $\|\cdot\|$, it is easy to see that $\|u_k\| \to \infty$. Finally, taking Remark 2.2.1 into account completes the proof of the theorem. $\qquad\square$

The following result is a special case of Theorem 2.4.1 with $F(t, t_1, \ldots, t_n) \equiv F(t_1, \ldots, t_n)$ and $\alpha_i = \beta_i = \gamma_i = \sigma_i = 1$, $i = 1, \ldots, n$.

Corollary 2.4.1. *Assume that*

$$\liminf_{\xi\to\infty} \frac{\max_{(t_1,\ldots,t_n)\in Q_1(\xi)} F(t_1, \ldots, t_n)}{\xi^{\underline{p}}} < \frac{\underline{m}\,\underline{p}}{2\overline{M}\overline{p}} \limsup_{t\to\infty} \frac{F(t, \ldots, t)}{\sum_{i=1}^{n} t^{p_i}}. \tag{2.4.10}$$

Then, for each $\lambda \in \Lambda_1$, the system

$$\begin{cases} -(\phi_{p_i}(u_i'))' = \lambda F_{u_i}(u_1, \ldots, u_n)h_i(u_i') & \text{in } (a, b), \\ u_i(a) - u_i'(a) = 0, \quad u_i(b) + u_i'(b) = 0, \end{cases} \quad i = 1, \ldots, n. \tag{2.4.11}$$

has an unbounded sequence of classical solutions in X, where

$$Q_1(\xi) = \left\{ (t_1, \ldots, t_n) \in \mathbb{R}^n \; : \; \sum_{i=1}^{n} |t_i| \leq \xi \sum_{i=1}^{n} \left(1 + (b - a)^{\frac{1}{q_i}} \right) \right\}$$

and

$$\Lambda_1 = \left(\frac{1}{\frac{m}{2} \frac{p}{\limsup_{t \to \infty}} \frac{(b-a)F(t,\ldots,t)}{\sum_{i=1}^{n} t^{p_i}}}, \right.$$

$$\left. \frac{1}{\overline{M} p (b-a) \liminf_{\xi \to \infty} \frac{\max_{(t_1,\ldots,n) \in Q_1(\xi)} F(t_1,\ldots,t_n)}{\xi^p}} \right).$$

The following result is a direct consequence of Corollary 2.4.1.

Corollary 2.4.2. *Assume that*

$$\liminf_{\xi \to \infty} \frac{\max_{(t_1,\ldots,t_n) \in Q_1(\xi)} F(t_1,\ldots,t_n)}{\xi^p} = 0$$

and

$$\limsup_{t \to \infty} \frac{F(t,\ldots,t)}{\sum_{i=1}^{n} t^{p_i}} = \infty.$$

Then, for each $\lambda \in (0, \infty)$, system (2.4.11) has an unbounded sequence of classical solutions in X.

We now present a version of Theorem 2.4.1 with $n = 1$. Let α, γ be two nonnegative constants, β, σ be two positive constants, and $p > 1$. Let $f : [a, b] \times \mathbb{R} \to \mathbb{R}$ be continuous and let $h : \mathbb{R} \to [0, \infty)$ be continuous and satisfy

$$0 < m := \inf_{t \in \mathbb{R}} h(t) \le M := \inf_{t \in \mathbb{R}} h(t) < \infty.$$

Define the function F by

$$F(x, t) = \int_0^t f(x, s) ds \quad \text{for each } (x, t) \in [a, b] \times \mathbb{R}, \tag{2.4.12}$$

and for any $\vartheta > 0$, let

$$W(\vartheta) = \left\{ t \in \mathbb{R} \; : \; |t| \le \vartheta \left(\sqrt[p]{\frac{\beta^{p-1}}{\alpha^{p-1}}} + (b-a)^{\frac{1}{q}} \right) \right\}, \tag{2.4.13}$$

where $1/p + 1/q = 1$.

The next two corollaries are consequences of Theorem 2.4.1.

Corollary 2.4.3. *Assume that*

$$\liminf_{\xi \to \infty} \frac{\int_a^b \sup_{t \in W(\xi)} F(x, t) dx}{\xi^p} < \frac{mp}{Mp \left(\frac{\alpha^{p-1}}{\beta^{p-1}} + \frac{\gamma^{p-1}}{\sigma^{p-1}} \right)} \limsup_{t \to \infty} \frac{\int_a^b F(x, t) dx}{t^p}$$

Then, for each $\lambda \in \Lambda_2$, the problem

$$\begin{cases} -(\phi_p(u'))' = \lambda f(x,u)h(u') & in\ (a,b), \\ \alpha u(a) - \beta u'(a) = 0, \quad \gamma u(b) + \sigma u'(b) = 0, \end{cases} \tag{2.4.14}$$

has an unbounded sequence of classical solutions in X, where

$$\Lambda_2 = \left(\frac{\frac{\alpha^{p-1}}{\beta^{p-1}} + \frac{\gamma^{p-1}}{\sigma^{p-1}}}{mp\,\limsup_{t\to\infty} \frac{\int_a^b F(x,t)dx}{t^p}}, \quad \frac{1}{Mp\,\liminf_{\xi\to\infty} \frac{\int_a^b \sup_{t\in W(\xi)} F(x,t)dx}{\xi^p}} \right).$$

Corollary 2.4.4. *Assume that*

$$\liminf_{\xi\to\infty} \frac{\int_a^b \sup_{t\in W(\xi)} F(x,t)dx}{\xi^p} = 0 \quad and \quad \limsup_{t\to\infty} \frac{\int_a^b F(x,t)dx}{t^p} = \infty.$$

Then, for each $\lambda \in (0,\infty)$, the problem (2.4.14) has an unbounded sequence of classical solutions in X.

Making use of part (c) of Theorem 1.1.5 and arguing as in the proof of Theorem 2.4.1, we can obtain the following result.

Theorem 2.4.2. *Assume that*

$$\liminf_{\xi\to 0^+} \frac{\int_a^b \sup_{(t_1,\ldots,t_n)\in Q(\xi)} F(x,t_1,\ldots,t_n)dx}{\xi^{\overline{p}}}$$

$$< \frac{m\,\underline{p}}{\overline{M\overline{p}}} \limsup_{t\to 0^+} \frac{\int_a^b F(x,t,\ldots,t)dx}{\sum_{i=1}^n \left(\frac{\alpha_i^{p_i-1}}{\beta_i^{p_i-1}} + \frac{\gamma_i^{p_i-1}}{\sigma_i^{p_i-1}}\right) t^{p_i}}. \tag{2.4.15}$$

Then, for each $\lambda \in \Lambda_3$, system (2.1.1) has a sequence of classical solutions in X converging uniformly to 0, where

$$\Lambda_3 = \left(\frac{1}{\underline{m}\,\underline{p}\limsup_{t\to 0^+} \frac{\int_a^b F(x,t,\ldots,t)dx}{\sum_{i=1}^n \left(\frac{\alpha_i^{p_i-1}}{\beta_i^{p_i-1}} + \frac{\gamma_i^{p_i-1}}{\sigma_i^{p_i-1}}\right) t^{p_i}}}, \right.$$

$$\left. \frac{1}{\overline{M\overline{p}}\liminf_{\xi\to 0^+} \frac{\int_a^b \sup_{(t_1,\ldots,t_n)\in Q(\xi)} F(x,t_1,\ldots,t_n)dx}{\xi^{\overline{p}}}} \right). \tag{2.4.16}$$

Proof. We again define the functionals $\Phi, \Psi : X \to \mathbb{R}$ by (2.2.1) and (2.2.2) and note that Lemma 2.2.2 ensures the regularity conditions in Theorem 1.1.5 are satisfied.

Define a sequence $\{\xi_k\}$ of positive numbers such that $\xi_k < 1$ for $k \in \mathbb{N}$, $\xi_k \to 0^+$ as $k \to \infty$, and

$$\lim_{k \to \infty} \frac{\int_a^b \sup_{(t_1,\ldots,t_n) \in Q(\xi_k)} F(x, t_1, \ldots, t_n) dx}{\xi_k^{\overline{p}}}$$

$$= \liminf_{\xi \to 0^+} \frac{\int_a^b \sup_{(t_1,\ldots,t_n) \in Q(\xi)} F(x, t_1, \ldots, t_n) dx}{\xi^{\overline{p}}}. \qquad (2.4.17)$$

Let $r_k = \frac{\xi_k^{\overline{p}}}{\overline{M}\overline{p}}$ for $k \in \mathbb{N}$ so that $\overline{M}\overline{p}r_k = \xi_k^{\overline{p}} < 1$. Then, by Lemma 2.2.3, it follows that

$$\max_{x \in [a,b]} \sum_{i=1}^n |u_i(x)| \leq (\overline{M}\overline{p}r_k)^{\frac{1}{\overline{p}}} \sum_{i=1}^n \left(\sqrt[p_i]{\frac{\beta_i^{p_i-1}}{\alpha_i^{p_i-1}}} + (b-a)^{\frac{1}{q_i}} \right)$$

$$= \xi_k \sum_{i=1}^n \left(\sqrt[p_i]{\frac{\beta_i^{p_i-1}}{\alpha_i^{p_i-1}}} + (b-a)^{\frac{1}{q_i}} \right) \qquad (2.4.18)$$

for $u = (u_1, \ldots, u_n) \in X$ with $\Phi(u) \leq r_k$.

Since $(\inf_X \Phi)^+ = 0$, from the definition δ, we have $\delta = \liminf_{r \to 0^+} \varphi(r)$. Using (2.4.15)–(2.4.18) and arguing as in showing that (2.4.5) in the proof of Theorem 2.4.1 holds, we can prove that $\delta < \infty$ and $\Lambda_3 \subseteq (0, 1/\delta)$. Let $\lambda \in \Lambda_3$ be fixed. By Theorem 1.1.5 (c), one of the following alternatives holds:

(c$_1$) either there is a global minimum of Φ which is a local minimum of $I_\lambda = \Phi - \lambda\Psi$, or

(c$_2$) there exists a sequence $\{(u_{1k}, \ldots, u_{nk})\}$ of pairwise distinct critical points of I_λ that converges weakly to a global minimum of Φ.

We now show that (c$_1$) does not hold. Since $\lambda \in \Lambda_3$, by (2.4.16), we see that there exist a real sequence $\{d_k\}$ of positive numbers, $\tau > 0$, and $N_1 \in \mathbb{N}$ such that $d_k \to 0^+$ as $k \to \infty$ and

$$\frac{1}{\lambda} < \tau < \frac{\int_a^b F(x, d_k, \ldots, d_k) dx}{\frac{1}{\underline{m}\ \underline{p}} \sum_{i=1}^n \left(\frac{\alpha_i^{p_i-1}}{\beta_i^{p_i-1}} + \frac{\gamma_i^{p_i-1}}{\sigma_i^{p_i-1}} \right) d_k^{p_i}} \qquad (2.4.19)$$

for each $k \geq N_1$. For $k \in \mathbb{N}$, let $w_k = (d_k, \ldots, d_k) \in X$, so (2.4.8) and (2.4.9) hold. From (2.4.8), (2.4.9), and (2.4.19),

$$I_\lambda(w_k) \leq \frac{1}{\underline{m}\ \underline{p}} \sum_{i=1}^n \left(\frac{\alpha_i^{p_i-1}}{\beta_i^{p_i-1}} + \frac{\gamma_i^{p_i-1}}{\sigma_i^{p_i-1}} \right) t^{p_i} - \lambda \int_a^b F(x, d_k, \ldots, d_k) dx$$

$$< (1 - \lambda\tau) \frac{1}{\underline{m}\ \underline{p}} \sum_{i=1}^n \left(\frac{\alpha_i^{p_i-1}}{\beta_i^{p_i-1}} + \frac{\gamma_i^{p_i-1}}{\sigma_i^{p_i-1}} \right) d_k^{p_i}$$

$$< 0 = I_\lambda(0)$$

for every $k \geq N_1$. Therefore, 0 is not a local minimum of I_λ, and since 0 is the only global minimum of Φ, this proves that (c_1) does not hold. Hence, there exists a sequence $\{(u_{1k}, \ldots, u_{nk})\}$ of pairwise distinct critical points of I_λ that converges weakly to 0. Now the embedding $X = X_1 \times \cdots \times X_n \hookrightarrow C^0([0,1]) \times \cdots \times C^0([0,1])$ is compact, so the critical points converge uniformly to zero. In view of Remark 2.2.1, this completes the proof of the theorem. $\qquad\square$

Remark 2.4.1. Results similar to Corollaries 2.4.1–2.4.4 can be obtained from Theorem 2.4.2; we omit those here.

As an example of our results we present the following example in which the form of the nonlinear function $f(t,x)$ is partly motivated by Example 3.1 in [97].

Example 2.4.1. Define the sequences $\{a_n\}$ and $\{b_n\}$ by $b_1 = 2$, $b_{n+1} = b_n^6$, and $a_n = b_n^4$ for $n \in \mathbb{N}$. For $(x,t) \in [0,1] \times \mathbb{R}$, let $f(x,t) = (3x^2 + 1)g(t)$ and $h(t) = 2 + \sin^3 t$, where

$$g(t) = \begin{cases} b_1^3\sqrt{1 - (1-t)^2} + 1, & t \in [0, b_1], \\ (a_n - b_n^3)\sqrt{1 - (a_n - 1 - t)^2} + 1, & t \in \cup_{n=1}^\infty [a_n - 2, a_n], \\ (b_{n+1}^3 - a_n)\sqrt{1 - (b_{n+1} - 1 - t)^2} + 1, & t \in \cup_{n=1}^\infty [b_{n+1} - 2, b_{n+1}], \\ 1, & \text{otherwise.} \end{cases}$$

Then, $f : [0,1] \times \mathbb{R} \to [1, \infty)$ and $h : \mathbb{R} \to [2, \infty)$ are positive continuous functions.

We claim that, for each $\lambda \in (0, \infty)$, the problem

$$\begin{cases} -(\phi_p(u'))' = \lambda f(x,u)h(u') & \text{in } (0,1), \quad p \geq 2, \\ u(0) - u'(0) = 0, \quad u(1) + u'(1) = 0, \end{cases} \tag{2.4.20}$$

has an unbounded sequence of classical solutions

In fact, with $a = 0$, $b = 1$, and $\alpha = \beta = \gamma = \sigma = 1$, it is clear that (2.4.20) is a special case of (2.4.14). Let $F(t,x)$ be defined by (2.4.12). Then,

$$F(x,t) = (3x^2 + 1)G(t),$$

where

$$G(t) = \int_0^t g(s)ds.$$

Moreover, for $x \in [0, 1]$, by simple computations,

$$
\begin{aligned}
G(a_n) = &\int_0^{a_n} 1 ds + b_1^3 \int_0^2 \sqrt{1 - (1 - s)^2} \, ds \\
&+ \sum_{i=1}^{n} \int_{a_i - 2}^{a_i} (a_i - b_i^3) \sqrt{1 - (a_i - 1 - s)^2} \, ds \\
&+ \sum_{i=1}^{n-1} \int_{b_{i+1}-2}^{b_{i+1}} (b_i^3 - a_i) \sqrt{1 - (b_{i+1} - 1 - s)^2} \, ds \\
= &\frac{\pi}{2} a_n + a_n
\end{aligned}
$$

and

$$
\begin{aligned}
G(b_n) = &\int_0^{b_n} 1 ds + b_1^3 \int_0^2 \sqrt{1 - (1 - s)^2} \, ds \\
&+ \sum_{i=1}^{n-1} \int_{a_i - 2}^{a_i} (a_i - b_i^3) \sqrt{1 - (a_i - 1 - s)^2} \, ds \\
&+ \sum_{i=1}^{n-1} \int_{b_{i+1}-2}^{b_{i+1}} (b_i^3 - a_i) \sqrt{1 - (b_{i+1} - 1 - s)^2} \, ds \\
= &\frac{\pi}{2} b_n^3 + b_n.
\end{aligned}
$$

Thus,

$$
\lim_{n \to \infty} \frac{G(a_n)}{a_n^2} = 0 \quad \text{and} \quad \lim_{n \to \infty} \frac{G(b_n)}{b_n^2} = \infty.
$$

Note that, for $W(\vartheta)$ defined by (2.4.13),

$$
W(\vartheta) = \{t \in \mathbb{R} \ : \ |t| \le 2\vartheta\}.
$$

Then, it is easy to see that

$$
\liminf_{\xi \to \infty} \frac{\int_0^1 \sup_{t \in W(\xi)} F(x, t) dx}{\xi^p} = \liminf_{\xi \to \infty} \frac{G(2\xi) \int_0^1 (3x^2 + 1) dx}{\xi^p}
$$
$$
= \liminf_{\xi \to \infty} \frac{2G(2\xi)}{\xi^p} = 0
$$

and

$$
\limsup_{t \to \infty} \frac{\int_0^1 F(x, t) dx}{t^p} = \limsup_{t \to \infty} \frac{G(t) \int_0^1 (3x^2 + 1) dx}{t^p} = \limsup_{t \to \infty} \frac{2G(t)}{t^p} = \infty.
$$

Thus, all the conditions of Corollary 2.4.4 are satisfied and so our claim is proved.

Chapter 3

Multi-point Problems

3.1 Introduction

Multi-point boundary value problems arise in a variety of applications and as a consequence and have been considered by a number of researchers in recent years. For results on these kinds of problems, we refer the reader to the papers [57,107–109,112,113,120,121,130–132,144,145,155,164,177–180, 218,219,222,280,281] as recent examples of work on these kinds of problems. In this chapter, we wish to obtain sufficient conditions for existence of one or more classical solutions to the system

$$\begin{cases} -(\phi_{p_i}(u_i'))' = \lambda F_{u_i}(x, u_1, \ldots, u_n), \ x \in (0,1), \\ u_i(0) = \sum_{j=1}^m a_j u_i(x_j), \ u_i(1) = \sum_{j=1}^m b_j u_i(x_j), \end{cases} \quad i = 1, \ldots, n. \quad (3.1.1)$$

Here, we have $p_i > 1$ and $\phi_{p_i}(t) = |t|^{p_i-2}t$ for $i = 1, \ldots, n$, λ is a positive parameter, $m, n \geq 1$ are integers, $a_j, b_j \in \mathbb{R}$ for $j = 1, \ldots, m$, and $0 < x_1 < x_2 < x_3 < \ldots < x_m < 1$. The function $F : [0,1] \times \mathbb{R}^n \to \mathbb{R}$ has the property that the mapping $(t_1, t_2, \ldots, t_n) \to F(x, t_1, t_2, \ldots, t_n)$ is in C^1 in $\mathbb{R}^n$ for all $x \in [0,1]$, F_{t_i} is continuous in $[0,1] \times \mathbb{R}^n$ for $i = 1, \ldots, n$, where F_{t_i} denotes the partial derivative of F with respect to t_i, and $F(x, 0, \ldots, 0) = 0$ for all $x \in [0,1]$. In order to prove our results, we make use of Theorems 1.1.1, 1.1.3, and 1.1.4 above.

When we attempt to show the existence of infinitely many solutions, we will employ the recent critical point theorem, Theorem 1.1.5 above, which appeared in [54,244]. This theorem and variations of it have frequently been used to obtain multiplicity results for nonlinear variational type problems.

In Section 5, we consider a similar system but with two parameters. To show that it possesses multiple solutions, we apply the three critical points theorem, Theorem 1.1.6 above. In the final section in this chapter, we consider the case where the boundary conditions depend on the derivative of the unknown function. As we will see, the approach used in the previous sections in this chapter will not work on such a problem.

3.2   Nontrivial Solutions of Multi-point Problems

In this section, we wish to obtain sufficient conditions to ensure that the problem (3.1.1) has at least one nontrivial classical solution. We again will use a variational approach and make use of the local minimum theorem above, namely Theorem 1.1.1. We begin by taking X to be the Cartesian product of n spaces of the form

$$X_i = \left\{ \xi \in W^{1,p_i}([0,1]) \ : \ \xi(0) = \sum_{j=1}^{m} a_j \xi(x_j), \quad \xi(1) = \sum_{j=1}^{m} b_j \xi(x_j) \right\},$$

for $i = 1, \ldots, n$, i.e., $X = X_1 \times \cdots \times X_n$, endowed with the norm

$$\|(u_1, \ldots, u_n)\| = \sum_{i=1}^{n} \|u_i\|_{p_i} \quad \text{for } u = (u_1, \ldots, u_n) \in X,$$

where

$$\|u_i\|_{p_i} = \left(\int_0^1 |u_i'(x)|^{p_i} dx \right)^{1/p_i}, \quad i = 1, \ldots, n.$$

Then, X is a reflexive Banach space, and we will let X^* denote to dual space of X.

By a *classical solution* of the system (3.1.1), we mean a function $u = (u_1, \ldots, u_n)$ such that, for $i = 1, \ldots, n$, $u_i \in C^1[0,1]$, $\phi_{p_i}(u_i') \in C^1[0,1]$, and $u_i(x)$ satisfies (3.1.1). A function $u = (u_1, \ldots, u_n) \in X$ is a *weak solution* of (3.1.1) if

$$\int_0^1 \sum_{i=1}^{n} \phi_{p_i}(u_i'(x))v_i'(x)dx - \lambda \int_0^1 \sum_{i=1}^{n} F_{u_i}(x, u_1(x), \ldots, u_n(x))v_i(x)dx = 0$$

for any $v = (v_1, \ldots, v_n) \in W_0^{1,p_1}([0,1]) \times W_0^{1,p_2}([0,1]) \times \cdots \times W_0^{1,p_n}([0,1])$. The following lemma shows the relationship between weak solutions of the problem (3.1.1) and classical solutions.

Lemma 3.2.1. ([130, Lemma 2.5]) *A weak solution of (3.1.1) coincides with a classical solution of (3.1.1).*

Without further mention, we will assume throughout this section that the following conditions hold:

(H1) Either $\underline{p} \geq 2$ or $\bar{p} < 2$, where $\underline{p} = \min\{p_1, \ldots, p_n\}$ and $\bar{p} = \max\{p_1, \ldots, p_n\}$.

(H2) $\sum_{j=1}^{m} a_j \neq 1$ and $\sum_{j=1}^{m} b_j \neq 1$.

Let

$$c = \max\left\{ \sup_{u_i \in X_i \setminus \{0\}} \frac{\max_{x \in [0,1]} |u_i(x)|^{p_i}}{\|u_i\|_{p_i}^{p_i}} \; : \; \text{for } i = 1, \ldots, n \right\}. \quad (3.2.1)$$

Since $p_i > 1$ for $i = 1, \ldots, n$, the embedding $X = X_1 \times \cdots \times X_n \hookrightarrow (C^0([0,1]))^n$ is compact, and so $c < \infty$. Moreover, if (H2) holds, from [108, Lemma 3.1],

$$\sup_{v \in X_i \setminus \{0\}} \frac{\max_{x \in [0,1]} |v(x)|}{\|v\|_{p_i}} \leq \frac{1}{2} \left(1 + \frac{\sum_{j=1}^{m} |a_j|}{|1 - \sum_{j=1}^{m} a_j|} + \frac{\sum_{j=1}^{m} |b_j|}{|1 - \sum_{j=1}^{m} b_j|} \right)$$

for $i = 1, \ldots, n$. Let

$$\sigma_n = \left[2^{p_n - 1} \left(x_1^{1-p_n} \left| 1 - \sum_{j=1}^{m} a_j \right|^{p_n} + (1 - x_m)^{1-p_n} \left| 1 - \sum_{j=1}^{m} b_j \right|^{p_n} \right) \right]^{1/p_n}. \quad (3.2.2)$$

Define

$$B_{1,n}(x) = \begin{cases} [x \sum_{j=1}^{m} a_j, x]^n, & \text{if } \sum_{j=1}^{m} a_j < 1, \\ [x, x \sum_{j=1}^{m} a_j]^n, & \text{if } \sum_{j=1}^{m} a_j > 1, \end{cases}$$

and

$$B_{2,n}(x) = \begin{cases} [x \sum_{j=1}^{m} b_j, x]^n, & \text{if } \sum_{j=1}^{m} b_j < 1, \\ [x, x \sum_{j=1}^{m} b_j]^n, & \text{if } \sum_{j=1}^{m} b_j > 1, \end{cases}$$

where $[\cdot, \cdot]^n = [\cdot, \cdot] \times \cdots \times [\cdot, \cdot]$.

For any $\gamma > 0$, the set $K(\gamma)$ by

$$K(\gamma) = \left\{ (t_1, \ldots, t_n) \in \mathbb{R}^n \; : \; \sum_{i=1}^{n} \frac{|t_i|^{p_i}}{p_i} \leq \gamma \right\}.$$

This set will be used in some of our hypotheses with appropriate choices of γ. For any constants $\nu \geq 0$ and $\tau > 0$ with

$$\nu \neq \sqrt[p_n]{c \prod_{i=1}^{n-1} p_i \, \sigma_n \tau},$$

where $\prod_{i=1}^{0} p_i = 1$, we define

$$
a_\tau(\nu) = \left[\int_0^1 \sup_{(t_1,\ldots,t_n) \in K\left(\frac{\nu^{p_n}}{\prod_{i=1}^{n} p_i}\right)} F(x, t_1, \ldots, t_n) dx \right.
$$
$$
\left. - \int_{\frac{x_1}{2}}^{\frac{1+x_m}{2}} F(x, 0, \ldots, 0, \tau) dx \right] \times \left[\frac{\nu^{p_n}}{c\prod_{i=1}^{n} p_i} - \frac{(\sigma_n \tau)^{p_n}}{p_n} \right]^{-1}.
$$

$$(3.2.3)$$

Our first existence result in this section is contained in the following theorem.

Theorem 3.2.1. *Assume that*

(C1) $F(x, t_1, \ldots, t_n) \geq 0$ *for each* $x \in [0, x_1/2] \cup [(1 + x_m)/2, 1]$ *and* $(t_1, \ldots, t_n) \in B_{1,n}(\tau) \cup B_{2,n}(\tau)$,

and there exist a nonnegative constant ν_1 *and two positive constants* ν_2 *and* τ *with*

$$
\nu_1 < \sqrt[p_n]{c \prod_{i=1}^{n-1} p_i} \; \sigma_n \tau < \nu_2 \tag{3.2.4}
$$

such that

(C2) $a_\tau(\nu_2) < a_\tau(\nu_1)$.

Then, for each $\lambda \in (1/a_\tau(\nu_1), \; 1/a_\tau(\nu_2))$, *the system* (3.1.1) *has at least one nontrivial classical solution* $u_0 = (u_{01}, \ldots, u_{0n}) \in X$ *such that*

$$
\nu_1 < \sqrt[p_n]{c \sum_{i=1}^{n} \prod_{\substack{j=1 \\ j \neq i}}^{n} p_j \|u_{0i}\|_{p_i}^{p_i}} < \nu_2. \tag{3.2.5}
$$

Proof. We will apply Theorem 1.1.1 to obtain our result, so for $u = (u_1, \ldots, u_n) \in X$, define the functionals Φ, Ψ and $I_\lambda : X \to \mathbb{R}$ by

$$
\Phi(u) = \sum_{i=1}^{n} \frac{\|u_i\|_{p_i}^{p_i}}{p_i}, \tag{3.2.6}
$$

$$
\Psi(u) = \int_0^1 F(x, u_1(x), \ldots, u_n(x)) dx. \tag{3.2.7}
$$

and

$$
I_\lambda = \Phi(u) - \lambda \Psi(u). \tag{3.2.8}
$$

As we noted in the previous chapter, Φ and Ψ are well defined and continuously differentiable functionals whose derivatives at the point $u = (u_1, \ldots, u_n) \in X$ are the functionals $\Phi'(u)$, $\Psi'(u) \in X^*$ given by

$$\Phi'(u)(v) = \int_0^1 \sum_{i=1}^n |u_i'(x)|^{p_i-2} u_i'(x) v_i'(x) dx \qquad (3.2.9)$$

and

$$\Psi'(u)(v) = \int_0^1 \sum_{i=1}^n F_{u_i}(x, u_1(x), \ldots, u_n(x)) v_i(x) dx \qquad (3.2.10)$$

for every $v = (v_1, \ldots, v_n) \in X$. In addition, Φ is coercive, Φ' admits a continuous inverse on X^* (see [130, Lemma 2.6]), $\Psi' : X \to X^*$ is a compact operator, and Φ is sequentially weakly lower semicontinuous (see [289, Proposition 25.20]). Let $w = (0, \ldots, 0, w_n(x))$ with

$$w_n(x) = \begin{cases} \tau\left(\sum_{j=1}^m a_j + \frac{2(1-\sum_{j=1}^m a_j)}{x_1} x\right), & \text{if } x \in [0, \frac{x_1}{2}), \\ \tau, & \text{if } x \in [\frac{x_1}{2}, \frac{1+x_m}{2}], \\ \tau\left(\frac{2-\sum_{j=1}^m b_j - x_m \sum_{j=1}^m b_j}{1-x_m} - \frac{2(1-\sum_{j=1}^m b_j)}{1-x_m} x\right), & \text{if } x \in (\frac{1+x_m}{2}, 1], \end{cases}$$

$r_1 = \frac{\nu_1^{p_n}}{c \prod_{i=1}^n p_i}$, and $r_2 = \frac{\nu_2^{p_n}}{c \prod_{i=1}^n p_i}$. Then, in view of (3.2.4), we have

$$\Phi(w) = \frac{(\sigma_n \tau)^{p_n}}{p_n} \text{ and } r_1 < \Phi(w) < r_2. \qquad (3.2.11)$$

For each $u = (u_1, \ldots, u_n) \in X$, (3.2.1) implies

$$\sup_{x \in [0,1]} |u_i(x)|^{p_i} \leq c\|u_i\|_{p_i}^{p_i} \quad \text{for } i = 1, \ldots, n.$$

Then, we have

$$\sup_{x \in [0,1]} \sum_{i=1}^n \frac{|u_i(x)|^{p_i}}{p_i} \leq c \sum_{i=1}^n \frac{\|u_i\|_{p_i}^{p_i}}{p_i}.$$

Hence,

$$\Phi^{-1}(-\infty, r_2) \subseteq \left\{ (u_1, \ldots, u_n) \in X : \sum_{i=1}^n \frac{|u_i(x)|^{p_i}}{p_i} \right.$$
$$\left. < \frac{\nu_2^{p_n}}{\prod_{i=1}^n p_i} \text{ for all } x \in [0, 1] \right\},$$

and so

$$\sup_{(u_1,\ldots,u_n)\in\Phi^{-1}(-\infty,r_2)}\Psi(u) = \sup_{(u_1,\ldots,u_n)\in\Phi^{-1}(-\infty,r_2)}\int_0^1 F(x,u_1(x),\ldots,u_n(x))dx$$

$$\leq \int_0^1 \sup_{(t_1,\ldots,t_n)\in K\left(\frac{\nu_2^{p_n}}{\prod_{i=1}^n p_i}\right)} F(x,t_1,\ldots,t_n)dx.$$

Note that

$$\tau\sum_{j=1}^m a_j \leq w_n(x) \leq \tau \quad \text{for each } x \in [0,x_1/2] \quad \text{if } \sum_{j=1}^m a_j < 1,$$

$$\tau \leq w_n(x) \leq \tau\sum_{j=1}^m a_j \quad \text{for each } x \in [0,x_1/2] \quad \text{if } \sum_{j=1}^m a_j > 1,$$

$$\tau\sum_{j=1}^m b_j \leq w_n(x) \leq \tau \quad \text{for each } x \in [(1+x_m)/2,1] \quad \text{if } \sum_{j=1}^m b_j < 1,$$

and

$$\tau \leq w_n(x) \leq \tau\sum_{j=1}^m b_j \quad \text{for each } x \in [(1+x_m)/2,1] \quad \text{if } \sum_{j=1}^m b_j > 1.$$

Then, (C1) ensures that

$$\int_0^{\frac{x_1}{2}} F(x,w_1(x),\ldots,w_n(x))dx + \int_{\frac{1+x_m}{2}}^1 F(x,w_1(x),\ldots,w_n(x))dx \geq 0.$$

Hence,

$$\Psi(w) = \int_0^1 F(x,w_1(x),\ldots,w_n(x))dx \geq \int_{\frac{x_1}{2}}^{\frac{1+x_m}{2}} F(x,0,\ldots,0,\tau)dx.$$

$$(3.2.12)$$

From (1.1.1), (3.2.3), (3.2.11), and (3.2.12), it follows that

$$\eta(r_1,r_2) \leq \frac{\sup_{u\in\Phi^{-1}(-\infty,r_2)}\Psi(u) - \Psi(w)}{r_2 - \Phi(w)}$$

$$\leq \frac{\int_0^1 \sup_{(t_1,\ldots,t_n)\in K\left(\frac{\nu_2^{p_n}}{\prod_{i=1}^n p_i}\right)} F(x,t_1,\ldots,t_n)dx - \Psi(w)}{r_2 - \Phi(w)}$$

$$\leq \left[\int_0^1 \sup_{(t_1,\ldots,t_n)\in K\left(\frac{\nu_2^{p_n}}{\prod_{i=1}^n p_i}\right)} F(x,t_1,\ldots,t_n)dx\right.$$

$$\left.- \int_{\frac{x_1}{2}}^{\frac{1+x_m}{2}} F(x,0,\ldots,0,\tau)dx\right] \times \left[\frac{\nu_2^{p_n}}{c\prod_{i=1}^n p_i} - \frac{(\sigma_n\tau)^{p_n}}{p_n}\right]^{-1}$$

$$= a_\tau(\nu_2).$$

Similarly, (1.1.2), (3.2.3), (3.2.11), and (3.2.12) imply

$$\rho(r_1, r_2) \geq \frac{\Psi(w) - \sup_{u \in \Phi^{-1}(-\infty, r_1)} \Psi(u)}{\Phi(w) - r_1}$$

$$\geq \frac{\Psi(w) - \int_0^1 \sup_{(t_1, \ldots, t_n) \in K\left(\frac{\nu_1^{p_n}}{\prod_{i=1}^n p_i}\right)} F(x, t_1, \ldots, t_n) dx}{\Phi(w) - r_1}$$

$$\geq \left[\int_{\frac{x_1}{2}}^{\frac{1+x_m}{2}} F(x, 0, \ldots, 0, \tau) dx \right.$$

$$\left. - \int_0^1 \sup_{(t_1, \ldots, t_n) \in K\left(\frac{\nu_1^{p_n}}{\prod_{i=1}^n p_i}\right)} F(x, t_1, \ldots, t_n) dx \right]$$

$$\times \left[\frac{(\sigma_n \tau)^{p_n}}{p_n} - \frac{\nu_1^{p_n}}{c \prod_{i=1}^n p_i} \right]^{-1}$$

$$= a_\tau(\nu_1).$$

Then, (C2) implies that

$$\eta(r_1, r_2) \leq a_\tau(\nu_2) < a_\tau(\nu_1) \leq \rho(r_1, r_2).$$

By Theorem 1.1.1, I_λ has at least one nontrivial critical point $u_0 = (u_{01}, \ldots, u_{0n})$ in X satisfying (3.2.4). Since weak solutions of (3.1.1) are the critical points of $I_\lambda(u)$, applying Lemma 3.2.1 completes the proof of the theorem. $\square$

Now we discuss some consequence of this theorem.

Corollary 3.2.1. *Assume that (C1) holds and there exists two positive constants ν and τ with*

$$\sqrt[p_n]{c \prod_{i=1}^{n-1} p_i} \, \sigma_n \tau < \nu \tag{3.2.13}$$

such that

$$\text{(C3)} \quad \frac{\int_0^1 \sup_{(t_1, \ldots, t_n) \in K\left(\frac{\nu^{p_n}}{\prod_{i=1}^n p_i}\right)} F(x, t_1, \ldots, t_n) dx}{\nu^{p_n}} < \frac{\int_{\frac{x_1}{2}}^{\frac{1+x_m}{2}} F(x, 0, \ldots, 0, \tau) dx}{(\sigma_n \tau)^{p_n} c \prod_{i=1}^{n-1} p_i}.$$

Then, for each

$$\lambda \in \left(\frac{\frac{(\sigma_n \tau)^{Pn}}{p_n}}{\int_{\frac{x_1}{2}}^{\frac{1+x_m}{2}} F(x,0,\ldots,0,\tau)dx} \right., $$

$$\left. \frac{\frac{\nu^{Pn}}{c\prod_{i=1}^{n} p_i}}{\int_0^1 \sup_{(t_1,\ldots,t_n)\in K\left(\frac{\nu^{Pn}}{\prod_{i=1}^{n} p_i}\right)} F(x,t_1,\ldots,t_n)dx} \right), $$

the system (3.1.1) *has at least one nontrivial classical solution* $u_0 = (u_{01},\ldots,u_{0n}) \in X$ *such that* $\sup_{x\in\Omega} \sum_{i=1}^{n} \frac{|u_{0i}(x)|^{p_i}}{p_i} < \frac{\nu^{Pn}}{\prod_{i=1}^{n} p_i}$.

Proof. Let $\nu_1 = 0$ and $\nu_2 = \nu$; then, (3.2.13) implies (3.2.4), and from (3.2.3) and (C3), we see that

$$a_\tau(\nu_2) = a_\tau(\nu)$$

$$< \frac{\left(1 - \frac{(\sigma_n \tau)^{Pn} c \prod_{i=1}^{n-1} p_i}{\nu^{Pn}}\right) \int_0^1 \sup_{(t_1,\ldots,t_n)\in K\left(\frac{\nu^{Pn}}{\prod_{i=1}^{n} p_i}\right)} F(x,t_1,\ldots,t_n)dx}{\frac{\nu^{Pn}}{c\prod_{i=1}^{n} p_i} - \frac{(\sigma_n \tau)^{Pn}}{p_n}}$$

$$= \frac{c\prod_{i=1}^{n} p_i}{\nu^{Pn}} \int_0^1 \sup_{(t_1,\ldots,t_n)\in K\left(\frac{\nu^{Pn}}{\prod_{i=1}^{n} p_i}\right)} F(x,t_1,\ldots,t_n)dx$$

$$< \frac{\int_{\frac{x_1}{2}}^{\frac{1+x_m}{2}} F(x,0,\ldots,0,\tau)dx}{\frac{(\sigma_n \tau)^{Pn}}{p_n}}$$

$$= a_\tau(0) = a_\tau(\nu_1),$$

i.e., (C2) holds. The conclusion then follows from Theorem 3.2.1.   □

Let $p > 1$. Define

$$Y = \left\{ \xi \in W^{1,p}([0,1]) \ : \ \xi(0) = \sum_{j=1}^{m} a_j\xi(x_j), \ \xi(1) = \sum_{j=1}^{m} b_j\xi(x_j) \right\}$$

and

$$d = \sup_{u\in Y\setminus\{0\}} \frac{\max_{x\in[0,1]} |u(x)|}{||u||_p},$$

where

$$||u||_p = \left(\int_0^1 |u'(x)|^p dx \right)^{1/p}.$$

Then we have that $d^p = c$ if $p_i = p$ and $n = 1$ in (3.2.1).

Let $f : [0,1] \times \mathbb{R} \to \mathbb{R}$ be an L^1-Carathéodory function and F be the function defined by $F(x,t) = \int_0^t f(x,s)ds$ for each $(x,t) \in [0,1] \times \mathbb{R}$. Let

$$
\sigma = \left[2^{p-1} \left(x_1^{1-p} \left| 1 - \sum_{j=1}^m a_j \right|^p + (1 - x_m)^{1-p} \left| 1 - \sum_{j=1}^m b_j \right|^p \right) \right]^{1/p}.
$$

For any given nonnegative constant ν and a positive constant τ with $\nu \neq d\sigma\tau$, define

$$
b_\tau(\nu) = \frac{\int_0^1 \sup_{|t| \leq \nu} F(x,t)dx - \int_{\frac{x_1}{2}}^{\frac{1+x_m}{2}} F(x,\tau)dx}{\frac{\nu^p}{d^p p} - \frac{(\sigma\tau)^p}{p}}. \tag{3.2.14}
$$

The following corollary is a consequence of Corollary 3.2.1.

Corollary 3.2.2. *Assume that there exist a nonnegative constant ν_1 and two positive constants ν_2 and τ with $\nu_1 < d\sigma\tau < \nu_2$ such that*

(C4) $f(x,t) \geq 0$ *for each* $x \in [0, x_1/2] \cup [(1+x_m)/2, 1]$ *and* $t \in B_{1,1}(\tau) \cup B_{2,1}(\tau)$,

(C5) $b_\tau(\nu_2) < b_\tau(\nu_1)$.

Then, for each $\lambda \in (1/b_\tau(\nu_1),\ 1/b_\tau(\nu_2))$, the problem

$$
\begin{cases} -(\phi_p(u'))' = \lambda f(x,u),\ t \in (0,1), \\ u(0) = \sum_{j=1}^m a_j u_i(x_j),\ u(1) = \sum_{j=1}^m b_j u_i(x_j), \end{cases}
$$

has at least one nontrivial classical solution $u_0 \in Y$ such that $\nu_1 < d\|u_0\|_p < \nu_2$.

We conclude this section with a special case of Corollary 3.2.2.

Corollary 3.2.3. *Let $h : [0,1] \to \mathbb{R}$ be a positive and essentially bounded function and $g : \mathbb{R} \to \mathbb{R}$ be a nonnegative continuous function such that $\lim_{t \to 0^+} \frac{g(t)}{t^{p-1}} = \infty$. Then, for each*

$$
\lambda \in \left(0,\ \frac{1}{pd^p \int_0^1 h(x)dx} \sup_{\nu > 0} \frac{\nu^p}{\int_0^\nu g(\xi)d\xi} \right),
$$

the problem

$$
\begin{cases} -(\phi_p(u'))' = \lambda h(x)g(u),\ t \in (0,1), \\ u(0) = \sum_{j=1}^m a_j u_i(x_j),\ u(1) = \sum_{j=1}^m b_j u_i(x_j), \end{cases}
$$

has at least one nontrivial classical solution in Y.

Proof. Since $f(x,t) = h(x)g(t)$, (C4) of Corollary 3.2.2 holds. Fix λ as in the conclusion of the corollary. Then there exists a positive constant ν such that

$$\lambda < \frac{1}{pd^p \int_0^1 h(x)dx} \frac{\nu^p}{\int_0^\nu g(\xi)d\xi}. \tag{3.2.15}$$

Moreover, the condition $\lim_{t \to 0+} \frac{g(t)}{t^{p-1}} = \infty$ implies that $\lim_{t \to 0+} \frac{\int_0^t g(\xi)d\xi}{t^p} = \infty$. Therefore, there exists $\tau^* > 0$ with $\tau^* < \nu/(d\sigma)$ such that

$$\frac{\sigma^p}{\lambda p} \frac{1}{\int_{\frac{x_1}{2}}^{\frac{1+x_m}{2}} h(x)dx} < \frac{\int_0^\tau g(\xi)d\xi}{\tau^p} \quad \text{for all } \tau \in (0, \tau^*). \tag{3.2.16}$$

From (3.2.15) and (3.2.16), it follows that

$$\int_0^\nu g(\xi)d\xi \int_0^1 h(x)dx < \frac{\nu^p}{(\sigma\tau)^p d^p} \int_0^\tau g(\xi)d\xi \int_{\frac{x_1}{2}}^{\frac{1+x_m}{2}} h(x)dx \quad \text{for } \tau \in (0, \tau^*).$$

Let $\nu_1 = 0$ and $\nu_2 = \nu$. In view of the fact that $F(x, \tau) = h(x) \int_0^\tau g(s)ds$. Then, (3.2.14) implies

$$b_\tau(\nu_2) = b_\tau(\nu) < \frac{\int_0^\tau g(\xi)d\xi \int_{\frac{x_1}{2}}^{\frac{1+x_m}{2}} h(x)dx \left(\frac{\nu^p}{(\sigma\tau)^p d^p} - 1\right)}{\frac{\nu^p}{d^p p} - \frac{(\sigma\tau)^p}{p}}$$

$$= \frac{\int_0^\tau g(\xi)d\xi \int_{\frac{x_1}{2}}^{\frac{1+x_m}{2}} h(x)dx}{\frac{(\sigma\tau)^p}{p}}$$

$$= b_\tau(\nu_1) \quad \text{for all } \tau \in (0, \tau^*),$$

i.e., (C5) of Corollary 3.2.2 holds for all $\tau \in (0, \tau^*)$. Since $\lim_{\tau \to 0+} b_\tau(\nu_1) = \infty$, the conclusion follows from Corollary 3.2.2. $\qquad\square$

3.3   Multiple Solutions of Multi-point Problems

In this section, we obtain sufficient conditions to ensure that there exist multiple solutions to the problem (3.1.1). To do this we formulate conditions under which the hypotheses of Theorem 1.1.3 are satisfied. Again in this section and without further reference, we assume that conditions (H1) and (H2) hold. We again define the constant c as in (3.2.1).

Our first multiplicity result for system (3.1.1) is the following.

Theorem 3.3.1. *Assume that there exist a function $w = (w_1, \ldots, w_n) \in X$ and a positive constant r such that*

(C6) $\sum_{i=1}^{n} \frac{||w_i||_{p_i}^{p_i}}{p_i} > 2r,$

(C7) $\dfrac{\int_0^1 \sup_{(t_1,\ldots,t_n)\in K(cr)} F(x,t_1,\ldots,t_n)dx}{r} < \dfrac{2}{3}\dfrac{\int_0^1 F(x,w_1(x),\ldots,w_n(x))dx}{\sum_{i=1}^{n}\frac{||w_i||_{p_i}^{p_i}}{p_i}},$

(C8)

$$\limsup_{|t_1|\to\infty,\ldots,|t_n|\to\infty} \frac{F(x,t_1,\ldots,t_n)}{\sum_{i=1}^{n}\frac{|t_i|^{p_i}}{p_i}}$$

$$< \frac{\int_0^1 \sup_{(t_1,\ldots,t_n)\in K(cr)} F(x,t_1,\ldots,t_n)dx}{cr}.$$

Then, for each

$$\lambda \in \left(\frac{3\sum_{i=1}^{n}\frac{||w_i||_{p_i}^{p_i}}{p_i}}{2\int_0^1 F(x,w_1(x),\ldots,w_n(x))dx}, \frac{r}{\int_0^1 \sup_{(t_1,\ldots,t_n)\in K(cr)} F(x,t_1,\ldots,t_n)dx} \right),$$

the system (3.1.1) *has at least three classical solutions.*

Proof. In order to apply Theorem 1.1.3 to our problem, we again define the functionals Φ, Ψ, and I_λ as in (3.2.6)–(3.2.8).

From (C6), we see that $\Phi(w) > 2r$. From (3.2.1), for $(u_1,\ldots,u_n) \in X$,

$$\sup_{x\in[0,1]} |u_i(x)|^{p_i} \le c||u_i||_{p_i}^{p_i} \quad \text{for } i = 1,\ldots,n,$$

where c is given in (3.2.1). Then, we have

$$\sup_{x\in[0,1]} \sum_{i=1}^{n} \frac{|u_i(x)|^{p_i}}{p_i} \le c \sum_{i=1}^{n} \frac{||u_i||_{p_i}^{p_i}}{p_i}, \tag{3.3.1}$$

and so

$$\Phi^{-1}(-\infty,r) = \left\{ u = (u_1,u_2,\ldots,u_n) \in X : \Phi(u) < r \right\}$$

$$= \left\{ u = (u_1,u_2,\ldots,u_n) \in X : \sum_{i=1}^{n} \frac{||u_i||_{p_i}^{p_i}}{p_i} < r \right\}$$

$$\subseteq \left\{ u = (u_1,u_2,\ldots,u_n) \in X : \sum_{i=1}^{n} \frac{|u_i(x)|^{p_i}}{p_i} \le cr \ \text{ for each } x \in [0,1] \right\}.$$

Thus,

$$\sup_{(u_1,\ldots,u_n)\in\Phi^{-1}(-\infty,r)} \Psi(u) = \sup_{(u_1,\ldots,u_n)\in\Phi^{-1}(-\infty,r)} \int_0^1 F(x,u_1(x),\ldots,u_n(x))dx$$

$$\le \int_0^1 \sup_{(t_1,\ldots,t_n)\in K(cr)} F(x,t_1,\ldots,t_n)dx.$$

Therefore, in view of (C7), it follows that

$$
\begin{aligned}
\frac{\sup_{u\in\Phi^{-1}(-\infty,r)}\Psi(u)}{r} &= \frac{\sup_{(u_1,\ldots,u_n)\in\Phi^{-1}(-\infty,r)}\int_0^1 F(x,u_1(x),\ldots,u_n(x))dx}{r} \\
&\leq \frac{\int_0^1 \sup_{(t_1,\ldots,t_n)\in K(cr)} F(x,t_1,\ldots,t_n)dx}{r} \\
&< \frac{2}{3}\frac{\int_0^1 F(x,w_1(x),\ldots,w_n(x))dx}{\sum_{i=1}^n \frac{||w_i||_{p_i}^{p_i}}{p_i}} \\
&= \frac{2}{3}\frac{\Psi(w)}{\Phi(w)},
\end{aligned}
$$

i.e., (a1) of Theorem 1.1.3 holds with $\overline{v} = w$.

From (C8), there exist constants $\eta,\ \vartheta \in \mathbb{R}$ with

$$
\eta < \frac{\int_0^1 \sup_{(t_1,\ldots,t_n)\in K(cr)} F(x,t_1,\ldots,t_n)dx}{r}
$$

such that

$$
cF(x,t_1,\ldots,t_n) \leq \eta \sum_{i=1}^n \frac{|t_i|^{p_i}}{p_i} + \vartheta \quad \text{for all } x\in[0,1] \text{ and } (t_1,\ldots,t_n)\in\mathbb{R}^n.
$$

Fix $(u_1,\ldots,u_n)\in X$. Then

$$
F(x,u_1(x),\ldots,u_n(x)) \leq \frac{1}{c}\left(\eta\sum_{i=1}^n \frac{|u_i(x)|^{p_i}}{p_i} + \vartheta\right) \quad \text{for all } x\in[0,1].
$$

$$(3.3.2)$$

To prove the coercivity of I_λ, assume that $\eta > 0$. Then, for

$$
\lambda \in \left(\frac{3}{2}\frac{\sum_{i=1}^n \frac{||w_i||_{p_i}^{p_i}}{p_i}}{\int_0^1 F(x,w_1(x),\ldots,w_n(x))dx}, \right.
$$

$$
\left. \frac{r}{\int_0^1 \sup_{(t_1,\ldots,t_n)\in K(cr)} F(x,t_1,\ldots,t_n)dx}\right),
$$

(3.3.1) and (3.3.2) yield

$$I_\lambda(u) = \sum_{i=1}^n \frac{||u_i||_{p_i}^{p_i}}{p_i} - \lambda \int_0^1 F(x, u_1(x), \ldots, u_n(x))dx$$

$$\geq \sum_{i=1}^n \frac{||u_i||_{p_i}^{p_i}}{p_i} - \frac{\lambda\eta}{c}\left(\sum_{i=1}^n \frac{1}{p_i}\int_0^1 |u_i(x)|^{p_i}dx\right) - \frac{\lambda\vartheta}{c}$$

$$\geq \sum_{i=1}^n \frac{||u_i||_{p_i}^{p_i}}{p_i} - \frac{\lambda\eta}{c}\left(c\sum_{i=1}^n \frac{||u_i||_{p_i}^{p_i}}{p_i}\right) - \frac{\lambda\vartheta}{c}$$

$$= \sum_{i=1}^n \frac{||u_i||_{p_i}^{p_i}}{p_i} - \lambda\eta\sum_{i=1}^n \frac{||u_i||_{p_i}^{p_i}}{p_i} - \frac{\lambda\vartheta}{c}$$

$$\geq \left(1 - \eta\frac{r}{\int_0^1 \sup_{(t_1,\ldots,t_n)\in K(cr)} F(x, t_1, \ldots, t_n)dx}\right)\sum_{i=1}^n \frac{||u_i||_{p_i}^{p_i}}{p_i} - \frac{\lambda\vartheta}{c}.$$

Therefore,

$$\lim_{||u||\to\infty} (\Phi(u) - \lambda\Psi(u)) = \infty.$$

On the other hand, if $\eta \leq 0$, then it is clear that $\lim_{||u||\to\infty} I_\lambda(u)) = \infty$. That is, both cases lead to the coercivity of the functional I_λ, so condition (a2) of Theorem 1.1.3 holds with $\overline{v} = w$. Thus, by Theorem 1.1.3, $I_\lambda(u)$ has at least three distinct critical points. Since weak solutions of the system (3.1.1) are the critical points of $I_\lambda(u)$ an application of Lemma 3.2.1 completes the proof of the theorem. $\square$

In our next theorem, we give sufficient conditions for the existence of three nonnegative solutions of the system (3.1.1).

Theorem 3.3.2. *Assume that*

(C9) $a_j, b_j \in [0, 1)$ *for* $j = 1, \ldots, m$ *with* $\sum_{j=1}^m a_j \in [0, 1)$ *and* $\sum_{j=1}^m b_j \in [0, 1)$,

(C10) $F_{t_i}(x, t_1, \ldots, t_n) \geq 0$ *for all* $(x, t_1, \ldots, t_n) \in [0, 1] \times [0, \infty)^n$ *and* $i = 1, \ldots, n$,

and there exist a function $w = (w_1, \ldots, w_n) \in X$ *and two positive constants* r_1 *and* r_2 *with* $2r_1 < \sum_{i=1}^n \frac{||w_i||_{p_i}^{p_i}}{p_i} < \frac{r_2}{2}$ *such that*

(C11) $\dfrac{\int_0^1 \sup_{(t_1,\ldots,t_n)\in K(cr_1)} F(x,t_1,\ldots,t_n)dx}{r_1} < \dfrac{2}{3}\dfrac{\int_0^1 F(x,w_1(x),\ldots,w_n(x))dx}{\sum_{i=1}^n \frac{||w_i||_{p_i}^{p_i}}{p_i}}$,

(C12) $\dfrac{\int_0^1 \sup_{(t_1,\ldots,t_n)\in K(cr_2)} F(x,t_1,\ldots,t_n)dx}{r_2} < \dfrac{1}{3}\dfrac{\int_0^1 F(x,w_1(x),\ldots,w_n(x))dx}{\sum_{i=1}^n \frac{||w_i||_{p_i}^{p_i}}{p_i}}$.

Then, for each

$$\lambda \in \left(\frac{3}{2} \frac{\sum_{i=1}^{n} \frac{||w_i||_{p_i}^{p_i}}{p_i}}{\int_0^1 F(x, w_1(x), \ldots, w_n(x)) dx}, \; \Theta_1 \right),$$

where

$$\Theta_1 = \min \left\{ \frac{r_1}{\int_0^1 \sup_{(t_1, \ldots, t_n) \in K(cr_1)} F(x, t_1, \ldots, t_n) dx}, \right.$$

$$\left. \frac{r_2}{2 \int_0^1 \sup_{(t_1, \ldots, t_n) \in K(cr_2)} F(x, t_1, \ldots, t_n) dx} \right\},$$

the system (3.1.1) has at least three nonnegative classical solutions $v^j = (v_1^j, \ldots, v_n^j)$, $j = 1, 2, 3$, such that

$$\sum_{i=1}^{n} \frac{|v_i^j(x)|^{p_i}}{p_i} \leq cr_2 \quad \text{for each } x \in [0, 1] \text{ and } j = 1, 2, 3.$$

To prove this theorem, we need the comparison result given in the following lemma.

Lemma 3.3.1. ([219, Lemma 2.1]) *Let (B1) hold. Assume that $y \in C^1[0, 1]$ satisfies $\phi_p(y') \in AC[0, 1]$ with $p > 1$ and*

$$\begin{cases} -(\phi_p(y'))' \geq 0, & t \in (0, 1), \\ y(0) = \sum_{j=1}^{m} a_j y(x_j), \; y(1) = \sum_{j=1}^{m} b_j y(x_j). \end{cases}$$

Then, $y(t) \geq 0$ for $t \in [0, 1]$.

Proof of Theorem 3.3.2. To prove this theorem, we will apply Theorem 1.1.4, so we first note that condition (i) is satisfied. Let $u^* = (u_1^*, \ldots, u_n^*)$ and $u^{**} = (u_1^{**}, \ldots, u_n^{**})$ be two local minima of I_λ; then, u^* and u^{**} are critical points of I_λ, and hence they are weak solutions of system (3.1.1). Thus, by Lemma 3.2.1, u^* and u^{**} are classical solutions of (3.1.1). Now $F(x, 0, \ldots, 0) = 0$ and (C10) imply $F(x, t_1, \ldots, t_n) \geq 0$ for all $(x, t_1, \ldots, t_n) \in [0, 1] \times [0, \infty)^n$. Then, for $i = 1, \ldots, n$, from (C9) and Lemma 3.3.1, we see that $u_i^*(x) \geq 0$ and $u_i^{**}(x) \geq 0$ on $[0, 1]$. This implies $su_i^* + (1 - s)u_i^{**} \geq 0$ on $[0, 1]$ for $i = 1, \ldots, n$. Thus, $F(x, su^* + (1 - s)u^{**}) \geq 0$, and so $\Psi(su^* + (1 - s)u^{**}) \geq 0$ for all $s \in [0, 1]$, i.e., (ii) of Theorem 1.1.4 holds.

Since we have $2r_1 < \sum_{i=1}^{n} \frac{||w_i||_{p_i}^{p_i}}{p_i} < \frac{r_2}{2}$, it follows that $2r_1 < \Phi(w) < \frac{r_2}{2}$. In view of (3.3.1), we see that

$$\Phi^{-1}(-\infty, r_1) = \{u = (u_1, u_2, \ldots, u_n) \in X \; : \; \Phi(u) < r_1\}$$

$$= \left\{u = (u_1, u_2, \ldots, u_n) \in X : \sum_{i=1}^{n} \frac{||u_i||_{p_i}^{p_i}}{p_i} < r_1\right\}$$

$$\subseteq \left\{u = (u_1, u_2, \ldots, u_n) \in X : \sum_{i=1}^{n} \frac{|u_i(x)|^{p_i}}{p_i} \leq cr_1 \;\; \text{for each } x \in [0, 1]\right\}.$$

Thus,

$$\sup_{(u_1,\ldots,u_n)\in\Phi^{-1}(-\infty,r_1)} \Psi(u) = \sup_{(u_1,\ldots,u_n)\in\Phi^{-1}(-\infty,r_1)} \int_0^1 F(x, u_1(x), \ldots, u_n(x))dx$$

$$\leq \int_0^1 \sup_{(t_1,\ldots,t_n)\in K(cr_1)} F(x, t_1, \ldots, t_n)dx.$$

Therefore, from (B3), it follows that

$$\frac{\sup_{u\in\Phi^{-1}(-\infty,r_1)}\Psi(u)}{r_1} = \frac{\sup_{(u_1,\ldots,u_n)\in\Phi^{-1}(-\infty,r_1)}\int_0^1 F(x, u_1(x), \ldots, u_n(x))dx}{r_1}$$

$$\leq \frac{\int_0^1 \sup_{(t_1,\ldots,t_n)\in K(cr_1)} F(x, t_1, \ldots, t_n)dx}{r_1}$$

$$< \frac{2}{3}\frac{\int_0^1 F(x, w_1(x), \ldots, w_n(x))dx}{\sum_{i=1}^{n} \frac{||w_i||_{p_i}^{p_i}}{p_i}}$$

$$= \frac{2}{3}\frac{\Psi(w)}{\Phi(w)},$$

so (b1) holds with $\overline{v} = w$.

Using (C12) and arguing as above, we have

$$\frac{\sup_{u\in\Phi^{-1}(-\infty,r_2)}\Psi(u)}{r_2} = \frac{\sup_{(u_1,\ldots,u_n)\in\Phi^{-1}(-\infty,r_2)}\int_0^1 F(x, u_1(x), \ldots, u_n(x))dx}{r_2}$$

$$\leq \frac{\int_0^1 \sup_{(t_1,\ldots,t_n)\in K(cr_2)} F(x, t_1, \ldots, t_n)dx}{r_2}$$

$$< \frac{1}{3}\frac{\int_0^1 F(x, w_1(x), \ldots, w_n(x))dx}{\sum_{i=1}^{n} \frac{||w_i||_{p_i}^{p_i}}{p_i}}$$

$$= \frac{1}{3}\frac{\Psi(w)}{\Phi(w)}.$$

Thus, (b2) holds with $\overline{v} = w$.

Therefore, by Theorem 1.1.4, $I_\lambda(u)$ has at least three distinct critical points, which are all nonnegative by Lemma 3.3.1. Taking into account the fact that the weak solutions of (3.1.1) are the critical points of $I_\lambda(u)$ and applying Lemma 3.2.1 and (3.3.1) completes the proof of the theorem. □

Next we present some easy consequences of our theorems in case the test function w is specified. Let

$$\sigma_n = \left[2^{p_n-1} \left(x_1^{1-p_n} \left| 1 - \sum_{j=1}^m a_j \right|^{p_n} + (1-x_m)^{1-p_n} \left| 1 - \sum_{j=1}^m b_j \right|^{p_n} \right) \right]^{1/p_n}$$

and define

$$B_{1,n}(x) = \begin{cases} [x\sum_{j=1}^m a_j, x]^n, & \text{if } \sum_{j=1}^m a_j < 1, \\ [x, x\sum_{j=1}^m a_j]^n, & \text{if } \sum_{j=1}^m a_j > 1, \end{cases}$$

and

$$B_{2,n}(x) = \begin{cases} [x\sum_{j=1}^m b_j, x]^n, & \text{if } \sum_{j=1}^m b_j < 1, \\ [x, x\sum_{j=1}^m b_j]^n, & \text{if } \sum_{j=1}^m b_j > 1, \end{cases}$$

where $[\cdot, \cdot]^n = [\cdot, \cdot] \times \cdots \times [\cdot, \cdot]$.

Corollary 3.3.1. *Assume there exist constants $\theta > 0$ and $\tau > 0$ such that*

(C13) $(\sigma_n \tau)^{p_n} > \dfrac{2\theta^{p_n}}{c\prod_{i=1}^{n-1} p_i}$,

(C14) $F(x, t_1, \ldots, t_n) \geq 0$ *for each $x \in [0, x_1/2] \cup [(1+x_m)/2, 1]$ and* $(t_1, \ldots, t_n) \in B_{1,n}(\tau) \cup B_{2,n}(\tau)$,

(C15) $\dfrac{\int_0^1 \sup_{(t_1,\ldots,t_n)\in K\left(\frac{\theta p_n}{\prod_{i=1}^n p_i}\right)} F(x,t_1,\ldots,t_n)\,dx}{\theta^{p_n}}$

$$< \frac{2}{3c(\sigma_n\tau)^{p_n}\prod_{i=1}^{n-1} p_i} \int_{\frac{x_1}{2}}^{\frac{1+x_m}{2}} F(x,0,\ldots,0,\tau)\,dx,$$

(C16) $\limsup_{|t_1|\to\infty,\ldots,|t_n|\to\infty} \dfrac{F(x,t_1,\ldots,t_n)}{\sum_{i=1}^n \frac{|t_i|^{p_i}}{p_i}}$

$$< \frac{\prod_{i=1}^n p_i}{\theta^{p_n}} \int_0^1 \sup_{(t_1,\ldots,t_n)\in K\left(\frac{\theta p_n}{\prod_{i=1}^n p_i}\right)} F(x,t_1,\ldots,t_n)\,dx.$$

Then, for each

$$\lambda \in \left(\frac{3}{2} \frac{(\sigma_n\tau)^{p_n}}{p_n \int_{\frac{x_1}{2}}^{\frac{1+x_m}{2}} F(x,0,\ldots,0,\tau)\,dx}, \right.$$

$$\left. \frac{\theta^{p_n}}{c\prod_{i=1}^n p_i \int_0^1 \sup_{(t_1,\ldots,t_n)\in K\left(\frac{\theta p_n}{\prod_{i=1}^n p_i}\right)} F(x,t_1,\ldots,t_n)\,dx} \right),$$

the system (3.1.1) has at least three classical solutions.

Proof. Under the conditions (C13)–(C16), the assumptions (C6)–(C8) of Theorem 3.3.1 are satisfied by choosing $w = (0, \ldots, 0, w_n(x))$ with

$$
w_n(x) = \begin{cases}
\tau\left(\sum_{j=1}^m a_j + \frac{2(1-\sum_{j=1}^m a_j)}{x_1}x\right), & \text{if } x \in [0, \frac{x_1}{2}), \\
\tau, & \text{if } x \in [\frac{x_1}{2}, \frac{1+x_m}{2}], \\
\tau\left(\frac{2-\sum_{j=1}^m b_j - x_m \sum_{j=1}^m b_j}{1-x_m} - \frac{2(1-\sum_{j=1}^m b_j)}{1-x_m}x\right), & \text{if } x \in (\frac{1+x_m}{2}, 1],
\end{cases}
$$
(3.3.3)

and $r = \frac{\theta p_n}{c\prod_{i=1}^n p_i}$. It is not hard to see that $w = (0, \ldots, 0, w_n) \in X$ and, in particular,

$$
\|w_n\|_{p_n}^{p_n} = (\sigma_n\tau)^{p_n}.
$$

Thus,

$$
\Phi(w) = \sum_{i=1}^n \frac{\|w_i\|_{p_i}^{p_i}}{p_i} = \frac{(\sigma_n\tau)^{p_n}}{p_n}.
$$

Then, (C13) implies (C6). On the other hand, since

$$
\tau\sum_{j=1}^m a_j \leq w_n(x) \leq \tau \quad \text{for each } x \in [0, x_1/2] \text{ if } \sum_{j=1}^m a_j < 1,
$$

$$
\tau \leq w_n(x) \leq \tau\sum_{j=1}^m a_j \quad \text{for each } x \in [0, x_1/2] \text{ if } \sum_{j=1}^m a_j > 1,
$$

$$
\tau\sum_{j=1}^m b_j \leq w_n(x) \leq \tau \quad \text{for each } x \in [(1+x_m)/2, 1] \text{ if } \sum_{j=1}^m b_j < 1,
$$

and

$$
\tau \leq w_n(x) \leq \tau\sum_{j=1}^m b_j \quad \text{for each } x \in [(1+x_m)/2, 1] \text{ if } \sum_{j=1}^m b_j > 1,
$$

condition (C14) ensures that

$$
\int_0^{\frac{x_1}{2}} F(x, w_1(x), \ldots, w_n(x))dx + \int_{\frac{1+x_m}{2}}^1 F(x, w_1(x), \ldots, w_n(x))dx \geq 0,
$$

and so

$$
\int_0^1 F(x, w_1(x), \ldots, w_n(x))dx \geq \int_{\frac{x_1}{2}}^{\frac{1+x_m}{2}} F(x, 0, \ldots, 0, \tau)dx.
$$

From this and (C15), it is easy to see that (C7) holds. Finally, note that (C16) implies (C8) holds. The conclusion then follows from Theorem 3.3.1.

$\square$

Corollary 3.3.2. *Assume that (C9) and (C10) hold and there exist three positive constants θ_1, θ_2, and τ with*

$$2\theta_1^{p_n} < (\sigma_n \tau)^{p_n} c \prod_{i=1}^{n-1} p_i < \frac{\theta_2^{p_n}}{2}$$

such that

(C17)
$$\frac{\int_0^1 \sup_{(t_1,\ldots,t_n)\in K\left(\frac{\theta_1^{p_n}}{\prod_{i=1}^n p_i}\right)} F(x,t_1,\ldots,t_n)\,dx}{\theta_1^{p_n}}$$
$$< \frac{2}{3c(\sigma_n \tau)^{p_n} \prod_{i=1}^{n-1} p_i} \int_{\frac{x_1}{2}}^{\frac{1+x_m}{2}} F(x,0,\ldots,0,\tau)\,dx,$$

(C18)
$$\frac{\int_0^1 \sup_{(t_1,\ldots,t_n)\in K\left(\frac{\theta_2^{p_n}}{\prod_{i=1}^n p_i}\right)} F(x,t_1,\ldots,t_n)\,dx}{\theta_2^{p_n}}$$
$$< \frac{1}{3c(\sigma_n \tau)^{p_n} \prod_{i=1}^{n-1} p_i} \int_{\frac{x_1}{2}}^{\frac{1+x_m}{2}} F(x,0,\ldots,0,\tau)\,dx.$$

Then, for each

$$\lambda \in \left(\frac{3}{2} \frac{(\sigma_n \tau)^{p_n}}{p_n \int_{\frac{x_1}{2}}^{\frac{1+x_m}{2}} F(x,0,\ldots,0,\tau)\,dx}, \; \Theta_2 \right),$$

where

$$\Theta_2 = \min \left\{ \frac{\theta_1^{p_n}}{c \prod_{i=1}^n p_i \int_0^1 \sup_{(t_1,\ldots,t_n)\in K\left(\frac{\theta_1^{p_n}}{\prod_{i=1}^n p_i}\right)} F(x,t_1,\ldots,t_n)\,dx}, \right.$$
$$\left. \frac{\theta_2^{p_n}}{c \prod_{i=1}^n p_i \int_0^1 \sup_{(t_1,\ldots,t_n)\in K\left(\frac{\theta_2^{p_n}}{\prod_{i=1}^n p_i}\right)} F(x,t_1,\ldots,t_n)\,dx} \right\},$$

the system (3.1.1) has at least three nonnegative classical solutions $v^j = (v_1^j,\ldots,v_n^j)$, $j = 1,2,3$, such that

$$\sum_{i=1}^n \frac{|v_i^j(x)|^{p_i}}{p_i} \leq \frac{\theta_2^{p_n}}{\prod_{i=1}^n p_i} \quad \text{for each } x \in [0,1] \text{ and } j = 1,2,3.$$

Proof. Let $w = (0,\ldots,0,w_n(x))$ with $w_n(x)$ defined by (3.3.3), $r_1 = \frac{\theta_1^{p_n}}{c \prod_{i=1}^n p_i}$, and $r_2 = \frac{\theta_2^{p_n}}{c \prod_{i=1}^n p_i}$. Then, under the conditions (C17) and (C18), it is easy to check that (C11) and (C12) of Theorem 3.3.2 hold. The conclusion then follows from Theorem 3.3.2; we omit the details. $\square$

3.4 Infinitely Many Solutions of Multi-point Problems

We now turn our attention to the question of existence of infinitely many solutions of the system (3.1.1). As we needed to do in the previous sections in this chapter, we will assume that condition (H2) holds without further mention. Our first result in this direction is the following.

Theorem 3.4.1. *Assume that*

(C20) $F(x, t_1, \ldots, t_n) \geq 0$ *for each* $x \in [0, x_1/2] \cup [(1 + x_m)/2, 1]$ *and* $(t_1, \ldots, t_n) \in \mathbb{R}^n$;

(C21) *there exists* $l \in \{1, \ldots, n\}$ *such that*

$$0 \leq \liminf_{\xi \to \infty} \frac{\int_0^1 \sup_{(t_1,\ldots,t_n) \in K(\xi^{p_l})} F(x, t_1, \ldots, t_n) dx}{\xi^{p_l}}$$

$$< \frac{p_l}{c\sigma_l^{p_l}} \limsup_{\xi \to \infty} \frac{\int_{\frac{x_1}{2}}^{\frac{1+x_m}{2}} F(x, 0, \ldots, \xi, \ldots, 0) dx}{\xi^{p_l}},$$

where, in $F(x, 0, \ldots, \xi, \ldots, 0)$, ξ *is the* $(l+1)$-*th argument. Then, for each* $\lambda \in \Lambda$, *the system* (3.1.1) *has an unbounded sequence of classical solutions, where*

$$\Lambda = \left(\frac{\sigma_l^{p_l}}{p_l \limsup_{\xi \to \infty} \frac{\int_{\frac{x_1}{2}}^{\frac{1+x_m}{2}} F(x,0,\ldots,\xi,\ldots,0) dx}{\xi^{p_l}}} , \right.$$

$$\left. \frac{1}{c \liminf_{\xi \to \infty} \frac{\int_0^1 \sup_{(t_1,\ldots,t_n) \in K(\xi^{p_l})} F(x,t_1,\ldots,t_n) dx}{\xi^{p_l}}} \right). \quad (3.4.1)$$

Proof. Our technique of proof to be used here is to apply Theorem 1.1.5 (b). Once again, we define the functionals Φ, Ψ, and I_λ as in (3.2.6)–(3.2.8) and whose Gâteaux derivatives are given in (3.2.9)–(3.2.10).

Let $\{\xi_k\}$ be a sequence of positive numbers such that $\xi_k \to \infty$ as $k \to \infty$ and

$$\lim_{k \to \infty} \frac{\int_0^1 \sup_{(t_1,\ldots,t_n) \in K(\xi_k^{p_l})} F(x, t_1, \ldots, t_n) dx}{\xi_k^{p_l}}$$

$$= \liminf_{\xi \to \infty} \frac{\int_0^1 \sup_{(t_1,\ldots,t_n) \in K(\xi^{p_l})} F(x, t_1, \ldots, t_n) dx}{\xi^{p_l}}. \quad (3.4.2)$$

For each $(u_1, \ldots, u_n) \in X$, we have

$$\sup_{x \in [0,1]} |u_i(x)|^{p_i} \leq c \|u_i\|_{p_i}^{p_i} \quad \text{for } i = 1, \ldots, n,$$

with c defined in (3.2.1). Hence, and so

$$\sup_{x \in [0,1]} \sum_{i=1}^{n} \frac{|u_i(x)|^{p_i}}{p_i} \leq c \sum_{i=1}^{n} \frac{\|u_i\|_{p_i}^{p_i}}{p_i}.$$

Let $r_k = c^{-1} \xi_k^{p_l}$ for $k \in \mathbb{N}$. Then, for $v = (v_1, \ldots, v_n) \in X$ with $\sum_{i=1}^{n} \frac{\|v_i\|_{p_i}^{p_i}}{p_i} < r_k$, we have

$$\sup_{x \in [0,1]} \sum_{i=1}^{n} \frac{|v_i(x)|^{p_i}}{p_i} \leq \xi_k^{p_l}.$$

Note that $0 \in \Phi^{-1}(-\infty, r_k)$ and $\Psi(0) \geq 0$ by (C20). Then,

$$\varphi(r_k) = \inf_{u \in \Phi^{-1}(-\infty, r_k)} \frac{(\sup_{v \in \Phi^{-1}(-\infty, r_k)} \Psi(v)) - \Psi(u)}{r_k - \Phi(u)}$$

$$\leq \frac{\sup_{v \in \Phi^{-1}(-\infty, r_k)} \Psi(v)}{r_k}$$

$$\leq \frac{c \int_0^1 \sup_{(t_1, \ldots, t_n) \in K(\xi_k^{p_l})} F(x, t_1, \ldots, t_n) dx}{\xi_k^{p_l}}.$$

Then, from (3.4.2) and (C21), we see that

$$\gamma := \liminf_{k \to \infty} \varphi(r_k) \leq c \liminf_{\xi \to \infty} \frac{\int_0^1 \sup_{(t_1, \ldots, t_n) \in K(\xi^{p_l})} F(x, t_1, \ldots, t_n) dx}{\xi^{p_l}} < \infty.$$

$$(3.4.3)$$

By (C21), (3.4.1), and (3.4.3), we have $\Lambda \subseteq (0, 1/\gamma)$. Let $\lambda \in \Lambda$ be fixed. By Theorem 1.1.5 (b), one of the following alternatives must hold:

(b$_1$) either $I_\lambda = \Phi - \lambda \Psi$ has a global minimum, or

(b$_2$) there exists a sequence $\{(u_{1k}, \ldots, u_{nk})\}$ of critical points of I_λ such that

$$\lim_{k \to \infty} \|(u_{1k}, \ldots, u_{nk})\| = \infty.$$

To show that (b$_1$) does not hold, take $\lambda \in \Lambda$. Then by (3.4.1), we have

$$\frac{1}{\lambda} < \frac{p_l}{\sigma_l^{p_l}} \limsup_{\xi \to \infty} \frac{\int_{\frac{x_1}{2}}^{\frac{1+x_m}{2}} F(x, 0, \ldots, \xi, \ldots, 0) dx}{\xi^{p_l}},$$

so there exists a sequence $\{d_k\}$ of positive numbers and a constant τ such that $d_k \to \infty$ as $k \to \infty$ and

$$\frac{1}{\lambda} < \tau < \frac{p_l}{\sigma_l^{p_l}} \frac{\int_{\frac{x_1}{2}}^{\frac{1+x_m}{2}} F(x, 0, \ldots, d_k, \ldots, 0) dx}{d_k^{p_l}} \tag{3.4.4}$$

for large $k \in \mathbb{N}$.

Let $\{w_k\}$ be the sequence in X defined by $w_k(x) = (0, \ldots, w_{lk}(x), \ldots, 0)$, where w_{lk} is the l-th argument of w_k and is defined by

$$w_{lk}(x) = \begin{cases} d_k \left(\sum_{j=1}^{m} a_j + \frac{2(1 - \sum_{j=1}^{m} a_j)}{x_1} x \right), & \text{if } x \in [0, \frac{x_1}{2}), \\ d_k, & \text{if } x \in [\frac{x_1}{2}, \frac{1+x_m}{2}], \\ d_k \left(\frac{2 - \sum_{j=1}^{m} b_j - x_m \sum_{j=1}^{m} b_j}{1 - x_m} - \frac{2(1 - \sum_{j=1}^{m} b_j)}{1 - x_m} x \right), & \text{if } x \in (\frac{1+x_m}{2}, 1]. \end{cases} \tag{3.4.5}$$

For any fixed $k \in \mathbb{N}$, it is easy to see that $w_k = (0, \ldots, w_{lk}, \ldots, 0) \in X$ and $\|w_{lk}\|_{p_l}^{p_l} = (\sigma_l d_k)^{p_l}$, so

$$\Phi(w_k) = \frac{(\sigma_l d_k)^{p_l}}{p_l}. \tag{3.4.6}$$

From (C20),

$$\Psi(w_k) \geq \int_{\frac{x_1}{2}}^{\frac{1+x_m}{2}} F(x, 0 \ldots, d_k, \ldots, 0) dx. \tag{3.4.7}$$

Now (3.4.4), (3.4.6), and (3.4.7) imply

$$\Phi(w_k) - \lambda \Psi(w_k) \leq \frac{(\sigma_l d_k)^{p_l}}{p_l} - \lambda \int_{\frac{x_1}{2}}^{\frac{1+x_m}{2}} F(x, 0, \ldots, d_k, \ldots, 0) dx$$

$$< \frac{(\sigma_l d_k)^{p_l}}{p_l} (1 - \lambda\tau) \tag{3.4.8}$$

for large $k \in \mathbb{N}$. Since $\lambda\tau > 1$ and $d_k \to \infty$, we see that $\lim_{k\to\infty}(I_\lambda(w_k)) = -\infty$, so I_λ is unbounded from below. Hence, alternative (b_1) does not hold.

Thus, (b_2) holds and so there is a sequence $\{(u_{1k}, \ldots, u_{nk})\}$ of critical points of I_λ such that $\lim_{k\to\infty} \|(u_{1k}, \ldots, u_{nk})\| = \infty$. Finally, since weak solutions of system (3.1.1) are the critical points of I_λ, applying Lemma 3.2.1 completes the proof of the theorem. $\square$

The following corollary is the special case of Theorem 3.4.1 where $F(x, t_1, \ldots, t_n) \equiv F(t_1, \ldots, t_n)$.

Corollary 3.4.1. *Assume that*

(C22) $F(t_1, \ldots, t_n) \geq 0$ for each $(t_1, \ldots, t_n) \in \mathbb{R}^n$;

(C23) there exists $l \in \{1, \ldots, n\}$ such that

$$\liminf_{\xi \to \infty} \frac{\sup_{(t_1, \ldots, t_n) \in K(\xi^{p_l})} F(t_1, \ldots, t_n)}{\xi^{p_l}}$$
$$< \frac{p_l(1 + x_m - x_1)}{2c\sigma_l^{p_l}} \limsup_{\xi \to \infty} \frac{F(0, \ldots, \xi, \ldots, 0)}{\xi^{p_l}}.$$

Then, for each

$$\lambda \in \left(\frac{2\sigma_l^{p_l}}{p_l(1 + x_m - x_1) \limsup\limits_{\xi \to \infty} \frac{F(0, \ldots, \xi, \ldots, 0)}{\xi^{p_l}}} \right.,$$

$$\left. \frac{1}{c \liminf\limits_{\xi \to \infty} \frac{\sup_{(t_1, \ldots, t_n) \in K(\xi^{p_l})} F(t_1, \ldots, t_n)}{\xi^{p_l}}} \right),$$

the system

$$\begin{cases} -(\phi_{p_i}(u_i'))' = \lambda F_{u_i}(u_1, \ldots, u_n), \ t \in (0, 1), \\ u_i(0) = \sum_{j=1}^m a_j u_i(x_j), \ u_i(1) = \sum_{j=1}^m b_j u_i(x_j), \end{cases} \quad i = 1, \ldots, n,$$

has an unbounded sequence of classical solutions.

The following two results are special cases of Corollary 3.4.1.

Corollary 3.4.2. *Assume that there exists a function $F : \mathbb{R}^n \to \mathbb{R}$ such that $F(x_1, \ldots, x_n) \geq 0$ in $\mathbb{R}^n$, $F(x_1, \ldots, x_n)$ is continuously differentiable in x_i, and $\partial F / \partial x_i = f_i$ for $i = 1, \ldots, n$. Suppose further that there exists $l \in \{1, \ldots, n\}$ such that*

$$\liminf_{\xi \to \infty} \frac{\sup_{(t_1, \ldots, t_n) \in K(\xi^{p_l})} F(t_1, \ldots, t_n)}{\xi^{p_l}} = 0$$

and

$$\limsup_{\xi \to \infty} \frac{F(0, \ldots, \xi, \ldots, 0)}{\xi^{p_l}} = \infty,$$

where, in $F(0, \ldots, \xi, \ldots, 0)$, ξ is the l-th argument and

$$K(\xi^{p_l}) = \left\{ (t_1, \ldots, t_n) \in \mathbb{R}^n : \sum_{i=1}^n \frac{|t_i|^{p_i}}{p_i} \leq \xi^{p_l} \right\}.$$

Then, the system

$$\begin{cases} -(\phi_{p_i}(u_i'))' = f_i(u_1, \ldots, u_n), \quad t \in (0, 1), \\ u_i(0) = \sum_{j=1}^m a_j u_i(x_j), \ u_i(1) = \sum_{j=1}^m b_j u_i(x_j), \end{cases} \quad i = 1, \ldots, n,$$

has an unbounded sequence of classical solutions.

Corollary 3.4.3. *Assume that* $g_i : \mathbb{R} \to \mathbb{R}$, $i = 1, \ldots, n$, *are continuously differentiable functions such that*

(C24) $g_i(t) \geq 0$ *for each* $i = 1, \ldots, n$ *and* $t \in \mathbb{R}$;

(C25) $\liminf_{\xi \to \infty} \dfrac{\sup_{(t_1, \ldots, t_n) \in K(\xi^{p_n})} \prod_{i=1}^{n} g_i(t_i)}{\xi^{p_n}}$
$$< \frac{p_n(1 + x_m - x_1)}{2c\sigma_n^{p_n}} \prod_{i=1}^{n-1} g_i(0) \limsup_{\xi \to \infty} \frac{g_n(\xi)}{\xi^{p_n}}.$$

Then, for each

$$\lambda \in \left(\frac{2\sigma_n^{p_n}}{p_n(1 + x_m - x_1) \prod_{i=1}^{n-1} g_i(0) \limsup_{\xi \to \infty} \frac{g_n(\xi)}{\xi^{p_n}}}, \right.$$

$$\left. \frac{1}{c \liminf_{\xi \to \infty} \frac{\sup_{(t_1, \ldots, t_n) \in K(\xi^{p_n})} \prod_{i=1}^{n} g_i(t_i)}{\xi^{p_n}}} \right),$$

the system

$$\begin{cases} -(\phi_{p_i}(u_i'))' = \lambda g_i'(u_i)(\prod_{j=1, j \neq i}^{n} g_j(u_j)), \ t \in (0,1), \\ u_i(0) = \sum_{j=1}^{m} a_j u_i(x_j), \ u_i(1) = \sum_{j=1}^{m} b_j u_i(x_j), \end{cases} \quad i = 1, \ldots, n, \quad (3.4.9)$$

has an unbounded sequence of classical solutions.

Next, we will make use of Theorem 1.1.5 (c) to obtain the following result.

Theorem 3.4.2. *In addition to condition (C20) assume that*

(C26) *there exists* $l \in \{1, \ldots, n\}$ *such that*

$$0 \leq \liminf_{\xi \to 0^+} \frac{\int_0^1 \sup_{(t_1, \ldots, t_n) \in K(\xi^{p_l})} F(x, t_1, \ldots, t_n) dx}{\xi^{p_l}}$$

$$< \frac{p_l}{c\sigma_l^{p_l}} \limsup_{\xi \to 0^+} \frac{\int_{\frac{x_1}{2}}^{\frac{1 + x_m}{2}} F(x, 0, \ldots, \xi, \ldots, 0) dx}{\xi^{p_l}},$$

where, in $F(x, 0, \ldots, \xi, \ldots, 0)$, ξ *is the* $(l+1)$*-th argument. Then, for each* $\lambda \in \Lambda_1$, *the system (3.1.1) has a sequence of classical solutions converging* *to zero, where*

$$\Lambda_1 = \left(\frac{\dfrac{\sigma_l^{p_l}}{p_l \limsup\limits_{\xi \to 0^+} \dfrac{\int_{\frac{x_1}{2}}^{\frac{1+x_m}{2}} F(x,0,...,\xi,...,0)dx}{\xi^{p_l}}}}{c \liminf\limits_{\xi \to 0^+} \dfrac{\int_0^1 \sup_{(t_1,...,t_n) \in K(\xi^{p_l})} F(x,t_1,...,t_n)dx}{\xi^{p_l}}} \right). \qquad (3.4.10)$$

Proof. Proceeding as in the proof of Theorem 3.4.1, let $\{\xi_k\}$ be a sequence of positive numbers such that $\xi_k \to 0^+$ as $k \to \infty$ and

$$\lim_{k \to \infty} \frac{\int_0^1 \sup_{(t_1,...,t_n) \in K(\xi_k^{p_l})} F(x,t_1,\ldots,t_n)dx}{\xi_k^{p_l}}$$

$$= \liminf_{\xi \to 0^+} \frac{\int_0^1 \sup_{(t_1,...,t_n) \in K(\xi^{p_l})} F(x,t_1,\ldots,t_n)dx}{\xi^{p_l}}.$$

From the fact that $\inf_X \Phi = 0$ and the definition of δ, we see that $\delta = \liminf_{r \to 0^+} \varphi(r)$. As in showing (3.4.3) in the proof of Theorem 3.4.1, we can prove that $\delta < \infty$ and $\Lambda_1 \subseteq (0, 1/\delta)$. Taking $\lambda \in \Lambda_1$, by Theorem 1.1.5 (c), one of the following must hold:

(c$_1$) either there is a global minimum of Φ which is a local minimum of $I_\lambda = \Phi - \lambda\Psi$, or

(c$_2$) there exists a sequence $\{(u_{1k},\ldots,u_{nk})\}$ of critical points of I_λ which converges weakly to a global minimum of Φ.

To show (c$_1$) does not hold, notice that (3.4.10) implies there exist $\tau > 0$ and a sequence $\{d_k\}$ of positive numbers such that $d_k \to 0^+$ as $k \to \infty$ and

$$\frac{1}{\lambda} < \tau < \frac{p_l}{\sigma_l^{p_l}} \frac{\int_{\frac{x_1}{2}}^{\frac{1+x_m}{2}} F(x,0,\ldots,d_k,\ldots,0)dx}{d_k^{p_l}} \qquad \text{for } k \in \mathbb{N} \text{ large enough.}$$

Let $\{w_k\}$ be a sequence in X defined by $w_k(x) = (0,\ldots,w_{lk}(x),\ldots,0)$, where $w_{lk}(x)$ is the l-th argument of w_k and is defined by (3.4.5) with the above d_k. Note that $\lambda\tau > 1$, Then, as in showing (3.4.8), we can obtain that

$$I_\lambda(w_k) = \Phi(w_k) - \lambda\Psi(w_k)$$
$$\leq \frac{(\sigma_l d_k)^{p_l}}{p_l} - \lambda \int_{\frac{x_1}{2}}^{\frac{1+x_m}{2}} F(x, 0, \ldots, d_k, \ldots, 0)dx$$
$$< \frac{(\sigma_l d_k)^{p_l}}{p_l}(1 - \lambda\tau) < 0$$

for every $k \in \mathbb{N}$ large enough. Then, since $\lim_{k\to\infty} I_\lambda(w_k) = I_\lambda(0) = 0$, we see that 0 is not a local minimum of I_λ. This, together with the fact that 0 is the only global minimum of Φ, shows that alternative (c_1) does not hold.

Thus, there exists a sequence $\{(u_{1k}, \ldots, u_{nk})\}$ of critical points of I_λ converges weakly to 0. Since the embedding $X = X_1 \times \cdots \times X_n \hookrightarrow C^0([0,1]) \times \cdots \times C^0([0,1])$ is compact, the sequence of critical points converges uniformly to zero. The fact that weak solutions of (3.1.1) are critical points of I_λ and applying Lemma 3.2.1 completes the proof of the theorem. $\qquad\square$

Results similar to Corollaries 3.4.1–3.4.3 can be obtained by applying Theorem 3.4.2.

As an example of our results in this section we give an example. It is motivated by [42, Example 4.1].

Example 3.4.1. Let $n \geq 1$ be an integer and

$$f(\xi) = \begin{cases} \dfrac{32(n+1)!^2\left[(n+1)!^2 - n!^2\right]}{\pi}\sqrt{\dfrac{1}{16(n+1)!^2} - \left(\xi - \dfrac{n!(n+2)}{2}\right)^2} + 1, \\ \qquad \text{if } \xi \in \cup_{n\in\mathbb{N}}[c_n, d_n], \\ 1, \qquad\qquad\qquad\qquad\qquad\qquad\qquad\qquad\qquad \text{otherwise,} \end{cases}$$

where

$$c_n = \frac{2n!(n+2)! - 1}{4(n+1)!} \quad \text{and} \quad d_n = \frac{2n!(n+2)! + 1}{4(n+1)!}.$$

Assume that $g_i(\xi) = \cos^2\xi$ for $i = 1, \ldots, n-1$, and g_n is a continuously differentiable function such that $g_n(\xi) \geq 0$ and $g_n'(\xi) = f(\xi)$ for $\xi \in \mathbb{R}$.

Then, by simple computations,

$$g_n(c_n) = g_n(0) + \int_0^{c_n} f(t)dt$$

$$= g_n(0) + \int_0^{c_n} 1dt$$

$$+ \frac{32(n+1)!^2 \left[(n+1)!^2 - n!^2\right]}{\pi}$$

$$\times \int_{t \in \cup_{i=1}^{n-1}[c_i, d_i]} \sqrt{\frac{1}{16(n+1)!^2} - \left(t - \frac{n!(n+2)}{2}\right)^2} \, dt$$

$$= g_n(0) + c_n + n!^2 - 1, \tag{3.4.11}$$

and

$$g_n(d_n) = g_n(0) + \int_0^{d_n} f(t)dt$$

$$= g_n(0) + \int_0^{d_n} 1dt$$

$$+ \frac{32(n+1)!^2 \left[(n+1)!^2 - n!^2\right]}{\pi}$$

$$\times \int_{t \in \cup_{i=1}^{n}[c_i, d_i]} \sqrt{\frac{1}{16(n+1)!^2} - \left(t - \frac{n!(n+2)}{2}\right)^2} \, dt$$

$$= g_n(0) + d_n + (n+1)!^2 - 1. \tag{3.4.12}$$

From (3.4.11) and (3.4.12), it is easy to see that

$$\lim_{n \to \infty} \frac{g_n(c_n)}{c_n^2} = 0, \quad \text{and} \quad \lim_{n \to \infty} \frac{g_n(d_n)}{d_n^2} = 4.$$

Note that there is no sequence $\{h_n\}$ such that $h_n \to \infty$ and $\lim_{n \to \infty} g(h_n)/h_n^2 > 4$. Then,

$$\liminf_{\xi \to \infty} \frac{g_n(\xi)}{\xi^2} = 0, \quad \text{and} \quad \limsup_{\xi \to \infty} \frac{g_n(\xi)}{\xi^2} = 4. \tag{3.4.13}$$

We claim that for each $\lambda \in (1/2, \infty)$, the system

$$\begin{cases} -u_i'' = \lambda g_i'(u_i)(\prod_{j=1, j\neq i}^n g_j(u_j)), \ t \in (0,1), \\ u_i(0) = \frac{1}{2}u_i(\frac{1}{2}), \ u_i(1) = \frac{1}{2}u_i(\frac{1}{2}), \end{cases} \quad i = 1, \ldots, n, \tag{3.4.14}$$

has an unbounded sequence of classical solutions.

In fact, with $m = 1$, $a_1 = b_1 = x_1 = 1/2$, and $p_i = 2$ for $i = 1, \ldots, n$, (3.4.14) is a special case of (3.4.9). From (3.4.13), we see that (C24) and

(C25) of Corollary 3.4.3 hold. From (3.2.2) we have $\sigma_n = \sqrt{2}$, and from (3.4.13)

$$\frac{2\sigma_n^{p_n}}{p_n(1 + x_m - x_1) \prod_{i=1}^{n-1} g_i(0) \limsup_{\xi \to \infty} \frac{g_n(\xi)}{\xi^{p_n}}} = \frac{1}{2}$$

and

$$\frac{1}{c \liminf_{\xi \to \infty} \frac{\sup_{(t_1, \ldots, t_n) \in K(\xi^{p_n})} \prod_{i=1}^{n} g_i(t_i)}{\xi^{p_n}}} = \infty.$$

The conclusion then follows directly from Corollary 3.4.3.

3.5 Two Parameter Systems

In this section, we consider a multipoint problem containing two parameters. This problem can be viewed as a perturbed form of the system (3.1.1) considered earlier in this chapter. For results related to the problem studied here, we refer the reader to [2, 5, 49, 108, 174, 175, 268].

Consider the multi-point boundary value system

$$\begin{cases} -(\phi_{p_i}(u_i'))' = \lambda F_{u_i}(x, u_1, \ldots, u_n) + \mu G_{u_i}(x, u_1, \ldots, u_n), & \text{in } (0,1), \\ u_i(0) = \sum_{j=1}^{m} a_j u_i(x_j), \quad u_i(1) = \sum_{j=1}^{m} b_j u_i(x_j), & \text{for } i = 1, \ldots, n, \end{cases}$$
$$(3.5.1)$$

where $p_i > 1$ and $\phi_{p_i}(t) = |t|^{p_i - 2} t$ for $i = 1, \ldots, n$, $\lambda, \mu > 0$ are parameters, $m, n \geq 1$ are integers, $a_j, b_j \in R$ for $j = 1, \ldots, m$, and $0 < x_1 \leq x_2 \leq x_3 \leq \cdots \leq x_m < 1$. Here, $F : [0,1] \times R^n \to R$ is a function such that the mapping $(t_1, t_2, \ldots, t_n) \to F(x, t_1, t_2, \ldots, t_n)$ is in C^1 in R^n for all $x \in [0,1]$, F_{t_i} is continuous in $[0,1] \times R^n$ for $i = 1, \ldots, n$, where F_{t_i} denotes the partial derivative of F with respect to t_i, and $F(x, 0, \ldots, 0) = 0$ for all $x \in [0,1]$. Similarly, $G : [0,1] \times R^n \to R$ is a function such that $(t_1, t_2, \ldots, t_n) \to G(x, t_1, t_2, \ldots, t_n)$ is in C^1 in R^n for all $x \in [0,1]$, and G_{t_i} is continuous in $[0,1] \times R^n$ for $i = 1, \ldots, n$.

In this section, we take X to be the Cartesian product of n spaces of the form

$$X_i = \left\{ \xi \in W^{1,p_i}([0,1]) \ : \ \xi(0) = \sum_{j=1}^{m} a_j \xi_j(x_j), \quad \xi(1) = \sum_{j=1}^{m} b_j \xi_j(x_j) \right\}$$

for $i = 1, \ldots, n$, i.e., $X = X_1 \times \cdots \times X_n$, endowed with the norm

$$\| (u_1, \ldots, u_n) \| = \sum_{i=1}^{n} \|u_i\|_{p_i},$$

where

$$\|u_i\|_{p_i} = \left(\int_0^1 |u_i'(x)|^{p_i} dx \right)^{1/p_i}, \quad i = 1, \dots, n.$$

Analogous to what we had earlier in this chapter, we say that a function $u = (u_1, \dots, u_n)$ is a *classical solution* of (3.5.1) if $u_i \in C^1[0,1]$, $\phi_{p_i}(u_i') \in C^1[0,1]$, and $u_i(t)$ satisfies (3.5.1) for $i = 1, \dots, n$. A function $u = (u_1, \dots, u_n) \in X$ is a *weak solution* of (3.5.1) if

$$\int_0^1 \sum_{i=1}^n \phi_{p_i}(u_i'(x))v_i'(x)dx - \lambda \int_0^1 \sum_{i=1}^n F_{u_i}(x, u_1(x), \dots, u_n(x))v_i(x)dx$$

$$- \mu \int_0^1 \sum_{i=1}^n G_{u_i}(x, u_1(x), \dots, u_n(x))v_i(x)dx = 0$$

for any $v = (v_1, \dots, v_n) \in W_0^{1,p_1}([0,1]) \times W_0^{1,p_2}([0,1]) \times \cdots \times W_0^{1,p_n}([0,1])$.

We will say that *Property S_d* holds if for some $d \in (0, \infty)$, there exist an open interval $\Lambda \subseteq [0, d)$ and a positive real number q such that:

> *for every $\lambda \in \Lambda$ and an arbitrary function $G : [0,1] \times R^n \to R$ such that the mapping $(t_1, t_2, \dots, t_n) \to G(x, t_1, t_2, \dots, t_n)$ is in C^1 in R^n for all $x \in [0,1]$ and G_{t_i} is continuous in $[0,1] \times R^n$ for $i = 1, \dots, n$, there is a $\delta > 0$ such that, for each $\mu \in [0, \delta]$, the system (3.5.1) admits at least three classical solutions in X whose norms are less than q.*

Our goal here is to determine sufficient conditions for Property S_d to hold for some $d \in [0, \infty)$.

The approach to be used here is to apply Ricceri's three critical points theorem, Theorem 1.1.6 above, (see [248], as well as [245] and [38] for related results) which equates the existence of critical points of the Euler functional to the existence of three solutions of the system (3.5.1).

Let $\phi_{p_i}^{-1}$ denote the inverse of ϕ_{p_i} for each $i = 1, \dots, n$. Then, $\phi_{p_i}^{-1}(t) = \phi_{q_i}(t)$ where $\frac{1}{p_i} + \frac{1}{q_i} = 1$. It is clear that ϕ_{p_i} is increasing on R,

$$\lim_{t \to -\infty} \phi_{p_i}(t) = -\infty \quad \text{and} \quad \lim_{t \to \infty} \phi_{p_i}(t) = \infty. \tag{3.5.2}$$

We will need the following four lemmas in proving our main results in this section. The first one is an easy consequence of (3.5.2).

Lemma 3.5.1. *For fixed* λ, $\mu \in R$, $u = (u_1, \ldots, u_n) \in (C([0,1]))^n$, *and* $i = 1, \ldots, n$, *define* $\alpha_i(t; u) : R \to R$ *by*

$$\alpha_i(t; u) = \int_0^1 \phi_{p_i}^{-1} \left(t - \lambda \int_0^\delta F_{u_i}(\xi, u_1(\xi), \ldots, u_n(\xi)) d\xi \right.$$

$$\left. - \mu \int_0^\delta G_{u_i}(\xi, u_1(\xi), \ldots, u_n(\xi)) d\xi \right) d\delta$$

$$+ \sum_{j=1}^m a_j u_i(x_j) - \sum_{j=1}^m b_j u_i(x_j).$$

Then, the equation

$$\alpha_i(t; u) = 0 \tag{3.5.3}$$

has a unique solution $t_{u,i}$.

A direct computation shows that the following lemma holds.

Lemma 3.5.2. *The function* $u = (u_1, \ldots, u_n)$ *is a solution of the system* (3.5.1) *if and only if* $u_i(x)$ *is a solution of the equation*

$$u_i(x) = \sum_{j=1}^m a_j u_i(x_j) + \int_0^x \phi_{p_i}^{-1} \left(t_{u,i} - \lambda \int_0^\delta F_{u_i}(\xi, u_1(\xi), \ldots, u_n(\xi)) d\xi \right.$$

$$\left. - \mu \int_0^\delta G_{u_i}(\xi, u_1(\xi), \ldots, u_n(\xi)) d\xi \right) d\delta$$

for $i = 1, \ldots, n$, *where* $t_{u,i}$ *is the unique solution of* (3.5.3).

In view of what proceeded in the earlier sections in this chapter, the following lemma should come as no surprise.

Lemma 3.5.3. *A weak solution to* (3.5.1) *coincides with a classical solution to* (3.5.1).

Proof. Let $u = (u_1, \ldots, u_n) \in X$ be a weak solution to (3.5.1). Then,

$$\int_0^1 \sum_{i=1}^n \phi_{p_i}(u_i'(x)) v_i'(x) dx - \lambda \int_0^1 \sum_{i=1}^n F_{u_i}(x, u_1(x), \ldots, u_n(x)) v_i(x) dx$$

$$- \mu \int_0^1 \sum_{i=1}^n G_{u_i}(x, u_1(x), \ldots, u_n(x)) v_i(x) dx = 0 \tag{3.5.4}$$

for every $(v_1, \ldots, v_n) \in W_0^{1,p_1}([0,1]) \times W_0^{1,p_2}([0,1]) \times \cdots \times W_0^{1,p_n}([0,1])$. Recall that, in one dimension, any weakly differentiable function is absolutely continuous, so that its classical derivative exists almost everywhere, and that the classical derivative coincides with the weak derivative. Integrating (3.5.4) by parts gives

$$\sum_{i=1}^{n} \int_0^1 [(\phi_{p_i}(u_i'(x)))' + \lambda F_{u_i}(x, u_1(x), \ldots, u_n(x))$$

$$+ \mu G_{u_i}(x, u_1(x), \ldots, u_n(x))]v_i(x)dx = 0,$$

and so for $i = 1, \ldots, n$,

$$(\phi_{p_i}(u_i'(x)))' + \lambda F_{u_i}(x, u_1(x), \ldots, u_n(x)) + \mu G_{u_i}(x, u_1(x), \ldots, u_n(x)) = 0$$

for almost every $x \in (0, 1)$. Then, by Lemmas 3.5.1 and 3.5.2, we see that

$$u_i(x) = \sum_{j=1}^{m} a_j u_i(x_j) + \int_0^x \phi_{p_i}^{-1}\left(t_{u,i} - \lambda \int_0^\delta F_{u_i}(s, u_1(s), \ldots, u_n(s))ds\right.$$

$$\left. - \mu \int_0^\delta G_{u_i}(s, u_1(s), \ldots, u_n(s))ds\right) d\delta$$

for $i = 1, \ldots, n$, where $t_{u,i}$ is the unique solution of (3.5.3). Hence, $u_i \in C^1([0, 1])$ and $\phi_{p_i}(u_i'(x)) \in C^1([0, 1])$ for $i = 1, \ldots, n$, i.e., $u = (u_1, \ldots, u_n)$ is a classical solution to the system (3.5.1). $\square$

Our final lemma introduces an operator that will be used in the proof of our main result. In the remainder of this section we will tacitly assume that conditions (H1) and (H2) hold throughout.

Lemma 3.5.4. *Let $T : X \to X^*$ be the operator defined by*

$$T(u_1, \ldots, u_n)(h_1, \ldots, h_n) = \int_0^1 \sum_{i=1}^{n} |u_i'(x)|^{p_i - 2} u_i'(x) h_i'(x)dx$$

for every $(u_1, \ldots, u_n)$, $(h_1, \ldots, h_n) \in X$. Then T admits a continuous inverse on X^*.

Proof. In the proof, we use $K_1, K_2, \ldots, K_8$ to denote appropriate positive constants. For any $u = (u_1, \ldots, u_n) \in X \setminus \{0\}$,

$$\lim_{||u|| \to \infty} \frac{\langle T(u), u \rangle}{||u||} = \lim_{||u|| \to \infty} \frac{\sum_{i=1}^n \int_0^1 |u_i'(x)|^{p_i} dx}{||u||}$$

$$= \lim_{||u|| \to \infty} \frac{\sum_{i=1}^n \left[\left(\int_0^1 |u_i'(x)|^{p_i} dx \right)^{1/p_i} \right]^{p_i}}{||u||}$$

$$\geq \lim_{||u|| \to \infty} \frac{\sum_{i=1}^n \left[\left(\int_0^1 |u_i'(x)|^{p_i} dx \right)^{1/p_i} \right]^{\underline{p}}}{||u||}$$

$$\geq \lim_{||u|| \to \infty} \frac{K_1 \left[\sum_{i=1}^n \left(\int_0^1 |u_i'(x)|^{p_i} dx \right)^{1/p_i} \right]^{\underline{p}}}{||u||}$$

$$= \lim_{||u|| \to \infty} \frac{K_1 [||u||]^{\underline{p}}}{||u||} = \infty.$$

Thus, T is coercive.

Now, for any $u = (u_1, \ldots, u_n) \in X$ and $v = (v_1, \ldots, v_n) \in X$, we have

$$\langle T(u) - T(v), u - v \rangle$$
$$= \sum_{i=1}^n \int_0^1 (\phi_{p_i}(u_i'(x)) - \phi_{p_i}(v_i'(x)))(u_i'(x) - v_i'(x)).$$

Then, by [253, Eq. (2.2)], we see that

$$\langle T(u) - T(v), u - v \rangle \geq \begin{cases} K_2 \sum_{i=1}^n \int_0^1 |u_i'(x) - v_i'(x)|^{p_i} dx, & \text{if } \underline{p} \geq 2, \\ K_3 \sum_{i=1}^n \int_0^1 \dfrac{|u_i'(x) - v_i'(x)|^2}{(|u_i'(x)| + |v_i'(x)|)^{2-p_i}} dx, & \text{if } \overline{p} < 2. \end{cases}$$

$$(3.5.5)$$

If $\underline{p} \geq 2$, we let $I_{u,v} = \{i : 1 \leq i \leq n \text{ and } ||u_i||_{p_i} + ||v_i||_{p_i} \neq 0\}$. Choose $M_1 > 1$ so that $M_1(||u_i||_{p_i} + ||v_i||_{p_i}) > 1$ for $i \in I_{u,v}$. Then,

$$K_2 \sum_{i=1}^n \int_0^1 |u_i'(x) - v_i'(x)|^{p_i} dx = K_2 \sum_{i=1}^n ||u_i' - v_i'||_{p_i}^{p_i}$$

$$= K_2 \sum_{I_{u,v}} \frac{1}{M_1^{p_i}} ||M_1(u_i' - v_i')||_{p_i}^{p_i}$$

$$\geq \frac{C_2}{M_1^{\overline{p}}} \sum_{I_{u,v}} ||M_1(u_i' - v_i')||_{p_i}^{\underline{p}}$$

$$\geq \frac{K_2 M_1^{\underline{p}}}{M_1^{\overline{p}}} \sum_{I_{u,v}} ||u_i' - v_i'||_{p_i}^{\underline{p}}.$$

Using the fact that $(q_1 + \cdots + q_n)^p \leq M_2(q_1^p + \cdots + q_n^p)$, for some $M_2 > 0$, it follows that

$$\langle T(u) - T(v), u - v \rangle \geq K_4 \|u - v\|^p.$$

Then, T is uniformly monotone. By [289, Theorem 26.A (d)], T^{-1} exists and is continuous on X^*.

If $\bar{p} < 2$, then, from Hölder's inequality, we obtain

$$\int_0^1 |u_i'(x) - v_i'(x)|^{p_i}\, dx$$

$$\leq \left(\int_0^1 \frac{|u_i'(x) - v_i'(x)|^2}{(|u_i'(x)| + |v_i'(x)|)^{2-p_i}}\, dx \right)^{p_i/2} \left(\int_0^1 (|u_i'(x)| + |v_i'(x)|)^{p_i}\, dx \right)^{(2-p_i)/2}$$

$$\leq K_5 \left(\int_0^1 \frac{|u_i'(x) - v_i'(x)|^2}{(|u_i'(x)| + |v_i'(x)|)^{2-p_i}}\, dx \right)^{p_i/2}$$

$$\times \left(\int_0^1 (|u_i'(x)|^{p_i} + |v_i'(x)|^{p_i})\, dx \right)^{(2-p_i)/2}$$

$$\leq K_6 \left(\int_0^1 \frac{|u_i'(x) - v_i'(x)|^2}{(|u_i'(x)| + |v_i'(x)|)^{2-p_i}}\, dx \right)^{p_i/2} (\|u\| + \|v\|)^{(2-p_i)p_i/2}. \quad (3.5.6)$$

For convenience, let

$$p^* = \begin{cases} \underline{p}, & \text{if } \|u\| + \|v\| \geq 1, \\ \bar{p}, & \text{if } \|u\| + \|v\| < 1. \end{cases}$$

Then from (3.5.5) and (3.5.6), it follows that

$$\langle T(u) - T(v), u - v \rangle$$

$$\geq K_7 \sum_{i=1}^n \frac{\left(\int_0^1 |u_i'(x) - v_i'(x)|^{p_i}\, dx \right)^{2/p_i}}{(\|u\| + \|v\|)^{2-p_i}}$$

$$\geq \frac{K_7}{(\|u\| + \|v\|)^{2-p^*}} \sum_{i=1}^n \left(\int_0^1 |u_i'(x) - v_i'(x)|^{p_i}\, dx \right)^{2/p_i}$$

$$\geq \frac{K_8}{(\|u\| + \|v\|)^{2-p^*}} \left(\sum_{i=1}^n \left(\int_0^1 |u_i'(x) - v_i'(x)|^{p_i}\, dx \right)^{1/p_i} \right)^2$$

$$= \frac{K_8 \|u - v\|^2}{(\|u\| + \|v\|)^{2-p^*}}. \quad (3.5.7)$$

Thus, T is strictly monotone. By [289, Theorem 26.A (d)], T^{-1} exists and is bounded. Given $g_1, g_2 \in X^*$, from (3.5.7), we have

$$\|T^{-1}(g_1) - T^{-1}(g_2)\| \leq \frac{1}{C_8}(\|T^{-1}(g_1)\| + \|T^{-1}(g_2)\|)^{2-p^*} \|g_1 - g_2\|_{X^*}.$$

So T^{-1} is locally Lipschitz continuous and hence continuous. This completes the proof of the lemma. □

We are now ready to present our main theorem in this section.

Theorem 3.5.1. *Assume there are a constant $r > 0$ and a function $w = (w_1, \ldots, w_n) \in X$ such that*

(C27) $\sum_{i=1}^{n} \frac{||w_i||_{p_i}^{p_i}}{p_i} > r;$

(C28)

$$M_1 := \left(r \prod_{i=1}^{n} p_i \right) \frac{\int_0^1 F(x, w_1(x), \ldots, w_n(x)) dx}{\sum_{i=1}^{n} \prod_{j=1, j \neq i}^{n} p_j ||w_i||_{p_i}^{p_i}}$$

$$- \int_0^1 \sup_{(t_1, \ldots, t_n) \in K(kr)} F(x, t_1, \ldots, t_n) dx > 0;$$

(C29) $\limsup\limits_{|t_1| \to \infty, \ldots, |t_n| \to \infty} \frac{F(x, t_1, \ldots, t_n)}{\sum_{i=1}^{n} \frac{|t_i|^{p_i}}{p_i}} < \frac{1}{kd}$ *uniformly with respect to $x \in [0,1]$ for some d satisfying $d > \frac{r}{M_1}$.*

Then, the Property S_d holds.

Proof. To apply Theorem 1.1.6 to our problem, we define the functionals Φ, $\Psi : X \to R$ as in (3.2.6) and (3.2.7), set $J = -\Psi$ and note that J satisfies the same smoothness and compactness properties that Ψ does.

From (C27), it follows that $0 < r < \Phi(w)$ as required in Theorem 1.1.7. Since for $(u_1, \ldots, u_n) \in X$,

$$\sup_{x \in [0,1]} |u_i(x)|^{p_i} \leq c||u_i||_{p_i}^{p_i}$$

for $i = 1, \ldots, n$ (see (3.2.1)), we have

$$\sup_{x \in [0,1]} \sum_{i=1}^{n} \frac{|u_i(x)|^{p_i}}{p_i} \leq c \sum_{i=1}^{n} \frac{||u_i||_{p_i}^{p_i}}{p_i} \qquad (3.5.8)$$

for each $u = (u_1, \ldots, u_n) \in X$. From (3.5.8), we obtain

$$
\begin{aligned}
\Phi^{-1}((-\infty, r]) &= \{(u_1, \ldots, u_n) \in X : \Phi(u_1, \ldots, u_n) \leq r\} \\
&= \left\{(u_1, \ldots, u_n) \in X : \sum_{i=1}^{n} \frac{\|u_i\|_{p_i}^{p_i}}{p_i} \leq r\right\} \\
&\subseteq \left\{(u_1, \ldots, u_n) \in X : \sum_{i=1}^{n} \frac{|u_i(x)|^{p_i}}{p_i} \leq cr\} \text{ for all } x \in [0, 1]\right\}.
\end{aligned}
$$

Then,

$$
\begin{aligned}
\sup_{(u_1, \ldots, u_n) \in \Phi^{-1}((-\infty, r])} & (-J(u_1, \ldots, u_n)) \\
&= \sup_{(u_1, \ldots, u_n) \in \Phi^{-1}((-\infty, r])} \int_0^1 F(x, u_1(x), \ldots, u_n(x)) dx \\
&\leq \int_0^1 \sup_{(t_1, \ldots, t_n) \in K(kr)} F(x, t_1, \ldots, t_n) dx.
\end{aligned}
$$

Therefore, from condition (C28), we have

$$
\begin{aligned}
\sup_{u \in \Phi^{-1}((-\infty, r])} (-J(u_1, \ldots, u_n)) &\leq \int_0^1 \sup_{(t_1, \ldots, t_n) \in K(kr)} F(x, t_1, \ldots, t_n) dx \\
&< r \frac{\int_0^1 F(x, w_1(x), \ldots, w_n(x)) dx}{\sum_{i=1}^{n} \frac{\|w_i\|_{p_i}^{p_i}}{p_i}} \\
&= r \frac{-J(w)}{\Phi(w)},
\end{aligned}
$$

which is also needed Theorem 1.1.7. Let $u_0 = 0$; then, $\Phi(u_0) = J(u_0) = 0$. Hence, all the conditions of Theorem 1.1.7 are satisfied, so condition (1.1.6) in Theorem 1.1.6 holds.

By (C29), there exist constants $\gamma, \vartheta \in R$ with $0 < \gamma < 1/d$ such that

$$
\begin{aligned}
cF(x, t_1, \ldots, t_n) \\
\leq \gamma \sum_{i=1}^{n} \frac{|t_i|^{p_i}}{p_i} + \vartheta \text{ for all } x \in [0, 1] \text{ and for all } (t_1, \ldots, t_n) \in R^n.
\end{aligned}
$$

For any $(u_1, \ldots, u_n) \in X$, we have

$$
F(x, u_1(x), \ldots, u_n(x)) \leq \frac{1}{c}\left(\gamma \sum_{i=1}^{n} \frac{|u_i(x)|^{p_i}}{p_i} + \vartheta\right) \text{ for all } x \in [0, 1].
$$

$$
\tag{3.5.9}
$$

So, for any fixed $\lambda \in (0, d)$, we obtain

$$\Phi(u_1, \ldots, u_n) + \lambda J(u_1, \ldots, u_n)$$

$$= \sum_{i=1}^{n} \frac{||u_i||_{p_i}^{p_i}}{p_i} - \lambda \int_0^1 F(x, u_1(x), \ldots, u_n(x))dx$$

$$\geq \sum_{i=1}^{n} \frac{||u_i||_{p_i}^{p_i}}{p_i} - \frac{\lambda\gamma}{c} \left(\sum_{i=1}^{n} \frac{1}{p_i} \int_0^1 |u_i(x)|^{p_i} dx \right) - \frac{\lambda\vartheta}{c}$$

$$\geq \sum_{i=1}^{n} \frac{||u_i||_{p_i}^{p_i}}{p_i} - \frac{\lambda\gamma}{c} \left(c \sum_{i=1}^{n} \frac{||u_i||_{p_i}^{p_i}}{p_i} \right) - \frac{\lambda\vartheta}{c}$$

$$\geq (1 - \gamma d) \sum_{i=1}^{n} \frac{||u_i||_{p_i}^{p_i}}{p_i} - \frac{\lambda\vartheta}{c}.$$

Thus,

$$\lim_{||(u_1,\ldots,u_n)|| \to \infty} (\Phi(u_1, \ldots, u_n) + \lambda J(u_1, \ldots, u_n)) = \infty,$$

i.e., condition (1.1.5) holds. Noting that the solutions of the equation $\Phi'(u_1, \ldots, u_n) + \lambda J'(u_1, \ldots, u_n) + \mu\Psi'(u_n, \ldots, u_n) = 0$ are exactly the weak solutions of (3.5.1), the conclusion follows from Theorem 1.1.6 $\qquad\square$

We now present some consequences of Theorem 3.5.1 in the case where the function w is specified.

For $i = 1, \ldots, n$, let

$$\sigma_i = \left[2^{p_i - 1} \left(x_1^{1-p_i} \left| 1 - \sum_{j=1}^{m} a_j \right|^{p_i} + (1 - x_m)^{1-p_i} \left| 1 - \sum_{j=1}^{m} b_j \right|^{p_i} \right) \right]^{1/p_i}.$$

(3.5.10)

Define

$$B_{1,n}(x) = \begin{cases} [x \sum_{j=1}^{m} a_j, x]^n, & \text{if } \sum_{j=1}^{m} a_j < 1, \\ [x, x \sum_{j=1}^{m} a_j]^n, & \text{if } \sum_{j=1}^{m} a_j > 1, \end{cases}$$

and

$$B_{2,n}(x) = \begin{cases} [x \sum_{j=1}^{m} b_j, x]^n, & \text{if } \sum_{j=1}^{m} b_j < 1, \\ [x, x \sum_{j=1}^{m} b_j]^n, & \text{if } \sum_{j=1}^{m} b_j > 1, \end{cases}$$

where $[\cdot, \cdot]^n = [\cdot, \cdot] \times \cdots \times [\cdot, \cdot]$.

Corollary 3.5.1. *Assume there exist constants $\theta > 0$ and $\tau > 0$ with $\sum_{i=1}^{n} \frac{(\tau\sigma_i)^{p_i}}{p_i} > \frac{\theta}{k \prod_{i=1}^{n} p_i}$ such that*

(C30) $F(x, t_1, \ldots, t_n) \geq 0$ *for each* $x \in [0, \frac{x_1}{2}] \cup [\frac{1+x_m}{2}, 1]$ *and* $(t_1, \ldots, t_n) \in$
$B_{1,n}(\tau) \cup B_{2,n}(\tau)$;

(C31) $M_2 := \dfrac{\theta}{\left(\sum_{i=1}^n \frac{(\tau\sigma_i)^{p_i}}{p_i}\right)\left(k\prod_{i=1}^n p_i\right)} \displaystyle\int_{\frac{x_1}{2}}^{\frac{1+x_m}{2}} F(x, \tau, \ldots, \tau)dx$

$\qquad\qquad - \displaystyle\int_0^1 \sup_{(t_1,\ldots,t_n)\in K\left(\frac{\theta}{\prod_{i=1}^n p_i}\right)} F(x, t_1, \ldots, t_n)dx > 0;$

(C32) $\limsup_{|t_1|\to\infty,\ldots,\ |t_n|\to\infty} \dfrac{F(x,t_1,\ldots,t_n)}{\sum_{i=1}^n \frac{|t_i|^{p_i}}{p_i}} < \frac{1}{kd}$ *uniformly with respect to*
$x \in [0, 1]$ *for some* d *satisfying* $d > \frac{\theta}{kM_2 \prod_{i=1}^n p_i}$.

Then, the Property S_d *holds.*

Proof. Set $w(x) = (w_1(x), \ldots, w_n(x))$ such that for $1 \leq i \leq n$,

$$
w_i(x) = \begin{cases}
\tau\left(\sum_{j=1}^m a_j + \frac{2(1-\sum_{j=1}^m a_j)}{x_1}x\right), & \text{if } x \in [0, \frac{x_1}{2}), \\[2mm]
\tau, & \text{if } x \in [\frac{x_1}{2}, \frac{1+x_m}{2}], \\[2mm]
\tau\left(\frac{2-\sum_{j=1}^m b_j - x_m\sum_{j=1}^m b_j}{1-x_m} - \frac{2(1-\sum_{j=1}^m b_j)}{1-x_m}x\right), & \text{if } x \in (\frac{1+x_m}{2}, 1],
\end{cases}
$$

and $r = \frac{\theta}{k\prod_{i=1}^n p_i}$. It is easy to see that $w = (w_1, \ldots, w_n) \in X$ and, in particular,

$$\|w_i\|_{p_i}^{p_i} = (\sigma_i\tau)^{p_i} \tag{3.5.11}$$

for $i = 1, \ldots, n$. Hence, from the assumption that $\sum_{i=1}^n \frac{(\tau\sigma_i)^{p_i}}{p_i} > \frac{\theta}{k\prod_{i=1}^n p_i}$
and (3.5.11), we have

$$\sum_{i=1}^n \frac{\|w_i\|_{p_i}^{p_i}}{p_i} > r,$$

which is condition (C27).

Since for $i = 1, \ldots, n$,

$$\tau\sum_{j=1}^m a_j \leq w_i(x) \leq \tau \quad \text{for each } x \in [0, \frac{x_1}{2}] \text{ if } \sum_{j=1}^m a_j < 1,$$

$$\tau \leq w_i(x) \leq \tau\sum_{j=1}^m a_j \quad \text{for each } x \in [0, \frac{x_1}{2}] \text{ if } \sum_{j=1}^m a_j > 1,$$

$$\tau\sum_{j=1}^m b_j \leq w_i(x) \leq \tau \quad \text{for each } x \in [\frac{1+x_m}{2}, 1] \text{ if } \sum_{j=1}^m b_j < 1,$$

and

$$\tau \le w_i(x) \le \tau \sum_{j=1}^{m} b_j \text{ for each } x \in [\frac{1+x_m}{2}, 1] \text{ if } \sum_{j=1}^{m} b_j > 1,$$

condition (C30) implies

$$\int_0^{\frac{x_1}{2}} F(x, w_1(x), \ldots, w_n(x))dx + \int_{\frac{1+x_m}{2}}^1 F(x, w_1(x), \ldots, w_n(x))dx \ge 0.$$

(3.5.12)

Moreover, from Assumption (C31) and (3.5.12), we have

$$\int_0^1 \sup_{(t_1,\ldots,t_n)\in K(cr)} F(x, t_1, \ldots, t_n)dx$$

$$< \frac{\theta}{(\sum_{i=1}^n \frac{(\tau\sigma_i)^{p_i}}{p_i})(c\prod_{i=1}^n p_i)} \int_{\frac{x_1}{2}}^{\frac{1+x_m}{2}} F(x, \tau, \ldots, \tau)dx$$

$$\le \frac{\theta}{c} \frac{\int_0^1 F(x, w_1(x), \ldots, w_n(x))dx}{\sum_{i=1}^n \prod_{j=1,j\ne i}^n p_j||w_i||_{p_i}^{p_i}}$$

$$= \left(r\prod_{i=1}^n p_i\right) \frac{\int_0^1 F(x, w_1(x), \ldots, w_n(x))dx}{\sum_{i=1}^n \prod_{j=1,j\ne i}^n p_j||w_i||_{p_i}^{p_i}},$$

where $K(cr) = \{(t_1, \ldots, t_n) : \sum_{i=1}^n \frac{|t_i|^{p_i}}{p_i} \le cr\}$, i.e., (C28) is satisfied. Since $M_2 \le M_1$, (C29) follows from (C32). Hence, Theorem 3.5.1 gives the desired conclusion. $\square$

As an example of Corollary 3.5.1, we have the following.

Example 3.5.1. Define the function $F : [0, 1] \times R^2 \to R$ by

$$F(x, t_1, t_2) = \begin{cases} 0, & \text{for } t_i < 0, i = 1, 2, \\ x^2 t_2^{50} e^{-t_2}, & \text{for } t_1 < 0, t_2 \ge 0, \\ x^2 t_1^{50} e^{-t_1}, & \text{for } t_1 \ge 0, t_2 < 0, \\ x^2 \sum_{i=1}^2 t_i^{50} e^{-t_i}, & \text{for } t_i \ge 0, i = 1, 2, \end{cases}$$

for $(x, t_1, t_2) \in [0, 1] \times R^2$. Now, $n = 2$ and $m = 1$, so choosing $p_1 = p_2 = 3$ and $a_1 = b_1 = x_1 = \frac{1}{2}$, we see that (H1) and (H2) hold. A calculation gives $k = \frac{27}{8}$ and $\sigma_1 = \sigma_2 = 4^{1/3}$, and so with $\theta = 3$ and $\tau = 50$, we see that

(C30) and (C32) are satisfied. Since

$$\int_0^1 \sup_{(t_1,t_2)\in K(\frac{\theta}{\Pi_{i=1}^2 p_i})} F(x,t_1,t_2)\,dx = \int_0^1 \sup_{(t_1,t_2)\in K(\frac{1}{3})} F(x,t_1,t_2)\,dx$$

$$\leq \int_0^1 \sup_{(t_1,t_2)\in K(\frac{1}{3})} x^2 \sum_{i=1}^2 t_i^{50} e^{-t_i}\,dx = \max_{(t_1,t_2)\in K(\frac{1}{3})} \sum_{i=1}^2 t_i^{50} e^{-t_i} \int_0^1 x^2\,dx$$

$$\leq \frac{2}{3} \max_{|t|\leq 1} t^{50} e^{-t} = \frac{2e}{3} < \frac{13}{1296}(50)^{47} e^{-50}$$

$$= \frac{\theta}{\left(\sum_{i=1}^2 \frac{(\tau\sigma_i)^{p_i}}{p_i}\right)\left(k\prod_{i=1}^2 p_i\right)} \int_{\frac{x_1}{2}}^{\frac{1+x_1}{2}} F(x,\tau,\tau)\,dx,$$

(C31) holds. Note that $\lim_{|t_1|\to\infty,|t_2|\to\infty} \frac{F(x,t_1,t_2)}{\sum_{i=1}^2 \frac{|t_i|^{p_i}}{p_i}} = 0$ uniformly for $x \in [0,1]$. For d satisfying

$$d > \frac{1}{\frac{81}{8}\left(\frac{13}{1296}(50)^{47}e^{-50} - \frac{2e}{3}\right)},$$

Corollary 3.5.1 can be applied to the system

$$\begin{cases} -(|u_1'|u_1')' = \lambda x^2 (u_1^+)^{49} e^{-u_1^+}(50 - u_1^+) + \mu G_{u_1}(x,u_1,u_2), & \text{in } (0,1), \\ -(|u_2'|u_2')' = \lambda x^2 (u_2^+)^{49} e^{-u_2^+}(50 - u_2^+) + \mu G_{u_2}(x,u_1,u_2), & \text{in } (0,1), \\ u_1(0) = u_1(1) = \frac{1}{2}u_1(\frac{1}{2}), \quad u_2(0) = u_2(1) = \frac{1}{2}u_2(\frac{1}{2}), \end{cases}$$

where $t^+ = \max\{t,0\}$ and $G : [0,1] \times R^2 \to R$ is an arbitrary function that is continuous in $[0,1]$ and C^1 in R^2.

The following result is an immediate consequence of Corollary 3.5.1

Corollary 3.5.2. *Assume that $F(x,t_1,\ldots,t_n) \equiv F(t_1,\ldots,t_n)$, and there exist constants $\theta > 0$ and $\tau > 0$ with $\sum_{i=1}^n \frac{(\tau\sigma_i)^{p_i}}{p_i} > \frac{\theta}{c\prod_{i=1}^n p_i}$ such that*

(C33) $F(t_1,\ldots,t_n) \geq 0$ *for each* $(t_1,\ldots,t_n) \in B_{1,n}(\tau) \cup B_{2,n}(\tau)$;

(C34) $M_3 := \dfrac{\theta(1+x_m-x_1)}{\left(\sum_{i=1}^n \frac{(\tau\sigma_i)^{p_i}}{p_i}\right)\left(2c\prod_{i=1}^n p_i\right)} F(\tau,\ldots,\tau)$

$\qquad - \max_{(t_1,\ldots,t_n)\in K\left(\frac{\theta}{\Pi_{i=1}^n p_i}\right)} F(t_1,\ldots,t_n) > 0$;

(C35) $\limsup_{|t_1|\to\infty,\ldots,\ |t_n|\to\infty} \dfrac{F(t_1,\ldots,t_n)}{\sum_{i=1}^n \frac{|t_i|^{p_i}}{p_i}} < \frac{1}{cd}$ *for some d satisfying $d >$*

$\dfrac{\theta}{cM_3 \prod_{i=1}^n p_i}$.

Then, the Property S_d holds.

Example 3.5.2. Let $p_1 = p_2 = 3$, $m = 2$, $x_1 = \frac{1}{3}$, $x_2 = \frac{2}{3}$ and $a_i = b_i = \frac{1}{3}$, i=1,2. We consider the problem

$$\begin{cases} -(|u_1'|u_1')' = \lambda(e^{-u_1}u_1^9(10 - u_1)) + \mu G_{u_1}(x, u_1, u_2), & \text{in } (0,1), \\ -(|u_2'|u_2')' = \lambda(e^{-u_2}u_2^{15}(16 - u_2)) + \mu G_{u_2}(x, u_1, u_2), & \text{in } (0,1), \\ u_1(0) = u_1(1) = \frac{1}{3}u_1(\frac{1}{3}) + \frac{1}{3}u_1(\frac{2}{3}), \quad u_2(0) = u_2(1) = \frac{1}{3}u_2(\frac{1}{3}) + \frac{1}{3}u_2(\frac{2}{3}), \end{cases}$$
$$(3.5.13)$$

where $G : [0,1] \times R^2 \to R$ is a function such that $G(\cdot, t_1, t_2)$ is continuous in $[0,1]$ for all $(t_1, t_2) \in R^2$ and $G(x, \cdot, \cdot)$ is C^1 in R^2 for every $x \in [0,1]$. Clearly, (H1) and (H2) hold. A simple calculation shows that $k = \frac{125}{8}$ and $\sigma_1 = \sigma_2 = \left(\frac{8}{3}\right)^{1/3}$. So, by choosing $\theta = 3$ and $\tau = 10$, (C33) and (C35) hold. Since $F(t_1, t_2) = t_1^{10}e^{-t_1} + t_2^{16}e^{-t_2}$, we have

$$\begin{aligned} \max_{(t_1,t_2)\in K(\frac{\theta}{\Pi_{i=1}^2 p_i})} F(t_1, t_2) &= \max_{(t_1,t_2)\in K(\frac{1}{3})} F(t_1, t_2) \\ &= \max_{(t_1,t_2)\in K(\frac{1}{3})} (t_1^{10}e^{-t_1} + t_2^{16}e^{-t_2}) \\ &\leq \left[\max_{|t_1|\leq 1} t_1^{10}e^{-t_1} + \max_{|t_2|\leq 1} t_2^{16}e^{-t_2}\right] \\ &= 2e \\ &< \frac{1}{125 \cdot 10^3}(10^{10}e^{-10} + 10^{16}e^{-10}) \\ &= \frac{\theta(1 + x_2 - x_1)}{\left(\sum_{i=1}^2 \frac{(\tau\sigma_i)^{p_i}}{p_i}\right)\left(2k\prod_{i=1}^2 p_i\right)}F(\tau, \tau), \end{aligned}$$

(C34) also holds. Note that $\lim_{|t_1|\to\infty, |t_2|\to\infty} \dfrac{F(t_1, t_2)}{\sum_{i=1}^2 \frac{|t_i|^{p_i}}{p_i}} = 0$. We see that for every real number d satisfying

$$d > \frac{1}{\frac{375}{8}\left(\frac{1}{125\cdot10^3}(10^{10}e^{-10} + 10^{16}e^{-10}) - 2e\right)},$$

Corollary 3.5.2 is applicable to the system (3.5.13).

We conclude this section with the case $n = 1$ in Corollary 3.5.2.

Corollary 3.5.3. *Let $f : R \to R$ be a continuous function. Assume that there exist four constant a_1, a_2, b_1, and b_2 with $a_1 + a_2 \neq 1$ and $b_1 + b_2 \neq 1$. Set $k = \frac{1}{4}\left(1 + \frac{|a_1|+|a_2|}{|1-a_1-a_2|} + \frac{|b_1|+|b_2|}{|1-b_1-b_2|}\right)$ and assume that there are constants $\theta > 0$ and $\tau > 0$ with $(\tau\sigma)^2 > \frac{\theta}{k}$ such that*

(C36) $f(t) \geq 0$ *for each $t \in B_{1,1}(\tau) \cup B_{2,1}(\tau)$;*

(C37) $\max_{t^2 \leq \theta} \int_0^t f(\xi)d\xi < \frac{\theta(1+x_2-x_1)}{2k(\tau\sigma)^2} \int_0^\tau f(\xi)d\xi$, *where*

$$\sigma = \left[2\left(x_1^{-1}|1 - a_1 - a_2|^2 + (1 - x_2)^{-1}|1 - b_1 - b_2|^2\right)\right]^{1/2};$$

(C38) $\limsup_{|t|\to\infty} \frac{\int_0^t f(\xi)d\xi}{t^2} \leq 0$.

Then, there exist a non-empty open interval $\Lambda \subseteq [0, d)$ for every

$$d > \frac{\theta}{2k\left(\frac{\theta(1+x_2-x_1)}{2k(\tau\sigma)^2} \int_0^\tau f(\xi)d\xi - \max_{t^2 \leq \theta} \int_0^t f(\xi)d\xi\right)}$$

and a positive real number q with the property that for every $\lambda \in \Lambda$ and an arbitrary continuous function $g : [0, 1] \times R \to R$, there is a $\delta > 0$ such that, for each $\mu \in [0, \delta]$, the problem

$$\begin{cases} -u'' = \lambda f(u) + \mu g(x, u), & in\ (0, 1), \\ u(0) = a_1 u(x_1) + a_2 u(x_2), & u(1) = b_1 u(x_1) + b_2 u(x_2), \end{cases}$$

admits at least three classical solutions belonging to $C^2([0, 1])$ whose norms in

$$X = \{\xi \in W^{1,2}([0, 1]) : \xi(0) = a_1\xi(x_1) + a_2\xi(x_2),\ \ \xi(1) = b_1\xi(x_1) + b_2\xi(x_2)\}$$

are less than q.

3.6　Existence by the Dual Action Principle

In this section we consider a multi-point boundary value problem in which the derivative of the unknown function u appears in the boundary condition. For simplicity, we consider the second-order differential system

$$\begin{cases} \frac{d}{dt}(\Phi_p(\dot{x}(t))) + \nabla F(t, x(t)) = 0, & t \in [0, T], \\ x(0) = \alpha x(\xi), & \dot{x}(T) = \beta \dot{x}(\eta), \end{cases} \tag{3.6.1}$$

where $x \in R^N$, $T > 0$, $0 \leq \xi < \eta \leq T$, $0 < \alpha < 1$, $0 < \beta < 1$, $p > 1$,

$$\Phi_p(x) = |x|^{p-2}x = \left(\sqrt{\sum_{i=1}^N x_i^2}\right)^{p-2} \begin{pmatrix} x_1 \\ \vdots \\ x_N \end{pmatrix}, \quad F : [0, T] \times R^N \to R \text{ satisfies}$$

$(t, x) \to F(t, x)$ is measurable in t for every $x \in R^N$ and continuously differentiable and convex in x for a.e. $t \in [0, T]$. The technique of proof to be used here is what is known as the dual least action principle. Our aim is to obtain sufficient conditions for the existence of solutions to the system (3.6.1).

In previous sections in this chapter, we defined our Banach space to be $X = X_1 \times \cdots \times X_n$, where each X_i a Sobolev space of the form

$$X_i = \left\{ \xi \in W^{1,p_i}([0,1]) \ : \ \xi(0) = \sum_{j=1}^{m} a_j \xi_j(x_j), \ \xi(1) = \sum_{j=1}^{m} b_j \xi_j(x_j) \right\}$$

for $i = 1, \ldots, n$. Unfortunately, this approach will not work in our case here because if in order to accommodate the boundary conditions, we would take

$$X = \{ x \in W^{1,p}([0,T]) \ : \ x(0) = \alpha x(\xi), \ \dot{x}(T) = \beta \dot{x}(\eta) \},$$

then X is no longer a Banach space.

To overcome this problem, we will convert the second-order system (3.6.1) into a system of two first-order equations which in a sense causes the boundary condition to "disappear;" see system (3.6.5) below. Using the dual least action principle, we will be able to obtain existence results for the first-order system (3.6.5), and this in turn will give the existence of solutions to the original system (3.6.1). There have not been many attempts to use variational methods to study multi-point boundary value problems with derivatives of the unknown function appearing in the boundary conditions.

To begin, we take $\Gamma_0(R^N)$ to be the set of convex lower semicontinuous functions

$$F : R^N \to (-\infty, \infty],$$

whose domain $D(F) = \{ u \in R^N \ : \ F(u) < \infty \}$ is not empty. We let $H : [0,T] \times R^{2N} \to R$, $(t,u) \to H(t,u)$ be a smooth Hamiltonian such that for each $t \in [0,T]$, $H(t, \cdot) \in \Gamma_0(R^{2N})$ is strictly convex and

$$\frac{H(t,u)}{|u|} \to \infty \quad \text{as} \quad |u| \to \infty.$$

The *Fenchel* (or *Legendre*) transform $H^*(t, \cdot)$ of $H(t, \cdot)$ is defined by

$$\begin{cases} H^*(t,v) = (v,u) - H(t,u), \\ v = \nabla H(t,u) \quad \text{or} \quad u = \nabla H^*(t,v). \end{cases} \tag{3.6.2}$$

Note that $\nabla H^*(t,v)$ does exist in view of [227, Proposition 2.4]. If for $u = (u_1; u_2) \in R^{2N}$ with $u_1, u_2 \in R^N$, $H(t,u)$ can be split into two parts $H(t,u) = H_1(t, u_1) + H_2(t, u_2)$, then (3.6.2) implies $H^*(t,v) = H_1^*(t, v_1) + H_2^*(t, v_2)$, $v = (v_1; v_2)$, $v_1, v_2 \in R^N$.

We will be making use of the following assumptions in this section:

(C39) there exists $l \in L^{2\max\{q,p-1\}}(0,T;R^N)$ such that for all $y \in R^N$ and a.e. $t \in [0,T]$, we have

$$F(t,y) \geq (l(t), |y|^{\frac{p-2}{2}} y);$$

(C40) there exists

$$k \in \left(0, \min\left\{\left(T\frac{1 + \Phi_p(\beta)}{1 - \Phi_p(\beta)} + T\frac{1 + \alpha}{1 - \alpha}\right)^{-1}, \right.\right.$$

$$\left.\left. \left(T\frac{1 + \Phi_p(\beta)}{1 - \Phi_p(\beta)} + T\frac{1 + \alpha}{1 - \alpha}\right)^{-\frac{p}{q}}\right\}\right)$$

and $\gamma \in L^{\max\{q,p-1\}}(0, T; R^N)$ such that

$$F(t, y) \le \frac{k^2}{p}|y|^p + \gamma(t) \quad \text{for } y \in R^N;$$

(C41) $\int_0^T F(t, y)dt \to \infty$ as $|y| \to \infty$, $y \in R^N$.

In order to transform the second-order system (3.6.1) into an equivalent system of two first-order equations, we let

$$u_1 = x \quad \text{and} \quad ku_2 = \Phi_p(\dot{x}),$$

where $k > 0$ is given in condition (C40). Then (3.6.1) is equivalent to

$$\begin{cases} \dot{u}_2 + \frac{1}{k}\nabla F(t, u_1) = 0, \\ -\dot{u}_1 + \Phi_q(ku_2) = 0, \\ u_1(0) = \alpha u_1(\xi), \quad u_2(T) = \Phi_p(\beta)u_2(\eta). \end{cases} \quad (3.6.3)$$

If we define $H : [0, T] \times R^{2N} \to R$ and $H_i : [0, T] \times R^N \to R$, $i = 1, 2$, by $H(t, u) = H_1(t, u_1) + H_2(t, u_2)$,

$$H_1(t, u_1) = \frac{1}{k}F(t, u_1), \quad \text{and} \quad H_2(t, u_2) = \frac{k^{q-1}}{q}|u_2|^q, \quad (3.6.4)$$

where $u = (u_1; u_2)$, then setting

$$J = \begin{pmatrix} 0 & I_N \\ -I_N & 0 \end{pmatrix},$$

(3.6.3) can be written as

$$\begin{cases} J\dot{u} + \nabla H(t, u) = 0, \\ u_1(0) = \alpha u_1(\xi), \quad u_2(T) = \Phi_p(\beta)u_2(\eta). \end{cases} \quad (3.6.5)$$

We need to define an appropriate space, so let

$$Y = \{u = (u_1; u_2) \in R^{2N} : u_1 \in W^{1,p}(0, T; R^N), \; u_2 \in W^{1,q}(0, T; R^N),$$
$$u_1(0) = \alpha u_1(\xi), \; u_2(T) = \Phi_p(\beta)u_2(\eta)\}$$

with the norm $\|u\|_Y = \|\dot{u}_1\|_{L^p} + \|\dot{u}_2\|_{L^q}$. Our first lemma concerns this norm in the space Y.

Lemma 3.6.1. *For $u \in Y$, the norm $\|\cdot\|_Y$ is equivalent to the usual norm* $\|u_1\|_{W^{1,p}} + \|u_2\|_{W^{1,q}}$.

Proof. Clearly,

$$\|u_1\|_{W^{1,p}} + \|u_2\|_{W^{1,q}} \geq \|\dot{u}_1\|_{L^p} + \|\dot{u}_2\|_{L^q},$$

so it suffices to show that

$$\|u_1\|_{W^{1,p}} + \|u_2\|_{W^{1,q}} \leq m(\|\dot{u}_1\|_{L^p} + \|\dot{u}_2\|_{L^q}) \quad \text{for some constant } m > 0.$$
$$(3.6.6)$$

By the Mean Value Theorem, for $u \in Y$, we have

$$u_1(t) = u_1(0) + \int_0^t \dot{u}_1(s)ds \text{ and } u_1(t) = u_1(\xi) + \int_\xi^t \dot{u}_1(s)ds \quad \text{for } t \in [0, T],$$

which, together with $u_1(0) = \alpha u_1(\xi)$, implies

$$u_1(t) = \frac{1}{1-\alpha}\left(\int_0^t \dot{u}_1(s)ds - \alpha \int_\xi^t \dot{u}_1(s)ds\right) \leq \frac{1+\alpha}{1-\alpha}\int_0^T |\dot{u}_1(s)|ds.$$

By Hölder's inequality,

$$\|u_1\|_{L^p} \leq \frac{1+\alpha}{1-\alpha}T\|\dot{u}_1\|_{L^p}. \tag{3.6.7}$$

Similarly, we have

$$u_2(t) = u_2(T) - \int_t^T \dot{u}_2(s)ds \text{ and } u_2(t) = u_2(\eta) + \int_\eta^t \dot{u}_2(s)ds \text{ for } t \in [0, T].$$

Then, from $u_2(T) = \Phi_p(\beta)u_2(\eta)$, we see that

$$u_2(t) = (1 - \Phi_p(\beta))^{-1}\left(-\int_t^T \dot{u}_2(s)ds - \Phi_p(\beta)\int_\eta^t \dot{u}_2(s)ds\right)$$
$$\leq \frac{1+\Phi_p(\beta)}{1-\Phi_p(\beta)}\int_0^T |\dot{u}_2(s)|ds.$$

Again by Hölder's inequality,

$$\|u_2\|_{L^q} \leq \frac{1+\Phi_p(\beta)}{1-\Phi_p(\beta)}T\|\dot{u}_2\|_{L^q}. \tag{3.6.8}$$

Then (3.6.6) follows from (3.6.7) and (3.6.8) and this completes the proof of the lemma. $\square$

Remark 3.6.1. *The space $(Y, \|\cdot\|_Y)$ is a reflexive Banach space since* $(Y, \|\cdot\|_Y)$ *is a closed subspace of the Banach space* $W^{1,p} \times W^{1,q}$.

Next, we construct a functional φ on the space Y by defining

$$\varphi(u) = \int_0^T \left[\frac{1}{2}(J\dot{u}, u) + H(t, u) \right] dt.$$

Clearly, $H(t, u)$ is continuously differentiable in u and strictly convex with respect to u. Thus, we can make the Fenchel transformation $H^*(t, \dot{v}) = \sup_{u \in R^{2N}} [(\dot{v}, u) - H(t, u)]$. By transform theory, there is only one u_v for v such that

$$(\dot{v}, u_v) - H(t, u_v) = \sup_{u \in R^{2N}} [(\dot{v}, u) - H(t, u)].$$

Therefore, $\dot{v} = \nabla H(t, u_v)$, $u_v = \nabla H^*(t, \dot{v})$, and $H(t, u_v) + H^*(t, \dot{v}) = (\dot{v}, u_v)$.

For any $l \in R$, define the space

$$X = \big\{ v = (v_1; v_2) \ : \ v_1 \in W^{1,q}(0, T; R^N), \ v_2 \in W^{1,p}(0, T; R^N),$$
$$v_1(T) = \Phi_p(\beta)v_1(\eta), \ v_2(0) = \alpha v_2(\xi), \ v_1(T) = lv_2(T), \ v_1(0) = lv_2(0) \big\}$$

with the norm $\|v\|_X = \|\dot{v}_1\|_{L^q} + \|\dot{v}_2\|_{L^p}$.

Lemma 3.6.2. *For $v \in X$, the norm $\| \cdot \|_X$ is equivalent to the usual one $\|v_1\|_{W^{1,q}} + \|v_2\|_{W^{1,p}}$.*

Proof. The proof is similar to that of Lemma 3.6.1, so we omit the details here.   □

Remark 3.6.2. *The space $(X, \| \cdot \|_X)$ is a reflexive Banach space since it is a closed subspace of $W^{1,q} \times W^{1,p}$.*

Lemma 3.6.3. *For every $v \in X$,*

$$\int_0^T (J\dot{v}, v)dt \geq -T \left(\frac{1 + \Phi_p(\beta)}{1 - \Phi_p(\beta)} + \frac{1 + \alpha}{1 - \alpha} \right) [(1/q)\|\dot{v}_1\|_{L^q}^q + (1/p)\|\dot{v}_2\|_{L^p}^p].$$

Proof. For $v = (v_1; v_2) \in X$, similar to (3.6.7) and (3.6.8), we have

$$\|v_1\|_{L^q} \leq \frac{1 + \Phi_p(\beta)}{1 - \Phi_p(\beta)} T \|\dot{v}_1\|_{L^q} \quad \text{and} \quad \|v_2\|_{L^p} \leq \frac{1 + \alpha}{1 - \alpha} T \|\dot{v}_2\|_{L^p}. \quad (3.6.9)$$

By Hölder's inequality,

$$\int_0^T (J\dot{v}, v)dt = \int_0^T (\dot{v}_2 v_1 - \dot{v}_1 v_2)dt \geq -\|\dot{v}_2\|_{L^p}\|v_1\|_{L^q} - \|\dot{v}_1\|_{L^q}\|v_2\|_{L^p}.$$

Then, from (3.6.9) and Young's inequality, it follows that

$$\int_0^T (J\dot{v}, v)dt \geq -\frac{1+\alpha}{1-\alpha}T\|\dot{v}_2\|_{L^p}\|\dot{v}_1\|_{L^q} - \frac{1+\Phi_p(\beta)}{1-\Phi_p(\beta)}T\|\dot{v}_1\|_{L^q}\|\dot{v}_2\|_{L^p}$$

$$\geq -T\left(\frac{1+\Phi_p(\beta)}{1-\Phi_p(\beta)} + \frac{1+\alpha}{1-\alpha}\right)\left[(1/p)\|\dot{v}_2\|_{L^p}^p + (1/q)\|\dot{v}_1\|_{L^q}^q\right].$$

This completes the proof of the lemma. $\qquad\square$

We define φ's dual action $\mathcal{K}$ on X by

$$\mathcal{K}(v) = \int_0^T \left[\frac{1}{2}(J\dot{v}(t), v(t)) + H^*(t, \dot{v}(t))\right] dt.$$

Let $\varepsilon_0 > 0$ satisfy

$$\min\{(k+\varepsilon_0)^{-q/p}, (k^{q-1}+\varepsilon_0)^{-p/q}\} > T\frac{1+\Phi_p(\beta)}{1-\Phi_p(\beta)} + T\frac{1+\alpha}{1-\alpha}. \qquad (3.6.10)$$

First, we consider the perturbed problem

$$J\dot{u} + \nabla H_\varepsilon(t, u) = 0, \quad u_1(0) = \alpha u_1(\xi), \quad u_2(T) = \Phi_p(\beta)u_2(\eta), \qquad (3.6.11)$$

where $0 < \varepsilon < \varepsilon_0$ and $H_\varepsilon(t, u) = H_{1\varepsilon}(t, u_1) + H_{2\varepsilon}(t, u_2)$ with

$$H_{i\varepsilon} : [0, T] \times R^N \rightarrow R, \quad (t, u_i) \mapsto \frac{\varepsilon|u_i|^p}{p} + H_i(t, u_i), \quad i = 1, 2.$$

We also define the perturbed dual action

$$\mathcal{K}_\varepsilon(v) = \int_0^T \left[\frac{1}{2}(J\dot{v}(t), v(t)) + H_{1\varepsilon}^*(t, \dot{v}_1(t)) + H_{2\varepsilon}^*(t, \dot{v}_2(t))\right] dt.$$

It is clear that $\mathcal{K}_\varepsilon(v)$ is continuously differentiable on X by [266, Lemma 3.2]. In the remainder of this section, we will write

$$u_\varepsilon = (u_{1\varepsilon}; u_{2\varepsilon}) \in Y \quad \text{and} \quad v_\varepsilon = (v_{1\varepsilon}; v_{2\varepsilon}) \in X \quad \text{for any } \varepsilon > 0.$$

Lemma 3.6.4. *If $\psi \in L^1[t_0, t_1]$ and*

$$\int_{t_0}^{t_1} \psi(t)\dot{\lambda}(t)dt = 0$$

for all $\lambda \in W = \{u \in C[t_0, t_1] : \dot{u}(t) \text{ exists for a.e. } t \in [t_0, t_1], u(t_0) = u(t_1) = 0\}$, then $\psi(t) = c$ for a.e. $t \in [t_0, t_1]$.

Proof. Let $c = \frac{1}{t_1-t_0}\int_{t_0}^{t_1} \psi(s)ds$, $\lambda(t) = \int_{t_0}^t (\psi(s) - c)ds \in W$, $t \in [t_0, t_1]$. Then

$$0 = \int_{t_0}^{t_1} \psi(t)\dot{\lambda}(t)dt = \int_{t_0}^{t_1} \psi(t)(\psi(t) - c)dt = \int_{t_0}^{t_1} (\psi(t) - c)^2 dt.$$

So $\psi(t) = c$ for a.e. $t \in [t_0, t_1]$. $\qquad\square$

Lemma 3.6.5. *If $v_\varepsilon \in X$ is a critical point of $\mathcal{K}_\varepsilon$, then the function $u_\varepsilon(t) = \nabla H_\varepsilon^*(t, \dot{v}_\varepsilon)$ is a solution of* (3.6.11).

Proof. If $v_\varepsilon \in X$ is a critical point of $\mathcal{K}_\varepsilon$, then $\langle \mathcal{K}_\varepsilon'(v_\varepsilon), h \rangle = 0$ for all $h \in X$, i.e.

$$0 = \langle \mathcal{K}_\varepsilon'(v_\varepsilon), h \rangle = \int_0^T \left[\frac{1}{2}(J\dot{v}_\varepsilon, h) - \frac{1}{2}(Jv_\varepsilon, \dot{h}) + (\nabla H_\varepsilon^*(t, \dot{v}_\varepsilon), \dot{h}) \right] dt.$$
(3.6.12)

Without loss of generality, we assume that

$$h \in Z := \{(h_1; h_2) \in X : h \in C^1(0, \xi), h(0) = 0, h(t) \equiv 0, t \in [\xi, T]\} \subseteq X.$$

So (3.6.12) means that

$$0 = \int_0^\xi \left[\frac{1}{2}(J\dot{v}_\varepsilon, h) - \frac{1}{2}(Jv_\varepsilon, \dot{h}) + (\nabla H_\varepsilon^*(t, \dot{v}_\varepsilon), \dot{h}) \right] dt. \qquad (3.6.13)$$

Define $w_1 \in C(0, \xi; R^{2N})$ by $w_1(t) = \int_0^t \frac{1}{2} J\dot{v}_\varepsilon(s)ds + \frac{1}{2}Jv_\varepsilon(0)$. Thus,

$$\int_0^\xi (w_1(t), \dot{h}(t))dt = \int_0^\xi \left(\int_0^t \frac{1}{2} J\dot{v}_\varepsilon(s)ds, \dot{h}(t) \right) dt + \int_0^\xi \left(\frac{1}{2}Jv_\varepsilon(0), \dot{h}(t) \right) dt.$$

By Fubini's theorem, since $h \in AC(0, T)$, one has

$$\int_0^\xi (w_1(t), \dot{h}(t))dt$$

$$= \int_0^\xi \left[\int_s^\xi \left(\frac{1}{2} J\dot{v}_\varepsilon(s), \dot{h}(t) \right) dt \right] ds + \left(\frac{1}{2}Jv_\varepsilon(0), h(\xi) - h(0) \right)$$

$$= \int_0^\xi \left(\frac{1}{2} J\dot{v}_\varepsilon(s), h(\xi) - h(s) \right) ds + \left(\frac{1}{2}Jv_\varepsilon(0), h(\xi) - h(0) \right)$$

$$= \left(\frac{1}{2}Jv_\varepsilon(\xi) - \frac{1}{2}Jv_\varepsilon(0), h(\xi) \right) - \int_0^\xi \left(\frac{1}{2} J\dot{v}_\varepsilon(s), h(s) \right) ds$$

$$+ \left(\frac{1}{2}Jv_\varepsilon(0), h(\xi) - h(0) \right).$$

Substituting this into (3.6.13), we have

$$\int_0^\xi \left(w_1(t) + \frac{1}{2}Jv_\varepsilon(t) - \nabla H_\varepsilon^*(t, \dot{v}_\varepsilon), \dot{h}(t) \right) dt$$

$$= \left(\frac{1}{2}Jv_\varepsilon(\xi) - \frac{1}{2}Jv_\varepsilon(0), h(\xi) \right) + \left(\frac{1}{2}Jv_\varepsilon(0), h(\xi) - h(0) \right) = 0$$

holds for all $h \in Z$.

By Lemma 3.6.4, $w_1(t) + \frac{1}{2}Jv_\varepsilon(t) - \nabla H_\varepsilon^*(t, \dot{v}_\varepsilon) = c_1$ a.e. on $[0, \xi]$, i.e.,

$$Jv_\varepsilon(t) = \nabla H_\varepsilon^*(t, \dot{v}_\varepsilon) + c_1, \text{ a.e. } t \in [0, \xi], \ c_1 = (c_{11}, c_{12}). \tag{3.6.14}$$

Similarly, we have

$$Jv_\varepsilon(t) = \nabla H_\varepsilon^*(t, \dot{v}_\varepsilon) + c_2, \text{ a.e. } t \in [\xi, \eta], \ c_2 = (c_{21}, c_{22}), \tag{3.6.15}$$

$$Jv_\varepsilon(t) = \nabla H_\varepsilon^*(t, \dot{v}_\varepsilon) + c_3, \text{ a.e. } t \in [\eta, T], \ c_3 = (c_{31}, c_{32}). \tag{3.6.16}$$

Next, we show that $c_1 = c_2 = c_3 = 0$. Observe that from (3.6.12),

$$0 = \int_0^T \left[\frac{1}{2}(J\dot{v}_\varepsilon, h) + \frac{1}{2}(Jv_\varepsilon, \dot{h}) - (Jv_\varepsilon, \dot{h}) + (\nabla H_\varepsilon^*, \dot{h}) \right] dt$$

$$= \int_0^T \left[\frac{1}{2}(J\dot{v}_\varepsilon, h) + \frac{1}{2}(Jv_\varepsilon, \dot{h}) - (Jv_\varepsilon - \nabla H_\varepsilon^*, \dot{h}) \right] dt.$$

Thus, from (3.6.14)–(3.6.16), we have

$$\int_0^T \left[\frac{1}{2}(J\dot{v}_\varepsilon, h) + \frac{1}{2}(Jv_\varepsilon, \dot{h}) \right] dt = \int_0^\xi (c_1, \dot{h})dt + \int_\xi^\eta (c_2, \dot{h})dt + \int_\eta^T (c_3, \dot{h})dt \tag{3.6.17}$$

for all $h \in X$. Define $w \in C(0, T; R^{2N})$ by $w(t) = \int_0^t \frac{1}{2}J\dot{v}_\varepsilon(s)ds + \frac{1}{2}Jv_\varepsilon(0)$. Similar to the above derivation, we have

$$\int_0^T (w(t), \dot{h}(t))dt = \left(\frac{1}{2}Jv_\varepsilon(T) - \frac{1}{2}Jv_\varepsilon(0), h(T) \right)$$

$$- \int_0^T \left(\frac{1}{2}J\dot{v}_\varepsilon(s), h(s) \right) ds + \left(\frac{1}{2}Jv_\varepsilon(0), h(T) - h(0) \right)$$

$$= - \int_0^T \left(\frac{1}{2}J\dot{v}_\varepsilon(s), h(s) \right) ds + \left(\frac{1}{2}Jv_\varepsilon(T), h(T) \right) - \left(\frac{1}{2}Jv_\varepsilon(0), h(0) \right). \tag{3.6.18}$$

By $v_\varepsilon \in X, h \in X$, we have

$$\left(\frac{1}{2}Jv_\varepsilon(T), h(T) \right) - \left(\frac{1}{2}Jv_\varepsilon(0), h(0) \right)$$

$$= \frac{1}{2}[v_{2\varepsilon}(T)h_1(T) - v_{1\varepsilon}(T)h_2(T)] - \frac{1}{2}[v_{2\varepsilon}(0)h_1(0) - v_{1\varepsilon}(0)h_2(0)]$$

$$= \frac{1}{2}[lv_{2\varepsilon}(T)h_2(T) - lv_{2\varepsilon}(T)h_2(T)] - \frac{1}{2}[lv_{2\varepsilon}(0)h_2(0) - lv_{2\varepsilon}(0)h_2(0)] = 0. \tag{3.6.19}$$

Substituting (3.6.18) and (3.6.19) into (3.6.17), we have

$$0 = \int_0^T (w(t) - \frac{1}{2}Jv_\varepsilon(t), \dot{h})dt + \int_0^\xi (c_1, \dot{h})dt + \int_\xi^\eta (c_2, \dot{h})dt + \int_\eta^T (c_3, \dot{h})dt$$

$$= -c_{11}h_1(0) - c_{12}h_2(0) + (c_{11} - c_{21})h_1(\xi) + (c_{12} - c_{22})h_2(\xi)$$

$$+ (c_{21} - c_{31})h_1(\eta) + (c_{22} - c_{32})h_2(\eta) + c_{31}h_1(T) + c_{32}h_2(T). \tag{3.6.20}$$

Without loss of generality, we assume $h_1(T) = h_2(T) = h_1(0) = h_2(0) = h_1(\eta) = h_2(\xi) = 0$. Now if $h_1(\xi) = 0$ and $h_2(\eta) = 1$, then (3.6.20) implies $c_{22} = c_{32}$; on the other hand, if $h_1(\xi) = 1$ and $h_2(\eta) = 0$, then (3.6.20) implies $c_{11} = c_{21}$. So (3.6.20) becomes

$$- c_{11}h_1(0) - c_{12}h_2(0) + (c_{12} - c_{22})h_2(\xi) + (c_{21} - c_{31})h_1(\eta)$$
$$+ c_{31}h_1(T) + c_{32}h_2(T) = 0. \quad (3.6.21)$$

If we take $h_1(T) = l$, $h_2(T) = 1$, $h_1(\eta) = \frac{l}{\Phi_p(\beta)}$, and $h_1(0) = h_2(0) = h_2(\xi) = 0$, (3.6.21) becomes $(c_{21} - c_{31})\frac{l}{\Phi_p(\beta)} + c_{31}l + c_{32} = 0$ for any $l \in R$. So $c_{32} = 0$. Now if we take $h_1(0) = l$, $h_2(0) = 1$, $h_2(\xi) = \frac{1}{\alpha}$, and $h_1(T) = h_2(T) = h_1(\eta) = 0$, (3.6.21) becomes $-c_{11}l - c_{12} + (c_{12} - c_{22})\frac{1}{\alpha} = 0$ for all $l \in R$. So $c_{11} = 0$. Then (3.6.21) becomes

$$- c_{12}h_2(0) + c_{12}h_2(\xi) - c_{31}h_1(\eta) + c_{31}h_1(T) = 0. \quad (3.6.22)$$

Now if $h_2(0) = 1$, $h_1(0) = l$, $h_2(\xi) = \frac{1}{\alpha}$, and $h_1(T) = h_1(\eta) = 0$, then $c_{12} = 0$ and (3.6.22) becomes $-c_{31}h_1(\eta) + c_{31}h_1(T) = 0$. Finally, if $h_1(\eta) = 1$ and $h_1(T) = \Phi_p(\beta)$, $c_{31} = 0$ and so $c_1 = c_2 = c_3 = 0$ and $Jv_\varepsilon = \nabla H_\varepsilon^*(t, \dot{v}_\varepsilon), t \in [0, T]$.

Since $u_\varepsilon(t) = \nabla H_\varepsilon^*(t, \dot{v}_\varepsilon)$, we obtain $\dot{u}_\varepsilon(t) = J\dot{v}_\varepsilon(t)$, $t \in [0, T]$ and $u_\varepsilon \in W^{1,p} \times W^{1,q}$ and $u_\varepsilon = Jv_\varepsilon$. By duality, $\dot{v}_\varepsilon = \nabla H_\varepsilon(t, u_\varepsilon)$, i.e., $J\dot{u}_\varepsilon = -\dot{v}_\varepsilon = -\nabla H(t, u_\varepsilon)$ a.e. on $[0, T]$. Moreover, by $v_\varepsilon \in X$, we have $u_{1\varepsilon}(0) = \alpha u_{1\varepsilon}(\xi)$, $u_{2\varepsilon}(T) = \Phi_p(\beta)u_{2\varepsilon}(\eta)$. Hence, u_ε is a solution of (3.6.11). This completes the proof of the lemma. □

The following lemma gives us a criteria for determining if a functional has a minimum.

Lemma 3.6.6. ([227, Theorem 1.1]) *If the functional φ is weakly lower semi-continuous on a reflexive Banach space X and has a bounded minimizing sequence, then φ has a minimum on X.*

Next, we show that system (3.6.11) has a solution.

Lemma 3.6.7. *Assume that (C39)–(C40) hold. Then $\mathcal{K}_\varepsilon$ has a minimum point $v_\varepsilon \in X$, i.e., (3.6.11) has at least one solution.*

Proof. To apply 3.6.6, we will first show that K_ε has a bounded minimizing sequence. By [227, Equation (4), p. 34] and [227, Remark 2, p. 32], we have

$$H_{1\varepsilon}^*(t, \dot{v}_1) \geq \frac{1}{q(k+\varepsilon)^{\frac{q}{p}}}|\dot{v}_1|^q - \frac{\gamma(t)}{k} \quad \text{and} \quad H_{2\varepsilon}^*(t, \dot{v}_2) = \frac{1}{p(k^{q-1}+\varepsilon)^{\frac{p}{q}}}|\dot{v}_2|^p.$$

This, together with Lemma 3.6.3, imply

$$\mathcal{K}_\varepsilon(v) \geq -T\left(\frac{1+\Phi_p(\beta)}{1-\Phi_p(\beta)} + \frac{1+\alpha}{1-\alpha}\right)[(1/q)\|\dot{v}_1\|_{L^q}^q + (1/p)\|\dot{v}_2\|_{L^p}^p]$$

$$+ \frac{1}{p(k^{q-1}+\varepsilon)^{p/q}}\|\dot{v}_2\|_{L^p}^p + \frac{1}{q(k+\varepsilon)^{q/p}}\|\dot{v}_1\|_{L^q}^q - \gamma^0 \qquad (3.6.23)$$

$$= \frac{\|\dot{v}_2\|_{L^p}^p}{p}b_2 + \frac{\|\dot{v}_1\|_{L^q}^q}{q}b_1 - \gamma^0,$$

where

$$b_1 = -T\left(\frac{1+\Phi_p(\beta)}{1-\Phi_p(\beta)} + \frac{1+\alpha}{1-\alpha}\right) + \frac{1}{(k+\varepsilon)^{q/p}},$$

$$b_2 = -T\left(\frac{1+\Phi_p(\beta)}{1-\Phi_p(\beta)} + \frac{1+\alpha}{1-\alpha}\right) + \frac{1}{(k^{q-1}+\varepsilon)^{p/q}},$$

and

$$\gamma^0 = 1/k\int_0^T \gamma(s)ds.$$

By (3.6.10), b_1, $b_2 > 0$. Let $\{v_k\}$ be a minimizing sequence for $\mathcal{K}_\varepsilon$. By (3.6.23), $\{\|\dot{v}_{1k}\|_{L^q}\}$ and $\{\|\dot{v}_{2k}\|_{L^p}\}$ are bounded. This, together with Lemma 3.6.2, implies that $\{v_k\}$ is bounded in X.

To show that $\mathcal{K}_\varepsilon$ is weakly lower semi-continuous on X, observe that the continuity of $H_{1\varepsilon}(t,\cdot)$ and the definition of $H_{1\varepsilon}^*$ imply that

$$\mathcal{K}_{\varepsilon,1} = \int_0^T H_{1\varepsilon}^*(t,\dot{v}_1(t))dt \quad \text{and} \quad \mathcal{K}_{\varepsilon,2} = \int_0^T H_{2\varepsilon}^*(t,\dot{v}_2(t))dt$$

are weakly lower semicontinuous in $W^{1,q}$ and $W^{1,p}$, respectively. Also,

$$\mathcal{K}_{\varepsilon,3}(v) = \frac{1}{2}\int_0^T [(\dot{v}_2(t), v_1(t)) - (\dot{v}_1(t), v_2(t))]dt$$

is weakly lower semicontinuous (in fact, even weakly continuous). Thus, $\mathcal{K}_\varepsilon = \mathcal{K}_{\varepsilon,1} + \mathcal{K}_{\varepsilon,2} + \mathcal{K}_{\varepsilon,3}$ is weakly lower semicontinuous on X. Since X is a reflexive Banach space, by Lemma 3.6.6, $\mathcal{K}_\varepsilon$ has a minimum $v_\varepsilon \in X$, which completes the proof of the lemma. $\qquad\square$

Our final lemma gives conditions under which solutions of (3.6.11) are bounded

Lemma 3.6.8. *Assume that (C39)–(C41) hold. If u_ε is a solution of the perturbed problem (3.6.11), then u_ε is bounded in Y.*

Proof. By (C30) and (3.6.4),

$$H_1(t, u_1) + \frac{|u_1|^p}{p} + \frac{p}{4k^2}|l(t)|^2 = \frac{1}{k}F(t, u_1) + \frac{|u_1|^p}{p} + \frac{p}{4k^2}|l(t)|^2$$

$$\geq \frac{1}{k}(l(t), |u_1|^{\frac{p-2}{2}}u_1) + \frac{|u_1|^{p-2}}{p}(u_1, u_1) + \frac{p}{4k^2}(l(t), l(t)) \geq 0.$$

Hence,

$$-\frac{p}{4k^2}|l(t)|^2 \leq H_1(t, u_1) + \frac{|u_1|^p}{p}.$$

Note that (C40) and (3.6.4) imply

$$H_1(t, u_1) + \frac{|u_1|^p}{p} \leq \frac{k+1}{p}|u_1|^p + \frac{\gamma(t)}{k}.$$

Therefore,

$$-\frac{p}{4k^2}|l(t)|^2 \leq H_1(t, u_1) + \frac{|u_1|^p}{p} \leq \frac{k+1}{p}|u_1|^p + \frac{\gamma(t)}{k}.$$

By [227, Proposition 2.2] and the fact that $(|l|^2 + \gamma)^{p-1} \in L^1(0, T; R)$, we have

$$|\nabla H_1(t, u_1)| \leq \left\{ q(k+1)^{\frac{q}{p}} \left[|u_1| + \frac{p|l(t)|^2}{4k^2} + \frac{\gamma(t)}{k} \right] + 1 \right\}^{p-1} + |u_1|^{p-1}$$

and

$$|\nabla H_2(t, u_2)| = (k|u_2|)^{q-1}.$$

It is then easy to verify that the function

$$\overline{H}_1 : R^N \to R, \ u_1 \mapsto \int_0^T H_1(t, u_1)dt$$

is continuously differentiable on $W^{1,p}$ and

$$\overline{H}_2 : R^N \to R, \ u_2 \mapsto \int_0^T H_2(t, u_2)dt$$

is continuously differentiable on $W^{1,q}$.

Now, by (C41), $\overline{H}_i$, $i = 1, 2$, has a minimum at some point $\underline{u}_i \in R^N$ for which

$$\int_0^T \nabla H_i(t, \underline{u}_i)dt = 0, \quad i = 1, 2,$$

so the problem

$$\dot{v}_i(t) = \nabla H_i(t, \underline{u}_i), \quad i = 1, 2, \tag{3.6.24}$$

has a unique solution $\underline{v}_1 \in W^{1,q}$, $\underline{v}_2 \in W^{1,p}$. By (3.6.24), $H_i^*(t, \underline{\dot{v}}_i(t)) = (\underline{\dot{v}}_i(t), \underline{u}_i) - H_i(t, \underline{u}_i)$, so $H_i^*(\cdot, \underline{\dot{v}}_i(\cdot)) \in L^1(0, T; R^N)$. From the obvious inequality $H_i(t, u_i) \leq H_{i\varepsilon}(t, u_i)$, we deduce that $H_{i\varepsilon}^*(t, u_i) \leq H_i^*(t, u_i)$, and from (3.6.23), we obtain

$$b_2 \frac{\|\dot{v}_{2\varepsilon}\|_{L^p}^p}{p} + b_1 \frac{\|\dot{v}_{1\varepsilon}\|_{L^q}^q}{q} - \gamma_1^0$$

$$\leq \mathcal{K}_\varepsilon(v_\varepsilon) \leq \mathcal{K}_\varepsilon(\underline{v}) = \mathcal{K}_{1\varepsilon}(\underline{v}) + \mathcal{K}_{2\varepsilon}(\underline{v})$$

$$\leq \int_0^T \left[\frac{1}{2}[(\underline{\dot{v}}_2(t), \underline{v}_1(t)) - (\underline{\dot{v}}_1(t), \underline{v}_2(t))] \right.$$

$$\left. + H_1^*(t, \underline{\dot{v}}_1(t)) + H_2^*(t, \underline{\dot{v}}_2(t)) \right] dt = c_4.$$

Therefore, $\|\dot{v}_{1\varepsilon}\|_{L^q} \leq c_5$ and $\|\dot{v}_{2\varepsilon}\|_{L^p} \leq c_6$, and from $J\dot{v}_\varepsilon = \dot{u}_\varepsilon$, we have

$$\|\dot{u}_{1\varepsilon}\|_{L^p} = \|\dot{v}_{2\varepsilon}\|_{L^p} \leq c_6, \quad \|\dot{u}_{2\varepsilon}\|_{L^q} = \|\dot{v}_{1\varepsilon}\|_{L^q} \leq c_5.$$

Hence, $\|u_\varepsilon\|_Y < \infty$. This proves the lemma. $\qquad\square$

Our first existence result is contained in the following theorem.

Theorem 3.6.1. *Assume that (C39)–(C41) hold. Then, problem (3.6.1) has at least one solution.*

Proof. By Lemmas 3.6.7 and 3.6.8, the perturbed problem (3.6.11) has a solution u_ε and this solution u_ε is bounded in Y. Because Y is a reflexive Banach space, there exists a sequence (ε_n) in $(0, \varepsilon_0]$ tending to 0 and some $u_1 \in W^{1,p}$ and $u_2 \in W^{1,q}$ such that $(u_{1\varepsilon_n})$ converges weakly to u_1 in $W^{1,p}$ and $(u_{2\varepsilon_n})$ converges weakly to u_2 in $W^{1,q}$. Moreover, as $\dot{v}_\varepsilon = -J\dot{u}_\varepsilon$, we have

$$v_\varepsilon(t) - v_\varepsilon(0) = -J(u_\varepsilon(t)) + Ju_\varepsilon(0),$$

so that (v_{ε_n}) converges weakly to

$$v(t) - v(0) = -J(u(t)) + Ju(0).$$

Clearly, Y is compactly embedded in $C[0, T]$. So, (u_{ε_n}) (resp. (v_{ε_n})) converges uniformly to u (resp. v) on $[0, T]$. Since $u_{\varepsilon_n} \in X$, we have $u \in X$. From (3.6.11) in integrated form

$$\begin{cases} u_{2\varepsilon_n}(t) - u_{2\varepsilon_n}(0) + \int_0^t [\varepsilon_n \Phi_p(u_{1\varepsilon_n}(s)) + \nabla H_1(s, u_{1\varepsilon_n}(s))] \, ds = 0, \\ -u_{1\varepsilon_n}(t) + u_{1\varepsilon_n}(0) + \int_0^t [\varepsilon_n \Phi_q(u_{2\varepsilon_n}(s)) + \nabla H_2(s, u_{2\varepsilon_n}(s))] \, ds = 0, \end{cases}$$

it follows that

$$\begin{cases} u_2(t) - u_2(0) + \int_0^t \nabla H_1(s, u_1(s))ds = 0, \\ -u_1(t) + u_1(0) + \int_0^t \nabla H_2(s, u_2(s))ds = 0, \end{cases}$$

i.e.,

$$Ju(t) - Ju(0) + \int_0^t \nabla H(s, u(s))ds = 0.$$

Hence, $J\dot{u} + \nabla H(t, u) = 0$ a.e. on $[0, T]$. Therefore, $u \in Y$ is a solution of (3.6.5), i.e., $u_1 \in W^{1,p}$ is a solution of (3.6.1). This completes the proof of the theorem. $\qquad\square$

Our next and final existence result in this section is the following.

Theorem 3.6.2. *Assume that* $\nabla F : [0, T] \times [0, \infty)^N \to [0, \infty)^N$ *and (C39)–(C41) hold. Then, problem* (3.6.1) *has at least one solution* u *with* $u_i(t) \geq 0$ *on* $[0, T]$ *for* $i = 1, 2, \ldots, N$.

Proof. By Theorem 3.6.1, problem (3.6.1) has at least one solution $u \in W^{1,p}$. Since $\nabla F : [0, T] \times [0, \infty)^N \to [0, \infty)^N$, we have $\frac{d}{dt}(\Phi_p(\dot{u}(t))) \leq 0$. Hence, $\dot{u}(t)$ is nonincreasing on $[0, T]$, which, together with $0 < \beta < 1$, implies that $\dot{u}(T) = \beta\dot{u}(\eta) \geq \beta\dot{u}(T)$, i.e., $\dot{u}(T) \geq 0$. Thus, $\dot{u}(t) \geq 0$ on $[0, T]$ and $u(t)$ is nondecreasing on $[0, T]$. Since $0 < \alpha < 1$, we have $u(0) = \alpha u(\xi) \geq \alpha u(0)$, so $u(0) \geq 0$. Since $u(t)$ is nondecreasing, $u(t) \geq 0$ on $[0, T]$. This proves the theorem. $\qquad\square$

Chapter 4

Impulsive Problems

4.1 Introduction

In this chapter, we will consider some types of boundary value problems that include impulsive conditions. Again, in keeping with the theme of this monograph, our results will involve using variational methods in their proofs. In Sections 4.2 and 4.3, we are concerned with showing the existence of nontrivial periodic solutions to second-order impulsive Hamiltonian system with periodic boundary conditions. Recent work on such problems can also be found in [80, 209, 255, 256, 302].

Hamiltonian systems are important because of their applicability to problems in fluid mechanics and gas dynamics and are a natural framework for modeling many other natural phenomena as well. Background information, basic results, and discussions of their applications can be found, for example, in [93, 240, 259]. Variational methods have been used to study existence and multiplicity of periodic solutions for Hamiltonian systems as can be seen, for example, in the papers [50–52, 78, 91, 92, 102, 118, 157, 162, 184, 227, 242, 260, 277, 295–297, 303] and the references therein.

Impulsive differential equations are appearing frequently in mathematical models, and for background, theory, and applications of impulsive differential equations, we refer reader to the recent monograph of Graef et al. [143] as well as [32, 33, 200, 252]. Variational methods and critical point theory have recently been used to study existence and multiplicity of solutions to impulsive problems; see for example, [30, 46, 233, 267, 292]. Chen and He [80] used variational methods and critical point theorems of Ricceri to obtain existence of three solutions for second-order impulsive Hamiltonian systems. Other results on second order Hamiltonian systems with impulsive effects can be found, for example, in [209, 255, 302].

4.2   Existence of Solutions

We consider the second-order impulsive Hamiltonian system with periodic boundary conditions

$$
\begin{cases}
-\ddot{u}(t) + A(t)u(t) = \lambda \nabla F(t, u(t)) + \nabla H(u(t)), & a.e.\ t \in [0, T], \\
\Delta(\dot{u}_i(t_j)) = I_{ij}(u_i(t_j)), & i = 1, 2, \ldots, N,\ j = 1, 2, \ldots, p, \\
u(0) - u(T) = \dot{u}(0) - \dot{u}(T) = 0,
\end{cases}
$$

$$(4.2.1)$$

where $N \geq 1$, $u = (u_1, \ldots, u_N)^T$ where T means transpose, $T > 0$, $\lambda > 0$ is a parameter, $A : [0, T] \to \mathbb{R}^{N \times N}$ is a continuous map from the interval $[0, T]$ to the set of $N \times N$ symmetric matrices, t_j, $j = 1, 2, \ldots, p$, $p \geq 2$, are the instants at which the impulses occur, $0 = t_0 < t_1 < \cdots < t_p < t_{p+1} = T$, and $\Delta(\dot{u}_i(t_j)) = \dot{u}_i(t_j^+) - \dot{u}_i(t_j^-) = \lim_{t \to t_j^+} \dot{u}_i(t) - \lim_{t \to t_j^-} \dot{u}_i(t)$.

The following conditions are assumed to hold throughout the remainder of this section. The functions $I_{ij} : \mathbb{R} \to \mathbb{R}$ are Lipschitz continuous with the Lipschitz constants $L_{ij} > 0$, i.e.,

$$|I_{ij}(s_1) - I_{ij}(s_2)| \leq L_{ij}|s_1 - s_2| \tag{4.2.2}$$

for every s_1, $s_2 \in \mathbb{R}$, and $I_{ij}(0) = 0$ for $i = 1, 2, \ldots, N$, $j = 1, 2, \ldots, p$. In addition, $F : [0, T] \times \mathbb{R}^N \to \mathbb{R}$ is measurable with respect to t for all $u \in \mathbb{R}^N$, continuously differentiable in u for almost every $t \in [0, T]$, $F(t, \mathbf{0}) = \mathbf{0}$ for $t \in [0, T]$, where $\mathbf{0} = (0, \ldots, 0)$, and satisfies the standard summability condition

$$\sup_{|\xi| \leq a} \max\{|F(\cdot, \xi)|,\ |\nabla F(\cdot, \xi)|\} \in L^1([0, T]) \tag{4.2.3}$$

for any $a > 0$. Also, the function $H : \mathbb{R}^N \to \mathbb{R}$ is continuously differentiable, ∇H is Lipschitz continuous with the Lipschitz constant $L > 0$, i.e.,

$$|\nabla H(\xi_1) - \nabla H(\xi_2)| \leq L|\xi_1 - \xi_2| \tag{4.2.4}$$

for every ξ_1, $\xi_2 \in \mathbb{R}^N$,

$$H(\mathbf{0}) = 0, \quad \text{and} \quad \nabla H(\mathbf{0}) = 0. \tag{4.2.5}$$

Notice that assuming $\nabla F : [0, T] \times \mathbb{R}^N \to \mathbb{R}$ is continuous implies that condition (4.2.3) is satisfied.

The main analytic tools to be used in this section are Theorems 1.1.1, 1.1.2, and 1.1.8 above. We begin by obtaining the existence of a nontrivial periodic solution of problem (4.2.1) by combining algebraic conditions on F and H.

Another result in this section is concerned with the existence of three periodic solutions. It is obtained by combining two algebraic conditions that guarantee the existence of two local minima of the Euler-Lagrange functional, and then applying the mountain pass theorem of Pucci and Serrin (see [239]) to ensure the existence of the third critical point. This approach of combining techniques to obtain multiple solutions of boundary value problems is somewhat unique.

The matrix A is assumed to satisfy the:

(D1) $A(t) = (a_{kl}(t))$, $k = 1, \ldots, N$, $l = 1, \ldots, N$, is a symmetric matrix with $a_{kl} \in L^{\infty}[0, T]$ for any $t \in [0, T]$;

(D2) There exists $\delta > 0$ such that $(A(t)\xi, \xi) \geq \delta|\xi|^2$ for any $\xi \in \mathbb{R}^N$ and a.e. $t \in [0, T]$, where $(\cdot, \cdot)$ denotes the inner product in $\mathbb{R}^N$.

We define the space

$$E = \{u : [0, T] \to \mathbb{R}^N \,|\, u \text{ is absolutely continuous,}$$
$$u(0) = u(T), \dot{u} \in L^2([0, T], \mathbb{R}^N)\}$$

with the inner product

$$\prec u, v \succ_E = \int_0^T [(\dot{u}(t), \dot{v}(t)) + (u(t), v(t))]dt$$

and the corresponding norm

$$\|u\|_E^2 = \int_0^T (|\dot{u}(t)|^2 + |u(t)|^2)dt, \quad \text{for all} \quad u \in E.$$

For every u, $v \in E$, we define

$$\prec u, v \succ = \int_0^T [(\dot{u}(t), \dot{v}(t)) + (A(t)u(t), v(t))]dt,$$

and note that conditions (D1) and (D2) ensure that this defines an inner product in E. Then E is a separable and reflexive Banach space with the norm

$$\|u\| = \prec u, u \succ^{\frac{1}{2}} \quad \text{for all} \quad u \in E.$$

Clearly, E is an uniformly convex Banach space.

A simple computation shows that $(A(t)\xi, \xi) = \sum_{k,l=1}^{N} a_{kl}(t)\xi_k\xi_l \leq \sum_{k,l=1}^{N} \|a_{kl}\|_{L^\infty} |\xi|^2$ for every $t \in [0, T]$ and $\xi \in \mathbb{R}^N$. This implies

$$\sqrt{m}\|u\|_E \leq \|u\| \leq \sqrt{M}\|u\|_E, \qquad (4.2.6)$$

where $m = \min\{1, \delta\}$ and $M = \max\{1, \sum_{k,l=1}^{N} \|a_{kl}\|_\infty\}$, which means the norm $\|\cdot\|$ is equivalent to the norm $\|\cdot\|_E$. Since $(E, \|\cdot\|)$ is compactly embedded in $C([0, T], \mathbb{R}^N)$ (see [227]), there exists a positive constant c such that

$$\|u\|_\infty \leq c\, \|\, u\, \|, \qquad (4.2.7)$$

where $\|u\|_\infty = \max_{t \in [0,T]} |u(t)|$ and $c = \sqrt{\frac{2}{m}} \max\{\frac{1}{\sqrt{T}}, \sqrt{T}\}$ (see [80]).

For $u \in E$, $\Delta\dot{u}(t) = \dot{u}(t^+) - \dot{u}(t^-) = 0$ does not necessarily hold for every $t \in (0, T)$, and the derivative $\dot{u}$ may possess discontinuities. This leads to the impulsive effects.

Next, we define what we mean by a solution of (4.2.1).

Definition 4.2.1. *A function* $u \in \{u \in E : \dot{u} \in (W^{1,2}(t_j, t_{j+1}))^N, j = 0, 1, 2, \ldots, p\}$ *is said to be a classical solution of the problem* (4.2.1) *if* u *satisfies the differential equation, the impulse relations, and the boundary conditions given in* (4.2.1).

Definition 4.2.2. *By a weak solution of the problem* (4.2.1)*, we mean any* $u \in E$ *such that*

$$\int_0^T [(\dot{u}(t), \dot{v}(t)) + (A(t)u(t), v(t)) - (\nabla H(u(t)), v(t))]dt$$

$$+ \sum_{j=1}^{p}\sum_{i=1}^{N} I_{ij}(u_i(t_j))v_i(t_j) - \lambda \int_0^T (\nabla F(t, u(t)), v(t))dt = 0 \quad (4.2.8)$$

for every $v \in E$.

An important relationship between a weak solution and a classical solution of (4.2.1) is given in the next lemma. The scalar version of it for second order impulsive Sturm-Liouville boundary value problems was proved in [257]. The proof for problem (4.2.1) is essentially the same so we omit the details.

Lemma 4.2.1. *If $u \in E$ is a weak solution of* (4.2.1), *then u is a classical solution of* (4.2.1).

Unless stated otherwise, in what follows, by a solution of (4.2.1) we mean a classical solution.

We will assume throughout the remainder of this section that

$$K := c^2 \left(2LT + \sum_{j=1}^{p} \sum_{i=1}^{N} L_{ij} \right) < 1. \tag{4.2.9}$$

The following lemma, which is somewhat interesting in its own right, will be used to prove our main results.

Lemma 4.2.2. *Let $J : E \to E^*$ be the operator defined by*

$$J(u)v = \int_0^T [(\dot{u}(t), \dot{v}(t)) + (A(t)u(t), v(t)) - (\nabla H(u(t)), v(t))]dt$$

$$+ \sum_{j=1}^{p} \sum_{i=1}^{N} I_{ij}(u_i(t_j))v_i(t_j)$$

for every u, $v \in E$. Then J admits a continuous inverse on E^*.

Proof. Since $-L|\xi|^2 \leq (\nabla H(\xi), \xi) \leq L|\xi|^2$ for every $\xi \in \mathbb{R}^N$, and $-L_{ij}|s|^2 \leq I_{ij}(s)s \leq L_{ij}|s|^2$ for every $s \in \mathbb{R}$ and all $i = 1, 2, \ldots, N$, $j = 1, 2, \ldots, p$, in view of (4.2.7), we have

$$J(u)u = \int_0^T [(\dot{u}(t), \dot{u}(t)) + (A(t)u(t), u(t)) - (\nabla H(u(t)), u(t))]dt$$

$$+ \sum_{j=1}^{p} \sum_{i=1}^{N} I_{ij}(u_i(t_j))u_i(t_j)$$

$$\geq \left(1 - c^2 LT - c^2 \sum_{j=1}^{p} \sum_{i=1}^{N} L_{ij} \right) \|u\|^2$$

$$> (1 - K)\|u\|^2.$$

Since $K < 1$, J is coercive. Now for any u, $v \in E$,

$$\langle J(u) - J(v), u - v \rangle = \int_0^T (\dot{u}(t) - \dot{v}(t), \dot{u}(t) - \dot{v}(t))dt$$

$$+ \sum_{j=1}^p \sum_{i=1}^N (I_{ij}(u_i(t_j)) - I_{ij}(v_i(t_j)))(u_i(t_j) - v_i(t_j))$$

$$- \int_0^T (\nabla H(u(t)) - \nabla H(v(t)), u(t) - v(t))dt$$

$$\geq \left(1 - c^2 LT - c^2 \sum_{j=1}^p \sum_{i=1}^N L_{ij} \right) \|u - v\|^2$$

$$> (1 - K)\|u - v\|^2,$$

so J is uniformly monotone. By [289, Theorem 26.A (d)], J^{-1} exists and is continuous on E^*. □

Our first existence result is contained in the following theorem. For a given function $w \in E$ and a given nonnegative constant r with

$$r \neq \frac{1}{2}(1 + K)\|w\|^2,$$

we set

$$a_w(r) := \frac{\int_0^T \max_{|\xi| \leq c(\frac{2r}{1-K})^{1/2}} F(t, \xi)dt - \int_0^T F(t, w(t))dt}{r - \frac{1}{2}(1 + K)\|w\|^2}.$$

Theorem 4.2.1. *Assume that there exist constants $r_1 \geq 0$ and $r_2 > 0$, and a function $w \in E$ such that*

(D7) $\left(\frac{2r_1}{1-K}\right)^{1/2} < \|w\| < \left(\frac{2r_2}{1+K}\right)^{1/2}$,
(D8) $a_w(r_2) < a_w(r_1)$.

Then, for each $\lambda \in \left(\frac{1}{a_w(r_1)}, \frac{1}{a_w(r_2)}\right)$, the problem (4.2.1) has a non-trivial periodic solution $u^* \in E$ such that

$$r_1 < \frac{1}{2}\|u^*\|^2 + \sum_{j=1}^p \sum_{i=1}^N \int_0^{u_i^*(t_j)} I_{ij}(s)ds - \int_0^T H(u^*(t))dt < r_2.$$

Remark 4.2.1. *In the above theorem, and in the results below, by u^* we mean the vector $(u_1^*, u_2^*, \ldots, u_N^*)$.*

Proof. Choose λ as in the conclusion of the theorem. In order to apply Theorem 1.1.1 to our problem, we take $X = E$ and define the functionals $\Phi, \Psi, I_\lambda : X \to \mathbb{R}$ by

$$\Phi(u) = \frac{1}{2}\|u\|^2 + \sum_{j=1}^{p}\sum_{i=1}^{N}\int_0^{u_i(t_j)} I_{ij}(s)ds - \int_0^T H(u(t))dt,$$

$$\Psi(u) = \int_0^T F(t, u(t))dt,$$

and

$$I_\lambda(u) = \Phi(u) - \lambda\Psi(u)$$

for all $u \in X$. It is well known that Ψ is a Gâteaux differentiable functional whose Gâteaux derivative at the point $u \in X$ is the functional $\Psi'(u) \in X^*$ given by

$$\Psi'(u)v = \int_0^T (\nabla F(t, u(t)), v(t))dt \qquad (4.2.10)$$

for every $v \in X$, and that $\Psi' : X \to X^*$ is a compact operator. Moreover, Φ is a Gâteaux differentiable functional whose Gâteaux derivative at the point $u \in X$ is the functional $\Phi'(u) \in X^*$ given by

$$\Phi'(u)v = \int_0^T [(\dot{u}(t), \dot{v}(t)) + (A(t)u(t), v(t)) - (\nabla H(u(t)), v(t))]dt$$

$$+ \sum_{j=1}^{p}\sum_{i=1}^{N} I_{ij}(u_i(t_j))v_i(t_j) \qquad (4.2.11)$$

for every $v = (v_1, v_2, \dots, v_N) \in X$. Furthermore, Φ is sequentially weakly lower semicontinuous (see [139]). From (4.2.4) and (4.2.5), we have $|H(\xi)| \leq L|\xi|^2$ for all $\xi \in \mathbb{R}^N$. From (4.2.2), (4.2.7), and the fact that $I_{ij}(0) = 0$, we have

$$\frac{1}{2}(1 - K)\|u\|^2 \leq \Phi(u) \leq \frac{1}{2}(1 + K)\|u\|^2 \qquad (4.2.12)$$

for $u \in X$. Condition (D7) together with (4.2.12) implies

$$r_1 < \Phi(w) < r_2.$$

From (4.2.7) and (4.2.12), for each $u \in X$,

$$\Phi^{-1}(-\infty, r_2) = \{u \in X : \ \Phi(u) < r_2\}$$

$$\subseteq \left\{u \in X : \ \frac{1}{2}(1 - K)\|u\|^2 < r_2\right\}$$

$$\subseteq \left\{u \in X : \ |u(t)| \leq c\left(\frac{2r_2}{1 - K}\right)^{1/2} \ \text{for each } t \in [0, T]\right\},$$

and it follows that

$$\sup_{u\in\Phi^{-1}(-\infty,r_2)} \Psi(u) = \sup_{u\in\Phi^{-1}(-\infty,r_2)} \int_0^T F(t,u(t))dt$$

$$\leq \int_0^T \max_{|\xi|\leq c(\frac{2r_2}{1-k})^{1/2}} F(t,\xi)dt.$$

Therefore,

$$\eta(r_1,r_2) \leq \frac{\sup_{u\in\Phi^{-1}(-\infty,r_2)} \Psi(u) - \Psi(w)}{r_2 - \Phi(w)}$$

$$\leq \frac{\int_0^T \max_{|\xi|\leq c\left(\frac{2r_2}{1-k}\right)^{1/2}} F(t,\xi)dt - \Psi(w)}{r_2 - \Phi(w)}$$

$$\leq \frac{\int_0^T \max_{|\xi|\leq c\left(\frac{2r_2}{1-k}\right)^{1/2}} F(t,\xi)dt - \int_0^T F(t,w(t))dt}{r_2 - \frac{1}{2}(1+K)\|w\|^2}$$

$$= a_w(r_2).$$

On the other hand, arguing as before,

$$\rho(r_1,r_2) \geq \frac{\Psi(w) - \sup_{u\in\Phi^{-1}(-\infty,r_1)} \Psi(u)}{\Phi(w) - r_1}$$

$$\geq \frac{\Psi(w) - \int_0^T \max_{|\xi|\leq c\left(\frac{2r_1}{1-k}\right)^{1/2}} F(t,\xi)dt}{\Phi(w) - r_1}$$

$$\geq \frac{\int_0^T F(t,w(t))dt - \int_0^T \max_{|\xi|\leq c\left(\frac{2r_1}{1-k}\right)^{1/2}} F(t,\xi)dt}{\frac{1}{2}(1+K)\|w\|^2 - r_1}$$

$$= a_w(r_1).$$

Hence, from condition (D7), we have $\eta(r_1,r_2) < \rho(r_1,r_2)$. Therefore, by Theorem 1.1.1, for each $\lambda \in \left(\frac{1}{a_w(r_1)}, \frac{1}{a_w(r_2)}\right)$, the functional I_λ admits at least one critical point $u^* \in X$ such that $r_1 < \Phi(u^*) < r_2$, that is, u^* is a nontrivial local minimum for I_λ in X.

Since weak solutions of problem (4.2.1) are precisely the solutions of the equation $I'_\lambda(u) = 0$ (see (4.2.8)–(4.2.11)), u^* is a weak solution of problem (4.2.1). In view of Lemma 4.2.1, this completes the proof of the theorem. $\square$

The following corollary provides a sufficient condition for applying Theorem 4.2.1 that does not require knowledge of two constants r_1, r_2 and a test function w satisfying (a_1) and (a_2).

Let

$$D = \frac{(T-t_p)^2}{t_1 t_p^2} + \frac{t_1}{3t_p^2}(t_p^2 + t_p T + T^2) + (t_p - t_1) + \frac{T-t_p}{t_p^2} + \frac{1}{3t_p^2}(T^3 - t_p^3) > 0,$$

and for a given nonnegative constant θ and a positive constant η, with

$$(1-K)\theta^2 \neq c^2(1+K)DM\eta^2,$$

let

$$b_\eta(\theta) := \frac{\int_0^T \max_{|\xi| \leq \theta} F(t,\xi)dt - \int_{t_1}^{t_p} F(t,\eta\varepsilon)dt}{\frac{1}{2}(1-K)\theta^2 - \frac{1}{2}c^2(1+K)DM\eta^2}$$

where $\varepsilon = (1,0,\ldots,0) \in \mathbb{R}^N$.

Corollary 4.2.1. *Assume there exist constants $\theta_1 \geq 0$, $\theta_2 > 0$, and $\eta > 0$ with $\frac{\theta_1}{c\sqrt{Dm}} < \eta < \frac{\theta_2}{c}\sqrt{\frac{1-K}{DM(1+K)}}$ such that*

(D9) $F(t,\xi) \geq 0$ *for each $t \in [0,t_1] \cup [t_p, T]$ and $|\xi| \leq \frac{\eta T}{t_p}$,*
(D10) $b_\eta(\theta_2) < b_\eta(\theta_1)$.

Then, for each $\lambda \in \left(\frac{1}{c^2} \frac{1}{b_\eta(\theta_1)}, \frac{1}{c^2} \frac{1}{b_\eta(\theta_2)}\right)$, the problem (4.2.1) has a nontrivial periodic solution $u^* \in E$ such that

$$\frac{1}{2}(1-K)\left(\frac{\theta_1}{c}\right)^2 < \frac{1}{2}\|u^*\|^2 + \sum_{j=1}^{p}\sum_{i=1}^{N}\int_0^{u_i^*(t_j)} I_{ij}(s)ds - \int_0^T H(u^*(t))dt$$

$$< \frac{1}{2}(1-K)\left(\frac{\theta_2}{c}\right)^2.$$

Proof. Choose $r_1 = \frac{1}{2}(1-K)(\frac{\theta_1}{c})^2$, $r_2 = \frac{1}{2}(1-K)(\frac{\theta_2}{c})^2$, and

$$w(t) = \begin{cases} (T + \frac{t_p - T}{t_1}t)\frac{\eta\varepsilon}{t_p}, & t \in [0,t_1), \\ \eta\varepsilon, & t \in [t_1, t_p], \\ \frac{\eta\varepsilon}{t_p}t, & t \in (t_p, T]. \end{cases} \quad (4.2.13)$$

Then $w \in E$ and $\|w\|_E^2 = D\eta^2$. By (4.2.6),

$$Dm\eta^2 \leq \|w\|^2 \leq DM\eta^2, \quad (4.2.14)$$

and this together with the condition on η implies (D7) is satisfied. Moreover, since $0 \le w(t) \le \frac{\eta T}{t_p}$ for each $t \in [0, T]$ and (D9) holds, we have

$$\int_0^{t_1} F\left(t, \left(T + \frac{t_1 - T}{t_p}t\right)\frac{\eta\varepsilon}{t_p}\right)dt + \int_{t_p}^{T} F\left(t, \frac{\eta\varepsilon}{t_p}t\right)dt \ge 0. \qquad (4.2.15)$$

Therefore, from (4.2.14) and (4.2.15) it follows that

$$a_w(r_2) = \frac{\int_0^T \max_{|\xi| \le c(\frac{2r_2}{1-K})^{1/2}} F(t, \xi)dt - \int_0^T F(t, w(t))dt}{r_2 - \frac{1}{2}(1 + K)\|w\|^2} \le c^2 b_\eta(\theta_2)$$

and

$$c^2 b_\eta(\theta_1) \le \frac{\int_0^T F(t, w(t))dt - \int_0^T \max_{|\xi| \le c(\frac{2r_1}{1-K})^{1/2}} F(t, \xi)dt}{\frac{1}{2}(1 + K)\|w\|^2 - r_1} = a_w(r_1).$$

Therefore, (D10) implies (D8) holds. Hence, by Theorem 4.2.1, the conclusion of the corollary follows. □

An easy consequence of Corollary 4.2.1 is the following existence result.

Corollary 4.2.2. *Assume there exist* $\theta > 0$ *and* $\eta > 0$ *with* $\eta < \frac{\theta}{c}\sqrt{\frac{1-K}{DM(1+K)}}$ *such that*

(D11) $\dfrac{\int_0^T \max_{|\xi| \le \theta} F(t, \xi)dt}{\theta^2} < \dfrac{1-K}{c^2(1+K)DM}\dfrac{\int_{t_1}^{t_p} F(t, \eta\varepsilon)dt}{\eta^2},$

     where $\varepsilon = (1, 0, \ldots, 0) \in \mathbb{R}^N$.

Then, for each $\lambda \in \left(\dfrac{(1+K)DM\eta^2}{2\int_{t_1}^{t_p} F(t, \eta\varepsilon)dt}, \dfrac{(1-K)\theta^2}{2c^2 \int_0^T \max_{|\xi| \le \theta} F(t, \xi)dt}\right)$, *problem* (4.2.1) *has a non-trivial periodic solution* $u^* \in E$ *such that*

$$0 < \frac{1}{2}\|u^*\|^2 + \sum_{j=1}^{p}\sum_{i=1}^{N}\int_0^{u_i^*(t_j)} I_{ij}(s)ds - \int_0^T H(u^*(t))dt < \frac{1}{2}(1 - K)\left(\frac{\theta}{c}\right)^2.$$

Proof. Choosing $\theta_1 = 0$ and $\theta_2 = \theta$, we have

$$b_\eta(\theta) < \frac{\left(1 - \frac{c^2(1+K)DM\eta^2}{(1-K)\theta^2}\right)\int_0^T \max_{|\xi|\le\theta} F(t,\xi)dt}{\frac{1}{2}(1-K)\theta^2 - \frac{1}{2}c^2(1+K)DM\eta^2}$$

$$= \frac{\int_0^T \max_{|\xi|\le\theta} F(t,\xi)dt}{\frac{1}{2}(1-K)\theta^2}$$

$$< \frac{1}{\frac{1}{2}c^2(1+K)DM} \frac{\int_{t_1}^{t_p} F(t,\eta\varepsilon)dt}{\eta^2}$$

$$= b_\eta(0).$$

In particular,

$$b_\eta(\theta) < \frac{\int_0^T \max_{|\xi|\le\theta} F(t,\xi)dt}{\frac{1}{2}(1-K)\theta^2}.$$

The conclusion then follows from Corollary 4.2.1. □

Next, we present an application of Theorem 1.1.2 that will be used to obtain multiple solutions to problem (4.2.1).

Theorem 4.2.2. *Assume there exist a constant $\bar{r} > 0$ and a function $\bar{w}$ with $\frac{2\bar{r}}{1-K} < \|\bar{w}\|^2$ such that:*

(D12)　$\int_0^T \max_{|\xi|\le c(\frac{2\bar{r}}{1-K})^{1/2}} F(t,\xi)dt < \int_0^T F(t,\bar{w}(t))dt;$

(D13)　$\limsup_{|\xi|\to\infty} \frac{F(t,\xi)}{|\xi|^2} \le 0$ *uniformly for $t \in [0,T]$.*

Then, for each $\lambda \in (\bar{\lambda}, \infty)$, where

$$\bar{\lambda} := \frac{\frac{1}{2}(1+K)\|\bar{w}\|^2 - \bar{r}}{\int_0^T F(t,\bar{w}(t))dt - \int_0^T \max_{|\xi|\le c(\frac{2\bar{r}}{1-K})^{1/2}} F(t,\xi)dt},$$

problem (4.2.1) admits at least one non-trivial periodic solution $\bar{u} \in E$ such that

$$\frac{1}{2}\|\bar{u}\|^2 + \sum_{j=1}^p \sum_{i=1}^N \int_0^{\bar{u}_i(t_j)} I_{ij}(s)ds - \int_0^T H(\bar{u}(t))dt > \bar{r}.$$

Proof. Choose λ as in the conclusion of the theorem. Taking X and the functionals Φ and Ψ as in the proof of Theorem 4.2.1, we see that all the regularity assumptions required in Theorem 1.1.2 are satisfied. By (D13), there is a constant ϵ and a function $h_\epsilon \in L^1([0,T])$ with $0 < \epsilon < \frac{1-K}{2\lambda c^2}$ such that

$$F(t,\xi) \leq \epsilon|\xi|^2 + h_\epsilon(t) \quad \text{for all } t \in [0,T], \ \xi \in \mathbb{R}^N. \tag{4.2.16}$$

From the definitions of Φ and Ψ, (4.2.7), (4.2.12) and (4.2.16), we obtain

$$
\begin{aligned}
I_\lambda(u) &\geq \frac{1}{2}(1-K)\|u\|^2 - \lambda\epsilon \int_0^T |u(t)|^2 dt - \lambda \int_0^T h_\epsilon(t)dt \\
&\geq \frac{1}{2}(1 - K - \lambda\epsilon c^2)\|u\|^2 - \lambda\|h_\epsilon\|_{L^1([0,T])}.
\end{aligned}
$$

Since $1 - K - \lambda\epsilon c^2 > 0$, the functional I_λ is coercive. Arguing as in the proof of Theorem 4.2.1 shows that

$$\rho_2(\bar{r}) \geq \frac{\int_0^T F(t,\bar{w}(t))dt - \int_0^T \max_{|\xi|\leq c(\frac{2\bar{r}}{1-K})^{1/2}} F(t,\xi)dt}{\frac{1}{2}(1+K)\|\bar{w}\|^2 - \bar{r}} > 0$$

by (D12) and (D13). By Theorem 1.1.2, the functional I_λ admits at least one local minimum $\bar{u} \in X$ such that

$$\frac{1}{2}\|\bar{u}\|^2 + \sum_{j=1}^p \sum_{i=1}^N \int_0^{\bar{u}_i(t_j)} I_{ij}(s)ds - \int_0^T H(\bar{u}(t))dt > \bar{r},$$

and the conclusion follows. $\qquad\qquad\square$

The following corollary provides a sufficient condition for applying Theorem 4.2.2 that does not require knowledge of a constant $\bar{r}$ and a test function $\bar{w}$ satisfying (D12) and (D13).

Corollary 4.2.3. *Assume that (D9) and (D13) hold and there exist constants $\bar{\theta} > 0$ and $\bar{\eta} > 0$ with $\frac{\bar{\theta}}{c\sqrt{Dm}} < \bar{\eta}$ such that*

(D14) $\int_0^T \max_{|\xi|\leq\bar{\theta}} F(t,\xi)dt < \int_{t_1}^{t_p} F(t,\bar{\eta}\varepsilon)dt$, *where $\varepsilon = (1,0,\ldots,0) \in \mathbb{R}^N$.*

Then, for each $\lambda \in (\bar{\lambda}',\infty)$, where

$$\bar{\lambda}' := \frac{\frac{1}{2}(1+K)DM\bar{\eta}^2 - \frac{1}{2}(1-K)(\frac{\bar{\theta}}{c})^2}{\int_{t_1}^{t_p} F(t,\bar{\eta}\varepsilon)dt - \int_0^T \max_{|\xi|\leq\bar{\theta}} F(t,\xi)dt},$$

problem (4.2.1) admits at least one non-trivial periodic solution $\bar{u} \in E$ such that

$$\frac{1}{2}\|\bar{u}\|^2 + \sum_{j=1}^{p}\sum_{i=1}^{N}\int_{0}^{\bar{u}_i(t_j)} I_{ij}(s)ds - \int_{0}^{T} H(\bar{u}(t))dt > \frac{1}{2}(1-K)\left(\frac{\bar{\theta}}{c}\right)^2.$$

Proof. Choose $\bar{r} = \frac{1}{2}(1-K)(\frac{\bar{\theta}}{c})^2$ and let $\bar{w}$ be as in (4.2.13) with η replaced by $\bar{\eta}$. The conclusion follows from an application of Theorem 4.2.2. $\square$

Next, we point out some results for which the function F is in factored form. To be precise, consider the problem

$$\begin{cases} -\ddot{u}(t) + A(t)u(t) = \lambda b(t)\nabla G(u(t)) + \nabla H(u(t)), & a.e.t \in [0,T], \\ \Delta(\dot{u}_i(t_j)) = I_{ij}(u_i(t_j)), & i = 1,2,\ldots,N, \ j = 1,2,\ldots,p, \\ u(0) - u(T) = \dot{u}(0) - \dot{u}(T) = 0, \end{cases}$$

$$(4.2.17)$$

where $b \in L^1([0,T])$, $b(t) \geq 0$ a.e. $t \in [0,T]$, $b \not\equiv 0$, $G \in C^1(\mathbb{R}^N, \mathbb{R})$, $G(0) = 0$, and each component of the vector $\nabla G : \mathbb{R}^N \to \mathbb{R}^N$ is a nonnegative continuous function.

Remark 4.2.2. Since the first term on the right hand side of the equation in (4.2.17) is nonnegative, any weak solution of (4.2.17) is nonnegative. To see this, assume that the set $\mathcal{A} = \{t \in [0,T] : u_0(t) < 0\}$ is non-empty and of positive measure. Let $\bar{v}(t) = \min\{0, u_0(t)\}$ for all $t \in [0,T]$. Clearly, $\bar{v} \in E$. Using the fact that u_0 is a weak solution of (4.2.17), we have

$$\int_{0}^{T} ((\dot{u}_0(t), \dot{\bar{v}}(t)) + (A(t)u_0(t), \bar{v}(t)) - (\nabla H(u_0(t)), \bar{v}(t)))dt$$

$$+ \sum_{j=1}^{p}\sum_{i=1}^{N} I_{ij}(u_{0i}(t_j))\bar{v}_i(t_j) = \lambda \int_{0}^{T} (b(t)\nabla G(u_0(t)), \bar{v}(t))dt.$$

Thus,

$$0 \leq \left(1 - c^2 LT - c^2 \sum_{j=1}^{p}\sum_{i=1}^{N} L_{ij}\right) \int_{\mathcal{A}} [(\dot{u}_0(t), \dot{u}_0(t)) + (A(t)u_0(t), u_0(t))]dt$$

$$\leq \int_{\mathcal{A}} ((\dot{u}_0(t), \dot{u}_0(t)) + (A(t)u_0(t), u_0(t)) - (\nabla H(u_0(t)), u_0(t)))dt$$

$$+ \sum_{j=1}^{p}\sum_{i=1}^{N} I_{ij}(u_{0i}(t_j))u_{0i}(t_j) \leq 0.$$

Now $c^2 LT + c^2 \sum_{j=1}^{p} \sum_{i=1}^{N} L_{ij} < 1$, so $u_0 = 0$ is in $\mathcal{A}$, which is a contradiction.

We will now present some existence results that are consequences of Corollaries 4.2.1, 4.2.2, and 4.2.3, respectively. For a given nonnegative constant θ and a positive constant η, with

$$(1 - K)\theta^2 \neq c^2(1 + K)DM\eta^2,$$

define

$$c_\eta(\theta) := \frac{\max_{|\xi| \le \theta} G(\xi) \int_0^T b(t)dt - G(\eta\varepsilon) \int_{t_1}^{t_p} b(t)dt}{\frac{1}{2}(1 - K)\theta^2 - \frac{1}{2}c^2(1 + K)DM\eta^2},$$

where $\varepsilon = (1, 0, \dots, 0) \in \mathbb{R}^N$.

Corollary 4.2.4. *Assume there exist constants $\theta_1 \ge 0$, $\theta_2 > 0$, and $\eta > 0$ with $\frac{\theta_1}{c\sqrt{Dm}} < \eta < \frac{\theta_2}{c}\sqrt{\frac{1-K}{DM(1+K)}}$ such that:*

(D16) $c_\eta(\theta_2) < c_\eta(\theta_1)$.

Then, for each $\lambda \in \left(\frac{1}{c^2} \frac{1}{c_\eta(\theta_1)}, \frac{1}{c^2} \frac{1}{c_\eta(\theta_2)} \right)$, the problem (4.2.17) has a nontrivial periodic solution $u^* \in E$ such that

$$\frac{1}{2}(1 - K)\left(\frac{\theta_1}{c}\right)^2 < \frac{1}{2}\|u^*\|^2 + \sum_{j=1}^{p} \sum_{i=1}^{N} \int_0^{u_i^*(t_j)} I_{ij}(s)ds - \int_0^T H(u^*(t))dt$$

$$< \frac{1}{2}(1 - K)\left(\frac{\theta_2}{c}\right)^2.$$

Corollary 4.2.5. *Assume there exist constants $\theta > 0$ and $\eta > 0$ with $\eta < \frac{\theta}{c}\sqrt{\frac{1-K}{DM(1+K)}}$ such that*

(D17) $\dfrac{\max_{|\xi| \le \theta} G(\xi) \int_0^T b(t)dt}{\theta^2} < \dfrac{1-K}{c^2(1+K)DM} \dfrac{G(\eta\varepsilon) \int_{t_1}^{t_p} b(t)dt}{\eta^2}$,   *where $\varepsilon = (1, 0, \dots, 0) \in \mathbb{R}^N$.*

Then, for each $\lambda \in \left(\frac{(1+K)DM\eta^2}{2G(\eta\varepsilon)\int_{t_1}^{t_p} b(t)dt}, \ \frac{(1-K)\theta^2}{2c^2 \max_{|\xi|\leq\theta} G(\xi)\int_0^T b(t)dt} \right)$, *the problem* (4.2.17) *has a non-trivial periodic solution* $u^* \in E$ *such that*

$$\frac{1}{2}\|u^*\|^2 + \sum_{j=1}^{p}\sum_{i=1}^{N}\int_0^{u_i^*(t_j)} I_{ij}(s)ds - \int_0^T H(u^*(t))dt < \frac{1}{2}(1-K)\left(\frac{\theta}{c}\right)^2.$$

Corollary 4.2.6. *Assume there exist constants* $\bar{\theta} > 0$ *and* $\bar{\eta} > 0$ *with* $\frac{\bar{\theta}}{c\sqrt{Dm}} < \bar{\eta}$ *such that*

(D18) $\max_{|\xi|\leq\bar{\theta}} G(\xi)\int_0^T b(t)dt < G(\bar{\eta}\varepsilon)\int_{t_1}^{t_p} b(t)dt$, *where* $\varepsilon = (1,0,\dots,$ $0) \in \mathbb{R}^N$;

(D19) $\limsup_{|\xi|\to\infty} \frac{G(\xi)}{|\xi|^2} \leq 0$.

Then, for each $\lambda \in (\hat{\lambda}, \infty)$, *where*

$$\hat{\lambda} := \frac{\frac{1}{2}(1+K)DM\bar{\eta}^2 - \frac{1}{2}(1-K)(\frac{\bar{\theta}}{c})^2}{G(\bar{\eta}\varepsilon)\int_{t_1}^{t_p} b(t)dt - \max_{|\xi|\leq\bar{\theta}} G(\xi)\int_0^T b(t)dt}, \tag{4.2.18}$$

problem (4.2.16) *has at least one positive periodic solution* $\bar{u} \in E$ *such that*

$$\frac{1}{2}\|\bar{u}\|^2 + \sum_{j=1}^{p}\sum_{i=1}^{N}\int_0^{\bar{u}_i(t_j)} I_{ij}(s)ds - \int_0^T H(\bar{u}(t))dt > \frac{1}{2}(1-K)\left(\frac{\bar{\theta}}{c}\right)^2.$$

One consequence of Corollary 4.2.5 is the following existence result.

Theorem 4.2.3. *Assume that*

$$\lim_{x\to 0^+} \frac{\max_{|\xi|\leq x} G(\xi)}{|x|^2} = \infty. \tag{4.2.19}$$

Then, for each $\lambda \in (0, \lambda^*)$, *where* $\lambda^* := \frac{1-K}{2c^2\int_0^T b(t)dt} \sup_{\theta>0} \frac{\theta^2}{\max_{|\xi|\leq\theta} G(\xi)}$, *the problem* (4.2.17) *has a non-trivial periodic solution* $u^* \in E$.

Proof. For fixed $\lambda \in (0, \lambda^*)$, there exists a positive constant θ such that

$$\lambda < \frac{1-K}{2c^2\int_0^T b(t)dt} \frac{\theta^2}{\max_{|\xi|\leq\theta} G(\xi)}.$$

Moreover, by (4.2.19), we can choose $\eta > 0$ satisfying $\eta < \frac{\theta}{c}\sqrt{\frac{1-K}{DM(1+K)}}$ such that

$$\frac{(1+K)DM}{2\lambda \int_{t_1}^{t_p} b(t)dt} < \frac{G(\eta\varepsilon)}{\eta^2},$$

where $\varepsilon = (1,0,\ldots,0) \in \mathbb{R}^N$. The conclusion then follows from Corollary 4.2.5. $\qquad\square$

The following examples illustrate some of our results.

Example 4.2.1. Take $N = 1$ and consider the problem

$$\begin{cases} -u''(t) + u(t) = \lambda b(t)g(u(t)) + h(u(t)), & a.e. \ t \in [0,3], \\ \Delta(u'(t_j)) = I_j(u(t_j)), & j = 1, 2, \\ u(0) - u(3) = u'(0) - u'(3) = 0, \end{cases} \qquad (4.2.20)$$

where $b(t) = e^t$ for every $t \in [0,3]$, $t_1 = 1$, $t_2 = 2$, $g(x) = 1 + 3x|x|e^{x^4} + 4x|x|^5 e^{x^4}$, $h(x) = \frac{1}{12}x^+$, $x^+ = \max\{x,0\}$, and $I_j(x) = \frac{1}{8}x$ for $j = 1,2$ for every $x \in \mathbb{R}$. It is easy to see that $G(x) = x + |x|^3 e^{x^4}$ and

$$\lim_{x \to 0^+} \frac{\max_{|\xi| \le x} G(\xi)}{x^2} = \infty.$$

Moreover, since $c = \sqrt{6}$, $L = \frac{1}{72}$, and $L_{1j} = \frac{1}{48}$ for $j = 1,2$, we see that $K = \frac{3}{4} < 1$. Hence, applying Theorem 4.2.3, for each $\lambda \in \left(0, \frac{1}{48(e^3-1)(1+e)}\right)$, problem (4.2.20) has a positive periodic solution.

Example 4.2.2. Let $N = 2$, $p = 2$, $T = 3$, $t_1 = 1$, and $t_2 = 2$. Let $A : [0,3] \to \mathbb{R}^{2\times 2}$ be the identity matrix, let $G(\xi_1,\xi_2) = \xi_1 + \xi_2 + \frac{1}{4}\xi_1^4 + \frac{1}{4}\xi_2^4$ for all $(\xi_1,\xi_2) \in \mathbb{R}^2$, $b \in L^1([0,3])$ be a positive function, $I_{ij}(s) = \frac{1}{96}s(1+e^{-s})$ for all $s \in \mathbb{R}$, for $i = 1,2$ and $j = 1,2$, and $H(\xi_1,\xi_2) = \frac{1}{72\sqrt{2}}(\frac{1}{2}\xi_1^2 + \xi_1\xi_2 + \frac{1}{2}\xi_2^2)$ for all $(\xi_1,\xi_2) \in \mathbb{R}^2$. It is clear that

$$\lim_{x \to 0^+} \frac{\max_{|\xi| \le x} G(\xi)}{x^2} = \infty.$$

Moreover, since $c = \sqrt{6}$, $L = \frac{1}{72\sqrt{2}}$ and $L_{ij} = \frac{1}{96}$ for $i = 1,2$, $j = 1,2$, we have $K = \frac{2+\sqrt{2}}{4\sqrt{2}} < 1$. Hence, applying Theorem 4.2.3, for each

$$\lambda \in \left(0, \frac{1 - \frac{2+\sqrt{2}}{4\sqrt{2}}}{12\int_0^T b(t)dt} \sup_{\theta > 0} \frac{\theta^2}{\max_{|\xi| \le \theta}\left(\xi_1 + \xi_2 + \frac{1}{4}\xi_1^4 + \frac{1}{4}\xi_2^4\right)}\right),$$

problem (4.2.17) has a positive periodic solution.

Our next theorem is for the existence of three positive periodic solutions to problem (4.2.17). It is based on Corollaries 4.2.5 and 4.2.6.

Theorem 4.2.4. *Let (D19) hold and assume there exist constants $\theta > 0$, $\eta > 0$, $\bar{\theta} > 0$, and $\bar{\eta} > 0$ with*

$$c\sqrt{\frac{DM(1+K)}{1-K}}\eta < \theta \leq \bar{\theta} < c\sqrt{Dm\bar{\eta}}$$

such that (D17) and (D18) hold. If

$$\frac{\max_{|\xi|\leq\theta} G(\xi) \int_0^T b(t)dt}{\theta^2} < \frac{1-K}{2c^2} \frac{\max_{|\xi|\leq\bar{\theta}} G(\xi) \int_0^T b(t)dt - G(\bar{\eta}\varepsilon) \int_{t_1}^{t_p} b(t)dt}{\frac{1}{2}(1-K)(\frac{\bar{\theta}}{c})^2 - \frac{1}{2}(1+K)DM\bar{\eta}^2},$$

$$(4.2.21)$$

where $\varepsilon = (1,0,\ldots,0) \in \mathbb{R}^N$, then, for each

$$\lambda \in \Lambda := \left(\max\left\{ \hat{\lambda}, \frac{(1+K)DM\eta^2}{2G(\eta\varepsilon)\int_{t_1}^{t_p} b(t)dt} \right\}, \frac{(1-K)(\frac{\theta}{c})^2}{2\max_{|\xi|\leq\theta} G(\xi)\int_0^T b(t)dt} \right)$$

where $\hat{\lambda}$ is given in (4.2.18), problem (4.2.17) has at least three positive periodic solutions.

Proof. First we observe that (4.2.21) implies $\Lambda \neq \emptyset$. Fix $\lambda \in \Lambda$. Using Corollary 4.2.5, we obtain the first positive periodic solution u^* as a local minimum of the functional I_λ with

$$\frac{1}{2}\|u^*\|^2 + \sum_{j=1}^p \sum_{i=1}^N \int_0^{u_i^*(t_j)} I_{ij}(s)ds - \int_0^T H(u^*(t))dt < \frac{1}{2}(1-K)\left(\frac{\theta}{c}\right)^2.$$

Corollary 4.2.6 guarantees a second positive periodic solution $\bar{u}$ with

$$\frac{1}{2}\|\bar{u}\|^2 + \sum_{j=1}^p \sum_{i=1}^N \int_0^{\bar{u}_i(t_j)} I_{ij}(s)ds - \int_0^T H(\bar{u}(t))dt > \frac{1}{2}(1-K)\left(\frac{\bar{\theta}}{c}\right)^2.$$

The mountain pass theorem of Pucci and Serrin ([239]) ensures the existence of a third positive periodic solution. $\square$

As a consequence of Theorem 4.2.4 we have the following result.

Theorem 4.2.5. *Assume that*

$$\limsup_{|x|\to 0^+} \frac{\max_{|\xi|\leq x} G(\xi)}{|x|^2} = \infty, \qquad (4.2.22)$$

$$\limsup_{|\xi|\to\infty} \frac{G(\xi)}{|\xi|^2} = 0, \tag{4.2.23}$$

and there are constants $\bar{\theta} > 0$ and $\bar{\eta} > 0$ with $\frac{\bar{\theta}}{c\sqrt{Dm}} < \bar{\eta}$ such that

$$\frac{\max_{|\xi|\leq\bar{\theta}} G(\xi) \int_0^T b(t)dt}{\bar{\theta}^2} < \frac{1-K}{c^2(1+K)DM} \frac{G(\bar{\eta}\varepsilon) \int_{t_1}^{t_p} b(t)dt}{\bar{\eta}^2}, \tag{4.2.24}$$

where $\varepsilon = (1,0,\ldots,0) \in \mathbb{R}^N$. Then, for each

$$\lambda \in \left(\frac{(1+K)DM\bar{\eta}^2}{2G(\bar{\eta}\varepsilon) \int_{t_1}^{t_p} b(t)dt}, \frac{(1-K)\bar{\theta}^2}{2c^2 \max_{|\xi|\leq\bar{\theta}} G(\xi) \int_0^T b(t)dt} \right),$$

(4.2.17) has at least three positive periodic solutions.

Proof. We can easily observe from (4.2.23) that (D19) is satisfied. Moreover, by choosing η small enough and $\theta = \bar{\theta}$, we see that (4.2.22) implies condition (D17) holds, and (4.2.24) implies (D18) and (4.2.21) hold. We thus have the conclusion of the theorem. $\qquad\square$

In conclusion, we would like to mention that we believe that this approach of combining techniques to obtain multiple solutions of boundary value problems, with or without impulses, will prove to be a valuable strategy.

4.3   Existence of Infinitely Many Solutions

Just as in Section 6 of Chapter 3 where we considered a perturbed (two parameter) form of the problem studied in the earlier sections in that chapter, here we consider a perturbed version of problem (4.2.1). Here we want to investigate the existence of infinitely many periodic solutions to the perturbed impulsive Hamiltonian system with periodic boundary conditions

$$\begin{cases} -\ddot{u}(t) + A(t)u(t) = \lambda \nabla F(t, u(t)) + \mu \nabla G(t, u(t)) \\ \qquad\qquad\qquad + \nabla H(u(t)), \quad a.e.\ t \in [0, T], \\ \Delta(\dot{u}_i(t_j)) = I_{ij}(u_i(t_j)), \quad i = 1, 2, \ldots, N,\ j = 1, 2, \ldots, p, \\ u(0) - u(T) = \dot{u}(0) - \dot{u}(T) = 0, \end{cases} \tag{4.3.1}$$

where again $u = (u_1, u_2, \ldots, u_N)^T$, $N \geq 1$, $p > 1$, $T > 0$, $\lambda > 0$ and $\mu \geq 0$ are parameters, $0 = t_0 < t_1 < \cdots < t_p < t_{p+1} = T$, $A : [0, T] \to \mathbb{R}^{N\times N}$ is a continuous map from the interval $[0, T]$ to the set of $N \times N$ symmetric matrices, and $\Delta(\dot{u}_i(t_j)) = \dot{u}_i(t_j^+) - \dot{u}_i(t_j^-) = \lim_{t \to t_j^+} \dot{u}_i(t) - \lim_{t \to t_j^-} \dot{u}_i(t)$. Here, each $I_{ij} : \mathbb{R} \to \mathbb{R}$ satisfies

$$|I_{ij}(s)| \le L_{ij}|s|$$

for every $s \in \mathbb{R}$ and $i = 1, 2, \ldots, N$, $j = 1, 2, \ldots, p$; $F, G : [0, T] \times \mathbb{R}^N \to \mathbb{R}$ are measurable with respect to t for all $u \in \mathbb{R}^N$, continuously differentiable in u for almost every $t \in [0, T]$, and satisfy the summability condition

$$\sup_{|x| \le \alpha} \max\{|F(\cdot, x)|, \ |\nabla F(\cdot, x)|, |G(\cdot, x)|, \ |\nabla G(\cdot, x)|\} \in L^1([0, T]) \quad (4.3.2)$$

for any $\alpha > 0$; $F(t, \mathbf{0}) = G(t, \mathbf{0}) = 0$ for all $t \in [0, T]$; and $H : \mathbb{R}^N \to \mathbb{R}$ is a continuously differentiable function for which there is a constant $L > 0$ such that

$$|H(x)| \le L|x|^2$$

for every $x \in \mathbb{R}^N$. Note that if $\nabla F, \nabla G : [0, T] \times \mathbb{R}^N \to \mathbb{R}$ are continuous, then clearly condition (4.3.2) is satisfied. It should also be understood that the matrix A satisfies conditions (D1) and (D2).

We need the concepts of classical and weak solutions as before.

Definition 4.3.1. *A function* $u \in \{u \in E : \dot{u} \in (W^{1,2}(t_j, t_{j+1}))^N, \ j = 0, 1, 2, \ldots, p\}$ *is said to be a classical solution of the problem* (4.3.1) *if* u *satisfies* (4.3.1).

Definition 4.3.2. *By a weak solution of the problem* (4.3.1), *we mean any* $u \in E$ *such that*

$$\int_0^T [(\dot{u}(t), \dot{v}(t)) + (A(t)u(t), v(t)) - (\nabla H(u(t)), v(t))]dt$$

$$+ \sum_{j=1}^p \sum_{i=1}^N I_{ij}(u_i(t_j))v_i(t_j) - \lambda \int_0^T (\nabla F(t, u(t)), v(t))dt$$

$$- \mu \int_0^T (\nabla G(t, u(t)), v(t))dt = 0$$

for every $v \in E$.

In view of what has proceeded, the following lemma should come as no surprise.

Lemma 4.3.1. *If* $u \in E$ *is a weak solution of* (4.3.1), *then* u *is a classical solution of* (4.3.1).

Set

$$D = \frac{(T - t_p)^2}{t_1 t_p^2} + \frac{t_1}{3t_p^2}(t_p^2 + t_p T + T^2) + (t_p - t_1) + \frac{T - t_p}{t_p^2} + \frac{1}{3t_p^2}(T^3 - t_p^3) > 0.$$

The constant K below is defined in (4.2.9). Our first main result in this section is as follows.

Theorem 4.3.1. *Assume that:*

(D21) $F(t, \xi) \geq 0$ *for each* $t \in [0, t_1] \cup [t_p, T]$, $|\xi| \in [0, \infty)$;

(D22)

$$\liminf_{\xi \to \infty} \frac{\int_0^T \max_{|x| \leq \xi} F(t, x) dt}{\xi^2}$$

$$< \frac{1 - K}{(1 + K)DMc^2} \limsup_{\xi \to \infty} \frac{\int_{t_1}^{t_p} F(t, \xi \varepsilon) dt}{\xi^2},$$

where $\varepsilon = (1, 0, \ldots, 0) \in \mathbb{R}^N$. *Then, for each* $\lambda \in (\lambda_1, \lambda_2)$ *with*

$$\lambda_1 := \frac{(1 + K)DM}{2 \limsup_{\xi \to \infty} \frac{\int_{t_1}^{t_p} F(t, \xi \varepsilon) dt}{\xi^2}}$$

and

$$\lambda_2 := \frac{(1 - K)(\frac{1}{c})^2}{2 \liminf_{\xi \to \infty} \frac{\int_0^T \max_{|x| \leq \xi} F(t, x) dt}{\xi^2}},$$

and for every arbitrary non-negative function $G : [0, T] \times \mathbb{R}^N \to \mathbb{R}$ *that is measurable with respect to t for all* $x \in \mathbb{R}^N$, *continuously differentiable in x for almost every* $t \in [0, T]$, *and satisfies*

$$G_\infty := \frac{2}{(1 - K)(\frac{1}{c})^2} \lim_{\xi \to \infty} \frac{\int_0^T \max_{|x| \leq \xi} G(t, x) dt}{\xi^2} < \infty, \qquad (4.3.3)$$

and for every $\mu \in [0, \mu_{G,\lambda})$, *where* $\mu_{G,\lambda} := \frac{1}{G_\infty}(1 - \frac{\lambda}{\lambda_2})$, *the problem* (4.3.1) *has an unbounded sequence of classical periodic solutions.*

Proof. We wish to apply Theorem 1.1.5, so take $\overline{\lambda} \in (\lambda_1, \lambda_2)$ and let G be a function satisfying condition (4.3.3). Since, $\overline{\lambda} < \lambda_2$, we have $\mu_{G,\overline{\lambda}} > 0$. Fix $\overline{\mu} \in [0, \mu_{G,\overline{\lambda}})$ and set $\nu_1 := \lambda_1$ and $\nu_2 := \frac{\lambda_2}{1 + \frac{\overline{\mu}}{\lambda} \lambda_2 G_\infty}$. If $G_\infty = 0$, then clearly $\nu_1 = \lambda_1$, $\nu_2 = \lambda_2$, and $\overline{\lambda} \in (\nu_1, \nu_2)$. If $G_\infty \neq 0$, since $\overline{\mu} < \mu_{G,\overline{\lambda}}$, we obtain $\frac{\overline{\lambda}}{\lambda_2} + \overline{\mu} G_\infty < 1$, and so $\frac{\lambda_2}{1 + \frac{\overline{\mu}}{\lambda} \lambda_2 G_\infty} > \overline{\lambda}$ and $\overline{\lambda} < \nu_2$. Since $\overline{\lambda} > \lambda_1 = \nu_1$, we see that $\overline{\lambda} \in (\nu_1, \nu_2)$.

Now set $Q(t, \xi) = F(t, \xi) + \frac{\overline{\mu}}{\lambda} G(t, \xi)$ for all $(t, \xi) \in [0, T] \times \mathbb{R}^N$. Take $X = E$ and consider the functionals $\Phi, \Psi, I_\lambda : X \to \mathbb{R}$ defined by

$$\Phi(u) = \frac{1}{2}\|u\|^2 + \sum_{j=1}^{p} \sum_{i=1}^{N} \int_0^{u_i(t_j)} I_{ij}(s)ds - \int_0^T H(u(t))dt,$$

$$\Psi(u) = \int_0^T Q(t, u(t))dt,$$

and

$$I(u) = \Phi(u) - \lambda\Psi(u)$$

for every $u \in X$. It is well known that Ψ is a Gâteaux differentiable functional whose Gâteaux derivative at the point $u \in X$ is the functional $\Psi'(u) \in X^*$ given by

$$\Psi'(u)v = \int_0^T \left(\nabla F(t, u(t)) + \frac{\overline{\mu}}{\lambda} \nabla G(t, u(t)), v(t) \right) dt \qquad (4.3.4)$$

for every $v \in X$. The functional Φ is also Gâteaux differentiable with Gâteaux derivative $\Phi'(u) \in X^*$ at the point $u \in X$ given by

$$\Phi'(u)v = \int_0^T [(\dot{u}(t), \dot{v}(t)) + (A(t)u(t), v(t)) - (\nabla H(u(t)), v(t))]dt$$
$$+ \sum_{j=1}^{p} \sum_{i=1}^{N} I_{ij}(u_i(t_j))v_i(t_j) \qquad (4.3.5)$$

for every $v \in X$.

Now, Φ is sequentially weakly lower semicontinuous. To see this, let $u_n \in X$ with $u_n \to u$ weakly in X, and using the sequential weakly lower semicontinuity of the norm, we have $\liminf_{n\to\infty} \|u_n\| \geq \|u\|$ and $u_n \to u$ uniformly on $[0, T]$. Hence, since H is continuous,

$$\liminf_{n\to\infty} \left(\frac{1}{2}\|u_n\|^2 + \sum_{j=1}^{p} \sum_{i=1}^{N} \int_0^{u_{n_i}(t_j)} I_{ij}(s)ds - \int_0^T H(u_n(t))dt \right)$$
$$\geq \frac{1}{2}\|u\|^2 + \sum_{j=1}^{p} \sum_{i=1}^{N} \int_0^{u_i(t_j)} I_{ij}(s)ds - \int_0^T H(u(t))dt,$$

i.e., $\liminf_{n\to\infty} \Phi(u_n) \geq \Phi(u)$. This implies Φ is sequentially weakly lower semicontinuous.

From the definition of Φ, since $(X, \|\cdot\|)$ is compactly embedded in $C([0, T], \mathbb{R}^N)$, we observe that Φ is strongly continuous. Since $-L|x|^2 \leq H(x) \leq L|x|^2$ for every $x \in \mathbb{R}^N$, and $-L_{ij}|s| \leq I_{ij}(s) \leq L_{ij}|s|$ for every $s \in \mathbb{R}$ for all $i = 1, 2, \ldots, N$ and $j = 1, 2, \ldots, p$, in view of (4.2.7), we see that

$$\frac{1}{2}(1 - K)\|u\|^2 \leq \Phi(u) \leq \frac{1}{2}(1 + K)\|u\|^2. \qquad (4.3.6)$$

This also shows that Φ is coercive. We want to verify that $\gamma < \infty$, where γ is defined in Theorem 1.1.5. Let $\{\xi_n\}$ be a real sequence such that $\xi_n \to \infty$ as $n \to \infty$ and

$$\lim_{n\to\infty} \frac{\int_0^T \max_{|x|\leq\xi_n} F(t,x)dt}{\xi_n^2} = \liminf_{\xi\to\infty} \frac{\int_0^T \max_{|x|\leq\xi} F(t,x)dt}{\xi^2}.$$

Set $r_n = \frac{1}{2}(1-K)(\frac{\xi_n}{c})^2$ for all $n \in \mathbb{N}$. From inequalities (4.2.7) and (4.3.6), for each $u \in X$, we have

$$\Phi^{-1}(-\infty, r_n] = \{u \in X : \Phi(u) < r_n\}$$
$$\subseteq \left\{u \in X : \frac{1}{2}(1-K)\|u\|^2 < r_n\right\}$$
$$\subseteq \{u \in X : |u(t)| \leq \xi_n \text{ for each } t \in [0,T]\}.$$

Hence, since $\Phi(0) = \Psi(0) = 0$, for large n,

$$\varphi(r_n) = \inf_{u\in\Phi^{-1}(-\infty,r_n)} \frac{\sup_{v\in\Phi^{-1}(-\infty,r_n]} \Psi(v) - \Psi(u)}{r_n - \Phi(u)}$$
$$\leq \frac{\sup_{v\in\Phi^{-1}(-\infty,r_n]} \Psi(v)}{r_n}$$
$$\leq \frac{\int_0^T \max_{|x|\leq\xi_n} Q(t,x)dt}{\frac{1}{2}(1-K)(\frac{\xi_n}{c})^2}$$
$$\leq \frac{\int_0^T \max_{|x|\leq\xi_n} F(t,x)dt}{\frac{1}{2}(1-K)(\frac{\xi_n}{c})^2} + \frac{\overline{\mu}}{\overline{\lambda}} \frac{\int_0^T \max_{|x|\leq\xi_n} G(t,x)dt}{\frac{1}{2}(1-K)(\frac{\xi_n}{c})^2}.$$

Moreover, from (D22), it follows that

$$\liminf_{\xi\to\infty} \frac{\int_0^T \max_{|x|\leq\xi} F(t,x)dt}{\xi^2} < \infty,$$

so

$$\lim_{n\to\infty} \frac{\int_0^T \max_{|x|\leq\xi_n} F(t,x)dt}{\xi_n^2} < \infty. \qquad (4.3.7)$$

Then, (4.3.3) and (4.3.7) imply

$$\lim_{n\to\infty} \frac{\int_0^T \max_{|x|\leq\xi_n} [F(t,x) + \frac{\overline{\mu}}{\overline{\lambda}}G(t,x)]dt}{\frac{1}{2}(1-K)(\frac{\xi_n}{c})^2} < \infty.$$

Therefore,

$$\gamma \leq \liminf_{n\to\infty} \varphi(r_n) \leq \lim_{n\to\infty} \frac{\int_0^T \max_{|x|\leq\xi_n} [F(t,x) + \frac{\overline{\mu}}{\overline{\lambda}}G(t,x)]dt}{\frac{1}{2}(1-K)(\frac{\xi_n}{c})^2} < \infty.$$
$$(4.3.8)$$

In view of (4.3.3), we see that

$$\liminf_{\xi\to\infty} \frac{\int_0^T \max_{|x|\le\xi} Q(t,x)dt}{\frac{1}{2}(1-K)(\frac{\xi}{c})^2} \le \liminf_{\xi\to\infty} \frac{\int_0^T \max_{|x|\le\xi} F(t,x)dt}{\frac{1}{2}(1-K)(\frac{\xi}{c})^2} + \frac{\overline{\mu}}{\lambda}G_\infty.$$

(4.3.9)

Moreover, since G is non-negative, we have

$$\limsup_{\xi\to\infty} \frac{\int_{t_1}^{t_p} Q(t,\xi\varepsilon)dt}{(1+K)DM\xi^2} \ge \limsup_{\xi\to\infty} \frac{\int_{t_1}^{t_p} F(t,\xi\varepsilon)dt}{(1+K)DM\xi^2}.$$

(4.3.10)

Therefore, from (4.3.8)–(4.3.10) and condition (D22), we have

$$\overline{\lambda} \in (\nu_1,\nu_2) \subseteq \left(\frac{1}{\limsup\limits_{\xi\to\infty} \frac{\int_{t_1}^{t_p} Q(t,\xi\varepsilon)dt}{\frac{1}{2}(1+K)DM\xi^2}}, \right.$$

$$\left. \frac{1}{\liminf\limits_{\xi\to\infty} \frac{\int_0^T \max_{|x|\le\xi_n} Q(t,x)dt}{\frac{1}{2}(1-K)(\frac{\xi_n}{c})^2}} \right) \subseteq \left(0,\frac{1}{\gamma}\right).$$

For the fixed $\overline{\lambda}$, the inequality (4.3.8) implies that condition (b) of Theorem 1.1.5 can be applied, and so either $I_{\overline{\lambda}}$ has a global minimum or there exists a sequence $\{u_n\}$ of weak solutions of the problem (4.3.1) such that $\lim_{n\to\infty}\|u_n\| = \infty$.

Next, we show that for the fixed $\overline{\lambda}$, the functional $I_{\overline{\lambda}}$ has no global minimum. To do this, we will show that the functional $I_{\overline{\lambda}}$ is unbounded from below. Since

$$\frac{1}{\overline{\lambda}} < 2\limsup_{\xi\to\infty} \frac{\int_{t_1}^{t_p} F(t,\xi\varepsilon)dt}{(1+K)DM\xi^2},$$

we can consider a real sequence $\{d_n\}$ and a positive constant τ such that $d_n \to \infty$ as $n \to \infty$ and

$$\frac{1}{\overline{\lambda}} < \tau < \frac{2\int_{t_1}^{t_p} F(t,d_n\varepsilon)dt}{(1+K)DMd_n^2}$$

(4.3.11)

for large $n \in \mathbb{N}$. Let $\{w_n\}$ be a sequence in X defined by

$$w_n(t) = \begin{cases} (T + \frac{t_p-T}{t_1}t)\frac{d_n\varepsilon}{t_p}, & t \in [0,t_1), \\ d_n\varepsilon, & t \in [t_1,t_p], \\ \frac{d_n\varepsilon}{t_p}t, & t \in (t_p,T]. \end{cases}$$

(4.3.12)

It is clear that $w_n \in X$ for all $n \in \mathbb{N}$, and $\|w_n\|_E^2 = Dd_n^2$. Therefore, from (4.2.6),

$$Dmd_n^2 \leq \|w\|^2 \leq DMd_n^2, \tag{4.3.13}$$

which together with (4.3.6) gives

$$\Phi(w_n) \leq \frac{1}{2}(1+K)DMd_n^2. \tag{4.3.14}$$

On the other hand, since G is non-negative, from the definition of Ψ and (D21), we see that

$$\Psi(w_n) \geq \int_{t_1}^{t_p} F(t, d_n\varepsilon)dt. \tag{4.3.15}$$

So (4.3.11), (4.3.14), and (4.3.15) imply

$$\begin{aligned}
I_{\overline{\lambda}}(w_n) &= \Phi(w_n) - \overline{\lambda}\Psi(w_n) \\
&\leq \frac{1}{2}(1+K)DMd_n^2 - \overline{\lambda}\int_{t_1}^{t_p} F(t, d_n\varepsilon)dt \\
&< (1 - \overline{\lambda}\tau)\frac{1}{2}(1+K)DMd_n^2 \to -\infty
\end{aligned}$$

as $n \to \infty$. Hence, the functional $I_{\overline{\lambda}}$ has no global minimum.

Therefore, applying Theorem 1.1.5, we conclude that there is a sequence $\{u_n\} \subset X$ of critical points of $I_{\overline{\lambda}}$ such that $\lim_{n\to\infty} \Phi(u_n) = \infty$, and from (4.3.6) it follows that $\lim_{n\to\infty} \|u_n\| = \infty$. In view of Definition 4.3.2, (4.3.4), and (4.3.5), we see that weak solutions of the problem (4.3.1) are exactly the solutions of the equation $\Phi'(u) - \lambda\Psi'(u) = 0$. Hence, by Lemma 4.3.1, the conclusion of the theorem follows. $\qquad\square$

Remark 4.3.1. Under the conditions

$$\liminf_{\xi\to\infty} \frac{\int_0^T \max_{|x|\leq\xi} F(t,x)dt}{\xi^2} = 0$$

and

$$\limsup_{\xi\to\infty} \frac{\int_{t_1}^{t_p} F(t, \xi\varepsilon)dt}{\xi^2} = \infty,$$

where $\varepsilon = (1, 0, \cdots, 0) \in \mathbb{R}^N$, Theorem 4.3.1 ensures that for every $\lambda > 0$ and for each $\mu \in [0, \frac{1}{G_\infty})$, the problem (4.3.1) admits infinitely many classical periodic solutions. Moreover, if $G_\infty = 0$, then the result holds for every $\lambda > 0$ and $\mu \geq 0$.

We now exhibit two examples for which the hypotheses of Theorem 4.3.1 are satisfied.

Example 4.3.1. Let $N = 1$, $p = 2$, $T = 3$, $t_1 = 1$, $t_2 = 2$, and define the sequences $\{a_n\}$ and $\{b_n\}$ by $b_1 = 2$, $b_{n+1} = b_n^6$, and $a_n = b_n^4$ for $n \in \mathbb{N}$. Let

$$
f(\xi) = \begin{cases} b_1^3 \sqrt{1 - (1 - \xi)^2} + 1, & \text{if } \xi \in [0, b_1], \\ (a_n - b_n^3)\sqrt{1 - (a_n - 1 - \xi)^2} + 1, & \text{if } \xi \in \bigcup_{n=1}^\infty [a_n - 2, a_n], \\ (b_{n+1}^3 - a_n)\sqrt{1 - (b_{n+1} - 1 - \xi)^2} + 1, & \text{if } \xi \in \bigcup_{n=1}^\infty [b_{n+1} - 2, b_{n+1}], \\ 1, & \text{otherwise,} \end{cases}
$$

$h(\xi) = \frac{1}{120} \frac{\xi(2+|\xi|)}{(1+|\xi|)^2}$ for every $\xi \in \mathbb{R}$, and $I_j(x) = \frac{1}{36}x(1 + \sin x)$ for $j = 1, 2$. Let $F(\xi) = \int_0^\xi f(x)dx$ and $H(\xi) = \int_0^\xi h(x)dx = \frac{1}{120}\frac{\xi^2}{1+|\xi|}$ for all $\xi \in \mathbb{R}$. Then, F is a C^1 function with $F' = f$. From the computation in Example 2.4.1, we have $F(a_n) = \frac{\pi}{2}a_n + a_n$ and $F(b_n) = \frac{\pi}{2}b_n^3 + b_n$, and so $\lim_{n\to\infty} \frac{F(a_n)}{a_n^2} = 0$ and $\lim_{n\to\infty} \frac{F(b_n)}{b_n^2} = \infty$. Therefore, $\liminf_{\xi\to\infty} \frac{F(\xi)}{\xi^2} = 0$ and $\limsup_{\xi\to\infty} \frac{F(\xi)}{\xi^2} = \infty$. Let $g(t, \xi) = e^{t - \xi^+}(\xi^+)^2(3 - \xi^+)$, where $\xi^+ = \max\{\xi, 0\}$, so that $G(t, \xi) = \int_0^\xi g(t, x)dx = e^{t-\xi^+}(\xi^+)^3$ for all $(t, \xi) \in [0, 3] \times \mathbb{R}$ and $G_\infty = 0$. We have $m = 1$, $M = 2$, $L = \frac{1}{120}$, $L_j = \frac{1}{18}$ for $j = 1, 2$, $c = \sqrt{6}$, and $K = \frac{29}{30}$. Hence, applying Theorem 4.3.1, for every $(\lambda, \mu) \in (0, \infty) \times [0, \infty)$, the problem

$$
\begin{cases} -u''(t) + u(t) = \lambda f(u(t)) + \mu g(t, u(t)) + h(u(t)), & \text{a.e. } t \in [0, 3], \\ \Delta(u'(t_j)) = I_j(u(t_j)), & j = 1, 2, \\ u(0) - u(3) = u'(0) - u'(3) = 0, \end{cases}
$$

admits an unbounded sequence of classical periodic solutions.

Example 4.3.2. Let $N = 3$, $p = 2$, $T = 3$, $t_1 = 1$, $t_2 = 2$, and $A : [0, 3] \to \mathbb{R}^{3\times3}$ be the identity matrix. Let $G : [0, 3] \times \mathbb{R}^3 \to \mathbb{R}$ be a nonnegative function that is measurable with respect to t for all $x = (x_1, x_2, x_3) \in \mathbb{R}^3$, continuously differentiable in x for almost every $t \in [0, 3]$, and satisfies

$$
\sup_{|x|\le a} \max\{|G(\cdot, x)|, |\nabla G(\cdot, x)|\} \in L^1([0, 3])
$$

for any $a > 0$, and

$$
\lim_{\xi\to\infty} \xi^{-2} \int_0^3 \sup_{|x|\le\xi} G(t, x)dx < \infty.
$$

Set $a_k := \frac{2k!(k+2)!-1}{4(k+1)!}$ and $b_k := \frac{2k!(k+2)!+1}{4(k+1)!}$ for every $k \in \mathbb{N}$. Define the function $f_1 : \mathbb{R} \to \mathbb{R}$ by

$$
f_1(s) = \begin{cases} \frac{32(k+1)!^2[k^3(k+1)!^4 - (k-1)^3 k!^4]}{\pi}\sqrt{\frac{1}{16(k+1)!^2} - (t - \frac{k!(k+2)}{2})^2} + 1, \\ \qquad\qquad\qquad\qquad\qquad \text{if } s \in \bigcup_{k\in\mathbb{N}}[a_k, b_k], \\ 1, \qquad\qquad\qquad\qquad\qquad\qquad \text{otherwise.} \end{cases}
$$

Now let $F(t, x_1, x_2, x_3) = \alpha(t) F_1(x_1) F_2(x_2, x_3)$ for all $(t, x_1, x_2, x_3) \in [0,3] \times \mathbb{R}^3$, where $\alpha \in L^1([0,3])$ is a positive function, $F_1(\xi) = \int_0^\xi f_1(s) ds$ for all $\xi \in \mathbb{R}$, and let $F_2 : \mathbb{R}^2 \to \mathbb{R}$ be a non-negative bounded and continuously differentiable function with $F_2(0,0) \neq 0$. Take $H(x) = \frac{1}{120} \frac{|x|^2}{1+|x|^2}$ for every $x \in \mathbb{R}^3$, $I_{i1}(t) = \frac{1}{360} t(1 + \sin t)$, and $I_{i2}(t) = \arctan \frac{t}{180}$ for every $t \in \mathbb{R}$ and $i = 1, 2, 3$. We then have

$$\lim_{k \to \infty} \frac{F_1(b_k) F_2(0,0)}{b_k^2} = 4 F_2(0,0)$$

and

$$\lim_{k \to \infty} \frac{F_1(a_k) F_2(a_k, a_k)}{a_k^2} = 0.$$

Therefore,

$$\liminf_{\xi \to \infty} \frac{\int_0^3 \max_{|x| \leq \xi} F(t, x) dt}{\xi^2}$$

$$= \|\alpha\|_{L^1([0,3])} \liminf_{\xi \to \infty} \frac{\max_{|x| \leq \xi} (F_1(x_1) F_2(x_2, x_3))}{\xi^2} = 0$$

and

$$\limsup_{\xi \to \infty} \frac{\int_{t_1}^{t_p} F(t, \xi \varepsilon) dt}{\xi^2}$$

$$= \|\alpha\|_{L^1([1,2])} \limsup_{\xi \to \infty} \frac{F_1(\xi) F_2(0,0)}{\xi^2} = 4 F_2(0,0) \|\alpha\|_{L^1([1,2])}.$$

We see that $m = 1$, $M = 3$, $c = \sqrt{6}$, $L = \frac{1}{120}$, $L_{i1} = \frac{1}{180}$ and $L_{i2} = \frac{1}{180}$ for $i = 1, 2, 3$, $D = \frac{7}{2}$, and $K = \frac{1}{2}$. Applying Theorem 4.3.1, for every $\lambda > \lambda_1 = \frac{7.875}{F_2(0,0)\|\alpha\|_{L^1([1,2])}}$ and for every μ in a convenient interval, the problem (4.3.1) has an unbounded sequence of classical periodic solutions.

Remark 4.3.2. Assumption (D22) in Theorem 4.3.1 can be replaced by the more general condition

(D22') there exist two sequence $\{\theta_n\}$ and $\{\eta_n\}$ with $(1 - K)(\frac{\eta_n}{c})^2 > (1 + K) DM\theta_n^2$ for every $n \in \mathbb{N}$ and $\lim_{n \to \infty} \theta_n = \infty$ such that

$$\lim_{n \to \infty} \frac{\int_0^T \max_{|x| \leq \eta_n} F(t, x) dt - \int_{t_1}^{t_p} F(t, \theta_n \varepsilon) dt}{\frac{1}{2}(1 - K)(\frac{\eta_n}{c})^2 - \frac{1}{2}(1 + K) DM\theta_n^2}$$

$$< \limsup_{|\xi| \to \infty} \frac{\int_{t_1}^{t_p} F(t, \xi \varepsilon) dt}{\frac{1}{2}(1 + K) DM\xi^2},$$

where $\varepsilon = (1, 0, \ldots, 0) \in \mathbb{R}^N$.

By choosing $\theta_n = 0$ for all $n \in \mathbb{N}$, (D22) follows from (D22'). Moreover, if we assume (D22') instead of (D22) and set $r_n = \frac{1}{2}(1 - K)(\frac{\eta_n}{c})^2$ for all $n \in \mathbb{N}$, by the same reasoning as in the proof of Theorem 4.3.1, we obtain

$$\varphi(r_n) \leq \frac{\sup_{v \in \Phi^{-1}(-\infty, r_n]} \Psi(v) - \int_0^T F(t, z_n(t)) dt}{r_n - \Phi(z_n)}$$

$$\leq \frac{\int_0^T \max_{|x| \leq \eta_n} Q(t, x) dt - \int_{t_1}^{t_p} F(t, \theta_n \varepsilon) dt}{\frac{1}{2}(1 - K)(\frac{\eta_n}{c})^2 - \frac{1}{2}(1 + K) D M \theta_n^2},$$

where

$$z_n(t) = \begin{cases} (T + \frac{t_p - T}{t_1} t) \frac{\theta_n \varepsilon}{t_p}, & t \in [0, t_1), \\ \theta_n \varepsilon, & t \in [t_1, t_p], \\ \frac{\theta_n \varepsilon}{t_p} t, & t \in (t_p, T]. \end{cases}$$

We have the same conclusion as in Theorem 4.3.1 with the interval (λ_1, λ_2) replaced by (λ_1', λ_2'), where

$$\lambda_1' := \frac{(1 + K) D M}{2 \limsup\limits_{\xi \to \infty} \frac{\int_{t_1}^{t_p} F(t, \xi \varepsilon) dt}{\xi^2}}$$

and

$$\lambda_2' := \left[\lim\limits_{n \to \infty} \frac{\int_0^T \max_{|x| \leq \eta_n} F(t, x) dt - \int_{t_1}^{t_p} F(t, \theta_n \varepsilon) dt}{\frac{1}{2}(1 - K)(\frac{\eta_n}{c})^2 - \frac{1}{2}(1 + K) D M \theta_n^2} \right]^{-1}.$$

Next, we point out a simple consequence of Theorem 4.3.1.

Corollary 4.3.1. *Assume that (D21) holds,*

(D23) $\quad \liminf_{\xi \to \infty} \frac{\int_0^T \max_{|x| \leq \xi} F(t, x) dt}{\xi^2} < \frac{1}{2}(1 - K) \frac{1}{c^2},$

and

(D24) $\quad \limsup_{\xi \to \infty} \frac{\int_{t_1}^{t_p} F(t, \xi \varepsilon) dt}{\xi^2} > \frac{1}{2}(1 + K) D M,$

where $\varepsilon = (1, 0, \cdots, 0) \in \mathbb{R}^N$. Then, for every arbitrary non-negative function $G : [0, T] \times \mathbb{R}^N \to \mathbb{R}$ that is measurable with respect to t for all $x \in \mathbb{R}^N$, continuously differentiable in x for almost every $t \in [0, T]$, and satisfies condition (4.3.3), and for every $\mu \in [0, \mu_{g,1})$, where

$$\mu_{g,1} := \frac{1}{G_\infty} \left(1 - \frac{1}{\lambda_2} \right)$$

and λ_2 is given in the statement of Theorem 4.3.1, the problem

$$\begin{cases} -\ddot{u}(t) + A(t)u(t) = \nabla F(t, u(t)) + \mu\nabla G(t, u(t)) + \nabla H(u(t)), & a.e.\ t \in [0, T], \\ \Delta(\dot{u}_i(t_j)) = I_{ij}(u_i(t_j)), & i = 1, 2, \dots, N,\ j = 1, 2, \dots, p, \\ u(0) - u(T) = \dot{u}(0) - \dot{u}(T) = 0, \end{cases}$$

has an unbounded sequence of classical periodic solutions.

By taking $\mu = 0$ in Corollary 4.3.1 will easily give us the following corollary.

Corollary 4.3.2. *Let $F : \mathbb{R}^N \to \mathbb{R}$ be a continuously differentiable function such that*

$$\liminf_{\xi \to \infty} \frac{\max_{|x| \le \xi} F(x)}{\xi^2} = 0 \ \text{ and } \ \limsup_{\xi \to \infty} \frac{F(\xi\varepsilon)}{\xi^2} = \infty,$$

where $\varepsilon = (1, 0, \cdots, 0) \in \mathbb{R}^N$. Then, the problem

$$\begin{cases} -\ddot{u}(t) + A(t)u(t) = \nabla F(u(t)) + \nabla H(u(t)), & a.e.\ t \in [0, T], \\ \Delta(\dot{u}_i(t_j)) = I_{ij}(u_i(t_j)), & i = 1, 2, \dots, N,\ j = 1, 2, \dots, p, \\ u(0) - u(T) = \dot{u}(0) - \dot{u}(T) = 0, \end{cases}$$

has an unbounded sequence of classical periodic solutions.

Arguing as in the proof of Theorem 4.3.1, but using conclusion (c) of Theorem 1.1.5 instead of (b), the following result holds.

Theorem 4.3.2. *Assume that (D21) and*

(D25)

$$\liminf_{\xi \to 0^+} \frac{\int_0^T \max_{|x| \le \xi} F(t, x)dt}{\xi^2}$$

$$< \frac{1 - K}{(1 + K)DMc^2} \limsup_{\xi \to 0^+} \frac{\int_{t_1}^{t_p} F(t, \xi\varepsilon)dt}{\xi^2}$$

holds, where $\varepsilon = (1, 0, \cdots, 0) \in \mathbb{R}^N$. Then, for each $\lambda \in (\lambda_3, \lambda_4)$, where

$$\lambda_3 := \frac{(1 + K)DM}{2\limsup\limits_{\xi \to 0^+} \dfrac{\int_{t_1}^{t_p} F(t, \xi\varepsilon)dt}{\xi^2}} \ \text{ and } \ \lambda_4 := \frac{(1 - K)(\frac{1}{c})^2}{2\liminf\limits_{\xi \to 0^+} \dfrac{\int_0^T \max_{|x| \le \xi} F(t, x)dt}{\xi^2}},$$

for every arbitrary non-negative function $G : [0, T] \times \mathbb{R}^N \to \mathbb{R}$ that is measurable with respect to t for all $x \in \mathbb{R}^N$, continuously differentiable in x for almost every $t \in [0, T]$, and satisfies

$$G_0 := \frac{2}{(1 - K)(\frac{1}{c})^2} \lim_{\xi \to 0^+} \frac{\int_0^T \max_{|x| \le \xi} G(t, x)dt}{\xi^2} < \infty, \tag{4.3.16}$$

and for every $\mu \in [0, \mu'_{G,\lambda})$, *where* $\mu'_{G,\lambda} := \frac{1}{G_0}\left(1 - \frac{\lambda}{\lambda_4}\right)$, *the problem* (4.3.1) *has a sequence of classical periodic solutions that converges uniformly to* 0 *in* $[0, T]$.

Proof. Fix $\overline{\lambda} \in (\lambda_3, \lambda_4)$ and let G be a function satisfying condition (4.3.16). Since $\overline{\lambda} < \lambda_4$, we have

$$\mu'_{G,\overline{\lambda}} := \frac{1}{G_0}\left(1 - \frac{\overline{\lambda}}{\lambda_4}\right) > 0.$$

Fix $\overline{\mu} \in [0, \mu'_{G,\overline{\lambda}})$ and set $\nu_3 := \lambda_3$ and $\nu_4 := \frac{\lambda_4}{1 + \frac{\overline{\mu}}{\lambda}\lambda_4 G_0}$. If $G_0 = 0$, clearly, $\nu_3 = \lambda_3$, $\nu_4 = \lambda_4$, and $\overline{\lambda} \in (\nu_3, \nu_4)$. If $G_0 \neq 0$, since $\overline{\mu} < \mu'_{G,\overline{\lambda}}$, we have $\frac{\overline{\lambda}}{\lambda_4} + \overline{\mu}G_0 < 1$, and so $\frac{\lambda_4}{1 + \frac{\overline{\mu}}{\lambda}\lambda_4 G_0} > \overline{\lambda}$ and $\overline{\lambda} < \nu_4$. Hence, since $\overline{\lambda} > \lambda_3 = \nu_3$, $\overline{\lambda} \in (\nu_3, \nu_4)$. Now, set $Q(t, \xi) = F(t, \xi) + \frac{\overline{\mu}}{\lambda}G(t, \xi)$ for all $(t, \xi) \in [0, T] \times \mathbb{R}^N$. Take X, Φ, Ψ, and $I_{\overline{\lambda}}$ as in the proof of Theorem 4.3.1.

To see that $\delta < \infty$, let $\{\xi_n\}$ be a sequence of positive numbers such that $\xi_n \to 0^+$ as $n \to \infty$ and

$$\lim_{n \to \infty} \frac{\int_0^T \max_{|x| \leq \xi_n} F(t, x)dt}{\xi_n^2} < \infty.$$

Setting $r_n = \frac{1}{2}(1 - 2LTc^2 - c^2 \sum_{j=1}^p \sum_{i=1}^N L_{ij})(\frac{\xi_n}{c})^2$ for all $n \in \mathbb{N}$, and arguing as in the proof of Theorem 4.3.1 shows that $\delta < \infty$.

Since

$$\frac{\int_0^T \max_{|x| \leq \xi} Q(t, x)dt}{\xi^2} \leq \frac{\int_0^T \max_{|x| \leq \xi} F(t, x)dt}{\xi^2} + \frac{\overline{\mu}}{\overline{\lambda}} \frac{\int_0^T \max_{|x| \leq \xi} G(t, x)dt}{\xi^2},$$

and (4.3.16) holds, we have

$$\liminf_{\xi \to 0^+} \frac{\int_0^T \max_{|x| \leq \xi} Q(t, x)dt}{\frac{1}{2}(1 - K)(\frac{\xi}{c})^2} \leq \liminf_{\xi \to 0^+} \frac{\int_0^T \max_{|x| \leq \xi} F(t, x)dt}{\frac{1}{2}(1 - K)(\frac{\xi}{c})^2} + \frac{\overline{\mu}}{\overline{\lambda}}G_0.$$

$$\text{(4.3.17)}$$

Moreover, since G is nonnegative, (c_1) implies

$$\limsup_{\xi \to 0^+} \frac{\int_{t_1}^{t_p} Q(t, \xi\varepsilon)dt}{\xi^2} \geq \limsup_{\xi \to 0^+} \frac{\int_{t_1}^{t_p} F(t, \xi\varepsilon)dt}{\xi^2}. \qquad \text{(4.3.18)}$$

Therefore, from (4.3.17) and (4.3.18),

$$\overline{\lambda} \in (\nu_3, \nu_4) \subseteq \left(\frac{(1 + K)DM}{2 \limsup_{\xi \to 0^+} \frac{\int_{t_1}^{t_p} Q(t, \xi\varepsilon)dt}{\xi^2}}, \frac{(1 - K)(\frac{1}{c})^2}{2 \liminf_{\xi \to 0^+} \frac{\int_0^T \max_{|x| \leq \xi} Q(t, x)dt}{\xi^2}}\right)$$

$$\subseteq \left(0, \frac{1}{\delta}\right).$$

We need to show that the functional $I_{\overline{\lambda}}$ does not have a local minimum at zero. To do this, let $\{d_n\}$ be a sequence of positive numbers and $\tau > 0$ be such that $d_n \to 0^+$ as $n \to \infty$ and

$$\frac{1}{\overline{\lambda}} < \tau < \frac{2 \int_{t_1}^{t_p} F(t, d_n \varepsilon) dt}{(1+K)DMd_n^2} \tag{4.3.19}$$

for large $n \in \mathbb{N}$. Let $\{w_n\}$ be a sequence in X defined as in (4.3.12). From (4.3.14), (4.3.15), and (4.3.19), we obtain

$$I_{\overline{\lambda}}(w_n) = \Phi(w_n) - \overline{\lambda}\Psi(w_n)$$

$$\leq \frac{1}{2}(1+K)DMd_n^2 - \overline{\lambda} \int_{t_1}^{t_p} F(t, d_n \varepsilon) dt$$

$$< (1 - \overline{\lambda}\tau)\frac{1}{2}(1+K)DMd_n^2$$

for large $n \in \mathbb{N}$. Since $I_{\overline{\lambda}}(0) = 0$ and (4.3.19) holds, the functional $I_{\overline{\lambda}}$ does not have a local minimum at zero. Hence, part (c) of Theorem 1.1.5 implies there is a sequence $\{u_n\}$ in X of critical points of $I_{\overline{\lambda}}$ such that $\|u_n\| \to 0$ as $n \to \infty$. Since the solutions of the equation $\Phi'(u) - \lambda\Psi'(u) = 0$ are exactly the weak solutions of the problem (4.3.1), by Lemma 4.3.1 we have the conclusion of the theorem. $\square$

We conclude this section with an example for which the hypotheses of Theorem 4.3.2 are satisfied.

Example 4.3.3. Consider the system

$$\begin{cases} -\ddot{u}(t) + A(t)u(t) = \lambda\nabla F(t, u(t)) + \mu\nabla G(t, u(t)) \\ \qquad\qquad\qquad + \nabla H(u(t)), \quad \text{a.e. } t \in [0,3], \\ \Delta(\dot{u}_i(t_j)) = I_{ij}(u_i(t_j)), \quad i = 1, 2, \ j = 1, 2, \\ u(0) - u(3) = \dot{u}(0) - \dot{u}(3) = 0, \end{cases} \tag{4.3.20}$$

where $A : [0,3] \to \mathbb{R}^{2\times 2}$ is the identity matrix, $N = p = 2$, $t_1 = 1$, $t_2 = 2$, and $F(t, x) = F(x)$ for every $(t, \xi) \in [0,3] \times \mathbb{R}^2$ is defined by

$$F(x) = \begin{cases} |x|^2(1 - \sin(\ln(|x|))), & \text{if } x = (x_1, x_2) \in (\mathbb{R} - \{0\})^2, \\ 0, & \text{if } x = (x_1, x_2) = (0,0). \end{cases}$$

In addition, let $H(x) = \frac{|x|^2}{108(1+|x|^4)}$, $G(t, x) = t|x|^{\frac{5}{4}}$ for $(t, x) \in [0,3] \times \mathbb{R}^2$, and $I_{ij}(s) = \frac{1}{288}s(1 + \sin s)$ for $s \in \mathbb{R}$, $i = 1, 2$, and $j = 1, 2$.

Since $m = 1$, $M = 2$, $c = \sqrt{6}$, $L = \frac{1}{108}$, $Lij = \frac{1}{288}$ for $i = 1, 2$, $j = 1, 2$, $D = \frac{7}{2}$, $K = \frac{5}{12}$,

$$\liminf_{\xi \to 0^+} \frac{\max_{|x| \leq \xi} F(x)}{\xi^2} = 0, \quad \limsup_{\xi \to 0^+} \frac{F(\xi, 0)}{\xi^2} = 2,$$

and $G_0 = 0$, we see that all the conditions of Theorem 4.3.2 are satisfied. Hence, for every $(\lambda, \mu) \in (\lambda_3, \infty) \times [0, \infty)$, where $\lambda_3 \approx 2.47$, system (4.3.20) has an unbounded sequence of classical periodic solutions that uniformly converges to 0 in E.

There are another number of similar results that can be obtained. For example, we can apply Theorem 4.3.2 to get results similar to Corollary 4.3.1. We leave their formulation to the reader.

4.4 Anti-periodic Solutions

In this section, we consider the existence of anti-periodic solutions to second order impulsive problems. At the same time, in order to show some of the possible variations and extensions of the types of problems considered in this monograph, we will do this in the setting of differential inclusions. Differential inclusions arise in a variety of models in such areas as engineering, biology, and game theory. For background, history, and applications of differential inclusions, we refer the reader to the recent monographs [33,143] and to the references therein. Anti-periodic boundary value problems are important because they arise in a number of important areas including the study of interpolation problems [100,106], wavelets [79], and various problems in physics [1, 12, 191, 236]. Anti-periodic problems for second-order and higher-order differential equations have been studied, for example, in [7, 8, 14, 85, 124–126, 160, 234].

We begin by considering the second-order impulsive differential inclusion with anti-periodic boundary conditions

$$\begin{cases} -\ddot{u}(t) + Mu(t) \in \lambda F(u(t)) + \mu G(t, u(t)), & a.e.\ t \in [0, T], \\ -\Delta(\dot{u}(t_j)) = I_j(\dot{u}(t_j)), & j = 1, 2, \dots, p, \\ u(0) = -u(T), & \dot{u}(0) = -\dot{u}(T), \end{cases} \qquad (4.4.1)$$

where $T > 0$, $M \geq 0$ is a constant, $\lambda > 0$ and $\mu \geq 0$ are parameters, $0 = t_0 < t_1 < \cdots < t_p < t_{p+1} = T$, and $\Delta(\dot{u}(t_j)) = \dot{u}(t_j^+) - \dot{u}(t_j^-) = \lim_{t \to t_j^+} \dot{u}(t) - \lim_{t \to t_j^-} \dot{u}(t)$. Moreover, we assume that $I_j \in C(R, R)$, $j = 1, 2, \dots, p$, F is a multi-valued function defined on $\mathbb{R}$ satisfying:

(D26) $F : \mathbb{R} \to 2^{\mathbb{R}}$ is upper semicontinuous with compact convex values;
(D27) $\min F, \max F : \mathbb{R} \to \mathbb{R}$ are Borel measurable;
(D28) $|\xi| \leq a(1 + |s|^{r-1})$ for all $s \in R$, $\xi \in F(s)$, $r > 1$ $a > 0$;

and G is a multi-valued function defined on $[0, T] \times \mathbb{R}$ satisfying:

(D29) $G(t, \cdot) : \mathbb{R} \to 2^{\mathbb{R}}$ is upper semicontinuous with compact convex values for a.e. $t \in [0, T]$,

(D30) $\min G, \max G : [0, T] \backslash \{t_1, t_2, \ldots, t_p\} \times \mathbb{R} \to \mathbb{R}$ are Borel measurable;

(D31) $|\xi| \leq a(1 + |s|^{r-1})$ for a.e. $t \in [0, T], s \in \mathbb{R}, \xi \in G(t, s), r > 1$.

We take as our Banach space the set $X = \{u \in W^{1,2}([0, T]) : u(0) = -u(T)\}$ endowed with the norm $\|u\|_X = \left(\int_0^T |u'(x)|^2 + M|u(x)|^2 dx \right)^{1/2}$ for all $u \in X$. We need to define what is meant by a solution of (4.4.1).

Definition 4.4.1. *A function $u : [0, T] \backslash \{t_1, t_2, \ldots, t_m\} \to \mathbb{R}$ is a solution of the impulsive differential inclusion (4.4.1) if $u \in C^1$, $\dot{u}$ is absolutely continuous, and there exists $u^* \in L^\gamma[0, T]$, for some $\gamma > 1$, such that*

$$-\ddot{u}(t) + Mu(t) = u^* \in \lambda F(u(t)) + \mu G(t, u(t))$$

and u satisfies the impulsive and boundary conditions in (4.4.1).

Our first theorem guarantees the existence of at least three solutions to the problem (4.4.1).

Theorem 4.4.1. *In addition to (D26)–(D28), assume that:*

(D32) *there exists $\tau > 0$ such that*

$$\frac{2\tau \int_0^T \min \int_0^{\frac{T}{2}-t} F(s) ds dt}{T \left[1 + \frac{MT^2}{12} \right] - \sum_{j=1}^{p} \int_0^{\frac{T}{2}-t_j} I_j(s) ds} > T \sup_{|u| < \frac{(2T\tau)^{1/2}}{2}} \min \int_0^u F(s) ds,$$

where $\frac{T}{2} \left(1 + \frac{MT^2}{12} \right) > \tau$;

(D33) $\min \int_0^u F(s) ds \leq b(1 + |u|^l)$ *for $u \in R, 1 < l < 2$;*

(D34) $I_j(0) = 0, I_j(s)s < 0, s \in R, j = 1, 2, \ldots, p$.

Then there exist a nonempty interval $[\alpha, \beta] \subset (0, \infty)$ and $r > 0$ such that for any $\lambda \in [\alpha, \beta]$ and any multi-valued function G satisfying (D29)–(D31), there exists $\delta > 0$ such that, for all $\mu \in [0, \delta]$, the problem (4.4.1) admits at least three solutions with norms in X less than r.

While the details of the proof of this theorem involves using a number of concepts from the study of differential inclusions and would take us away from the original intent of this work, we do wish to point out that the proof does rely on the variational results in Theorems 1.1.6 and 1.1.7 above.

It is also possible to obtain sufficient conditions for the existence if infinitely many solutions to problem (4.4.1). The proof of this theorem makes use of the critical points result in Theorem 1.1.5 above.

Theorem 4.4.2. *In addition to conditions (D26)–(D28), assume that*

(D35)
$$\liminf_{\xi \to \infty} \frac{\sup_{|u| \leq \xi} \min \int_0^u F(s)ds}{\xi^2}$$

$$< \frac{2}{T^2} \limsup_{\xi \to \infty} \frac{\int_0^T \min \int_0^{\xi(\frac{T}{2}-t)} F(s)dsdt}{\frac{\xi^2}{2}T\left[1 + \frac{MT^2}{12}\right] - \sum_{j=1}^p \int_0^{\xi(\frac{T}{2}-t_j)} I_j(s)ds};$$

(D36) $I_j(0) = 0$, $I_j(s)s < 0$, *for* $s \in \mathbb{R}$ *and* $j = 1, 2, \ldots, p$.

Then, for every $\lambda \in (\lambda_1, \lambda_2)$, *where*

$$\lambda_1 := \left(\limsup_{\xi \to \infty} \frac{\int_0^T \min \int_0^{\xi(\frac{T}{2}-t)} F(s)dsdt}{\frac{\xi^2}{2}T\left[1 + \frac{MT^2}{12}\right] - \sum_{j=1}^p \int_0^{\xi(\frac{T}{2}-t_j)} I_j(s)ds}\right)^{-1}$$

and

$$\lambda_2 := \left(\liminf_{\xi \to \infty} \frac{\sup_{|u| \leq \xi} \min \int_0^u F(s)ds}{\left(\frac{2\xi^2}{T^2}\right)}\right)^{-1},$$

and for every multi-valued function G *satisfying (D29)–(D31) and*

(D37) $\int_0^T \min \int_0^u G(t,s)dsdt \geq 0$ *for all* $u \in \mathbb{R}$,

(D38) $G_\infty := \lim_{\xi \to \infty} \frac{\sup_{|u| \leq \xi} \min \int_0^u G(t,s)ds}{\xi^2} < \infty$,

the problem (4.4.1) has an unbounded sequence of solutions in X *for every* $\mu \in [0, \mu_{G,\lambda})$, *where* $\mu_{G,\lambda} := \frac{1}{G_\infty}\frac{2}{T^2}\left(1 - \lambda\frac{T^2}{2}\liminf_{\xi \to \infty} \frac{\sup_{|u| \leq \xi} \min \int_0^u F(s)ds}{\xi^2}\right)$ *and* $\mu_{G,\lambda} = \infty$ *if* $G_\infty = 0$.

Chapter 5

Partial Differential Equations

5.1 Introduction

In this chapter, we will examine the existence of solutions for several types of partial differential equations. In particular, we study the existence of multiple solutions for the Kirchhoff-type and biharmonic problems.

5.2 A Kirchhoff-type Problem Involving Two Parameters

In this section, we are interested in establishing the existence of at least three weak solutions to the Kirchhoff-type problem

$$\begin{cases} -K\left(\int_\Omega |\nabla u(x)|^2 dx\right)\Delta u(x) = \lambda f(x,u) + \mu g(x,u), & \text{in } \Omega, \\ u = 0, & \text{on } \partial\Omega, \end{cases} \tag{5.2.1}$$

where λ and μ are two positive parameters, $K : [0,\infty) \to \mathbb{R}$ is a continuous function, $\Omega \subset \mathbb{R}^N$ is a nonempty bounded open set with a smooth boundary $\partial\Omega$, $f, g : \Omega \times \mathbb{R} \to \mathbb{R}$ are L^1-Carathéodory functions, and f is non-negative. Here, a *weak solution* of (5.2.1) is any function $u \in W_0^{1,2}(\Omega)$ such that

$$K\left(\int_\Omega |\nabla u(x)|^2 dx\right)\int_\Omega \nabla u(x)\nabla v(x)dx - \lambda\int_\Omega f(x,u(x))v(x)dx$$

$$-\mu\int_\Omega g(x,u(x))v(x)dx = 0$$

for every $v \in W_0^{1,2}(\Omega)$.

The problem (5.2.1) is related to the stationary analogue of the equation

$$u_{tt} - K\left(\int_\Omega |\nabla u(x)|^2 dx\right)\Delta u(x) = g(x,t)$$

proposed by Kirchhoff [190] as an extension of the classical D'Alembert's wave equation for free vibrations of elastic strings. Kirchhoff's model takes into account the changes in length of the string produced by transverse vibrations. Similar nonlocal problems also model several physical and biological systems where u describes a process that depends on the average of itself, for example, the population density. The reader is referred to [15, 19, 20, 61, 87, 88, 98, 190, 211, 235, 237, 249] and the references therein for previous work on this subject. In particular, these papers discuss the historical development of the problem as well as describe situations that can be realistically modeled by (5.2.1) with a nonconstant K.

In this section, we will present some results for the existence of three weak solutions of (5.2.1). Our analysis is based on Theorem 1.1.15. This theorem and its various variations have been frequently used to obtain multiplicity theorems for nonlinear problems of a variational nature. The reader may refer to [38, 40, 47, 53, 55, 57–60, 63, 96, 130, 224, 245, 248, 251] for related works in the literature.

Throughout this section, we always assume, and without further mention, that the following conditions hold:

(E1) there exists a continuous function $h : [0, \infty) \to \mathbb{R}$ satisfying $h(tK(t^2)) = t$ for all $t \geq 0$;

(E2) there exists a constant $m > 0$ such that $\inf_{t \geq 0} K(t) \geq m$.

It was pointed out in [249, Remark 3] that when the function K is nondecreasing in $[0, \infty)$ and $K(0) > 0$, condition (E1) is satisfied.

In what follows, let X be the Sobolev space $W_0^{1,2}(\Omega)$ endowed with the norm

$$||u|| = \left(\int_\Omega |\nabla u(x)|^2 dx \right)^{1/2}.$$

Then, X is a real reflexive Banach space. For each $u \in X$, define the functionals $\Phi, \Psi : X \to \mathbb{R}$ by

$$\Phi(u) = \frac{1}{2}\tilde{K}(||u||^2) \tag{5.2.2}$$

and

$$\Psi(u) = \int_\Omega F(x, u(x))dx, \tag{5.2.3}$$

where

$$\tilde{K}(t) = \int_0^t K(s)ds, \quad t \geq 0,$$

and

$$F(x, t) = \int_0^t f(x, \xi)d\xi, \quad (x, t) \in \Omega \times \mathbb{R}.$$

Standard arguments can be used to show that the functionals Φ and Ψ are well defined and continuously Gâteaux differentiable and whose Gâteaux derivatives at the point $u \in X$ are the functionals $\Phi'(u)$, $\Psi'(u) \in X^*$ given by

$$\Phi'(u)(v) = K\left(\int_\Omega |\nabla u(x)|^2 dx\right) \int_\Omega \nabla u(x) \nabla v(x) dx \qquad (5.2.4)$$

and

$$\Psi'(u)(v) = \int_\Omega f(x, u(x)) v(x) dx \qquad (5.2.5)$$

for every $v \in X$. Moreover, Φ is sequentially weakly lower semicontinuous and Ψ is sequentially weakly upper semicontinuous.

Lemma 5.2.1 below shows that Φ' admits a continuous inverse on X^*, while Lemma 5.2.2 asserts that Ψ' is compact. Proofs of these lemmas are actually embedded in the proof of Theorem 1 in [249].

Lemma 5.2.1. *The operator* $\Phi' : X \to X^*$ *defined by* (5.2.4) *admits a continuous inverse on* X^*.

Lemma 5.2.2. *The operator* $\Psi' : X \to X^*$ *defined by* (5.2.5) *is compact.*

From (E2) and (5.2.2), we have

$$\Phi(u) \geq \frac{m}{2}||u||^2 \quad \text{for all } u \in X. \qquad (5.2.6)$$

Recall that X is compactly embedded in

$$\begin{cases} L^q(\Omega) \text{ for all } q \in [1, 2N/(N-2)), & \text{if } N > 2, \\ L^q(\Omega) \text{ for all } q \geq 1, & \text{if } N = 2, \\ C^0(\overline{\Omega}), & \text{if } N = 1. \end{cases} \qquad (5.2.7)$$

Let

$$k = \sup_{u \in X \setminus \{0\}} \frac{\sup_{x \in \Omega} |u(x)|}{||u||}. \qquad (5.2.8)$$

If $N = 1$, we have $k \leq |\Omega|^{\frac{1}{2}}/2$, where $|\Omega|$ is the Lebesgue measure of the set Ω.

We first present an existence result for the case $N = 1$.

Theorem 5.2.1. *Let* $N = 1$. *Assume that there exist a constant* $r > 0$ *and a function* $w \in X$ *such that*

(E3) $\widetilde{K}(||w||^2) > 2r;$

(E4) $\int_\Omega \sup_{|t| \le k\sqrt{\frac{2r}{m}}} F(x,t)dx < \frac{2r \int_\Omega F(x,w(x))dx}{\widetilde{K}(\|w\|^2)}$;

(E5) $\limsup_{|t| \to \infty} \frac{F(x,t)}{t^2} \le 0$ *for almost every* $x \in \Omega$.

Then, for each compact interval

$$[a,b] \subset \Lambda_r := \left(\frac{\widetilde{K}(\|w\|^2)}{2\int_\Omega F(x,w(x))dx}, \frac{r}{\int_\Omega \sup_{|t| \le k\sqrt{\frac{2r}{m}}} F(x,t)dx} \right),$$

there exists $\rho > 0$ *with the following property: for every* $\lambda \in [a,b]$ *and every* L^1-*Carathéodory function* g, *there exists* $\delta > 0$ *such that, for each* $\mu \in [0,\delta]$, *the problem* (5.2.1) *has at least three distinct weak solutions in* X *whose norms are less than* ρ.

Proof. Let Φ and Ψ be defined by (5.2.2) and (5.2.3), respectively. Fix any $\lambda \in \Lambda_r$. Our aim is to apply Theorem 1.1.15 to our problem. To this end, first note from Lemmas 5.2.1 and 5.2.2 that Φ and Ψ satisfy the regularity assumptions required in Theorem 1.1.15. Moreover, from (E3), $0 < r < \Phi(w)$. Also, using (5.2.6) and (5.2.8), we obtain

$$\Phi^{-1}((-\infty, r]) = \{u \in X \; : \; \Phi(u) \le r\}$$
$$\subseteq \left\{ u \in X \; : \; \|u\| \le \sqrt{\frac{2r}{m}} \right\}$$
$$\subseteq \left\{ u \in X \; : \; |u(x)| \le k\sqrt{\frac{2r}{m}} \text{ for all } x \in \Omega \right\},$$

and then, from (5.2.3), we have

$$\sup_{\Phi(u) \le r} \Psi(u) \le \int_\Omega \sup_{|t| \le k\sqrt{\frac{2r}{m}}} F(x,t)dx,$$

which, in conjunction with (E4), implies that

$$\frac{\sup_{\Phi(u) \le r} \Psi(u)}{r} < \frac{\Psi(w)}{\Phi(w)}.$$

Now, fix $\epsilon \in (0, 2m/|\Omega|^2)$. In view of (E5), there exists $M \in L^1(\Omega)$ such that

$$F(x,t) \le \epsilon t^2 + M(x) \quad \text{for all } t \in \mathbb{R} \text{ and almost every } x \in \Omega.$$

Then,

$$F(x, u(x)) \le \epsilon |u(x)|^2 + M(x) \quad \text{for all } u \in X \text{ and almost every } x \in \Omega.$$

Since $N = 1$, we have (see (5.2.7) or (5.2.8))

$$\sup_{x \in \Omega} |u(x)| \leq \frac{1}{2} |\Omega|^{1/2} ||u|| \quad \text{for } u \in X.$$

Then,

$$F(x, u(x)) \leq \frac{1}{4} \epsilon |\Omega| \, ||u||^2 + M(x) \quad \text{for all } u \in X \text{ and almost every } x \in \Omega. \tag{5.2.9}$$

From (5.2.3), (5.2.6), and (5.2.9), it follows that

$$\Phi(u) - \lambda \Psi(u) \geq \frac{m}{2} ||u||^2 - \frac{1}{4} \epsilon |\Omega|^2 ||u||^2 - \int_\Omega M(x) dx$$

$$= \left(\frac{m}{2} - \frac{1}{4} \epsilon |\Omega|^2 \right) ||u||^2 - \int_\Omega M(x) dx.$$

By the choice of ϵ, we see that $m/2 - \epsilon |\Omega|^2/4 > 0$. Thus, $\Phi - \lambda \Psi$ is coercive. Hence, all the assumptions Theorem 1.1.15 are satisfied. In addition, if we let

$$\Gamma(u) = \int_\Omega \left(\int_0^{u(x)} g(x, \xi) d\xi \right) dx \quad \text{for all } u \in X,$$

then

$$\Gamma'(u)(v) = \int_\Omega g(x, u(x)) v(x) dx \quad \text{for every } v \in X.$$

As in Lemma 5.2.2, we see that Γ' is compact. Since the critical points of the functional $\Phi - \lambda \Psi - \mu \Gamma$ are exactly the weak solutions of (5.2.1), the conclusion then readily follows from Theorem 1.1.15. $\square$

The following result addresses the case when $N \geq 2$.

Theorem 5.2.2. *Let $N \geq 2$. Assume that there exist a constant $r_1 > 0$ and a function $\overline{w} \in X$ such that*

(E6) $\tilde{K}(||\overline{w}||^2) > 2r_1$;

(E7) *there exist constants $b_1 > 0$, $\theta > 0$, and $0 < s < 2$ satisfying $\theta > 2$ if $N = 2$, and $2 < \theta < 2N/(N-2)$ if $N > 2$, such that*

$$|F(x, t)| \leq b_1 (1 + |t|^s) \quad \text{for all } x \in \Omega \text{ and } t \in R, \tag{5.2.10}$$

$$\limsup_{|t| \to 0} \frac{\sup_{x \in \Omega} |F(x, t)|}{|t|^\theta} < \infty, \tag{5.2.11}$$

and

$$\int_\Omega F(x, \overline{w}(x)) dx > 0. \tag{5.2.12}$$

Then, for each compact interval

$$[a, b] \subset \Lambda_r := \left(\frac{\widetilde{K}(\|\overline{w}\|^2)}{2 \int_\Omega F(x, \overline{w}(x)) dx}, \frac{r}{\int_\Omega \sup_{\widetilde{K}(\|u\|^2) \leq 2r} F(x, u) dx} \right),$$

where $r \in (0, r_1)$ is small enough, there exists $\rho > 0$ with the following property: for every $\lambda \in [a, b]$ and every L^1-Carathéodory function g, there exists $\delta > 0$ such that, for each $\mu \in [0, \delta]$, the problem (5.2.1) has at least three distinct weak solutions in X whose norms are less than ρ.

Remark 5.2.1. From the proof given below (see (5.2.15)), we can see that the above interval Λ_r is well defined for sufficiently small $r > 0$.

Proof. As in the proof of Theorem 5.2.1, let Φ and Ψ be defined by (5.2.2) and (5.2.3), respectively. We will apply Theorem 1.1.15 to prove Theorem 5.2.2. Again by Lemmas 5.2.1 and 5.2.2, Φ and Ψ satisfy the regularity assumptions required in Theorem 1.1.15. From (5.2.11), there exist $b_2 > 0$ and $\eta \in (0, 1)$ satisfying

$$|F(x, t)| \leq b_2 |t|^\theta \quad \text{for all } x \in \Omega \text{ and } |t| \leq \eta.$$

This, together with (5.2.10), implies that

$$|F(x, t)| \leq b_3 |t|^\theta \quad \text{for all } x \in \Omega \text{ and } |t| \leq \eta, \tag{5.2.13}$$

where

$$b_3 = \max \left\{ b_2, \sup_{|t| > \eta} \frac{b_1(1 + |t|^s)}{|t|^\theta} \right\} = \max \left\{ b_2, \frac{b_1(1 + \eta^s)}{\eta^\theta} \right\}.$$

For any $r \in (0, r_1)$, from (5.2.6), we have

$$\Phi^{-1}((-\infty, r]) = \{u \in X : \Phi(u) \leq r\}$$

$$\subseteq \left\{ u \in X : \|u\|^2 \leq \frac{2r}{m} \right\} = \left\{ u \in X : \|u\| \leq \sqrt{\frac{2r}{m}} \right\}.$$

Then, since $N \geq 2$, from (5.2.3), (5.2.7), and (5.2.13), it follows that

$$\sup_{u \in \Phi^{-1}((-\infty, r])} \Psi(u) \leq \sup_{u \in \Phi^{-1}((-\infty, r])} \int_\Omega |F(x, u(x))| dx$$

$$\leq b_3 \sup_{u \in \Phi^{-1}((-\infty, r])} \int_\Omega |u(x)|^\theta dx$$

$$\leq b_4 \sup_{u \in \Phi^{-1}((-\infty, r])} \|u\|^\theta$$

$$\leq b_4 \left(\frac{2r}{m} \right)^{\theta/2}$$

for a suitable constant $b_4 > 0$. Since $\theta > 2$, we have

$$\lim_{r \to 0} \frac{\sup_{u \in \Phi^{-1}(]-\infty,r])} \Psi(u)}{r} \leq 0. \tag{5.2.14}$$

By (5.2.12), $\Psi(\overline{w}) > 0$. Then, from (5.2.14), there exists $0 < r < r_1$ small enough such that

$$\frac{\sup_{u \in \Phi^{-1}(]-\infty,r])} \Psi}{r} < \frac{\Psi(\overline{w})}{\Phi(\overline{w})}. \tag{5.2.15}$$

Moreover, from (E6), $0 < r < \Phi(\overline{w})$. For any $\lambda > 0$, by (5.2.3), (5.2.6), (5.2.10), and Hölder's inequality,

$$\Phi(u) - \lambda\Psi(u) \geq \frac{1}{2}\tilde{K}(||u||^2) - \lambda b_1 \int_\Omega (1 + |u(x)|^s)dx$$

$$\geq \frac{m}{2}||u||^2 - \lambda b_1 |\Omega| - |\Omega|^{(2-s)/2}\left(\int_\Omega |u(x)|^2\right)^{s/2}.$$

Then, by Poincaré's inequality, we see that

$$\Phi(u) - \lambda\Psi(u) \geq \frac{m}{2}||u||^2 - c||u||^s - \lambda b_1 |\Omega|,$$

where $c > 0$ is a constant independent of u. Since $s < 2$, we have

$$\lim_{||u|| \to \infty} (\Phi(u) - \lambda\Psi(u)) = \infty,$$

i.e., $\Phi - \lambda\Psi$ is coercive. Hence, all the assumptions Theorem 1.1.15 are satisfied. In addition, if we let

$$\Gamma(u) = \int_\Omega \left(\int_0^{u(x)} g(x,\xi)d\xi\right) dx \quad \text{for all } u \in X,$$

then, as in the proof of Theorem 5.2.1, we see that Γ is a C^1 functional with compact derivative. Since the critical points of the functional $\Phi - \lambda\Psi - \mu\Gamma$ are exactly the weak solutions of (5.2.1), the conclusion then readily follows from Lemma Theorem 1.1.15. $\quad\square$

In the sequel, we present some applications of Theorem 5.2.1. To this end, fix $x_0 \in \Omega$ and choose τ and σ with $0 < \sigma < \tau$ such that $B(x_0, \sigma) \subset B(x_0, \tau) \subseteq \Omega$, where $B(x_0, \rho)$ denotes the open ball with center at x_0 and radius ρ.

Corollary 5.2.1. *Let* $N = 1$*. Assume that (E5) holds and there exist two constants* $c > 0$ *and* $d \neq 0$ *such that*

(E8) $\widetilde{K}\left(\frac{2d^2}{\tau-\sigma}\right) > m(\frac{c}{k})^2$;

(E9) $F(x,t) \geq 0$ for $(x,t) \in \left(B(x_0,\tau) \setminus B(x_0,\sigma)\right) \times S_d$, where

$$S_d = \begin{cases} [d,0], & d < 0, \\ [0,d], & d > 0. \end{cases} \tag{5.2.16}$$

(E10) $\int_\Omega \sup_{|t| \leq c} F(x,t)dx < m\left(\frac{c}{k}\right)^2 \frac{\int_{B(x_0,\sigma)} F(x,d)dx}{\widetilde{K}\left(\frac{2d^2}{\tau-\sigma}\right)}$.

Then, for each compact interval

$$[a,b] \subset \left(\frac{\widetilde{K}\left(\frac{2d^2}{\tau-\sigma}\right)}{2\int_{B(x_0,\sigma)} F(x,d)dx}, \frac{mc^2}{2k^2 \int_\Omega \sup_{|t| \leq c} F(x,t)dx}\right),$$

there exists $\rho > 0$ with the following property: for every $\lambda \in [a,b]$ and every L^1-Carathéodory function g, there exists $\delta > 0$ such that, for each $\mu \in [0,\delta]$, the problem (5.2.1) has at least three distinct weak solutions in X whose norms are less than ρ.

Proof. Let

$$w(x) = \begin{cases} 0, & \text{if } x \in \Omega \setminus B(x_0,\tau), \\ \frac{d}{\tau-\sigma}(\tau - \text{dist}(x-x_0)), & \text{if } x \in B(x_0,\tau) \setminus B(x_0,\sigma), \\ d, & \text{if } x \in B(x_0,\sigma), \end{cases}$$

and $r = \frac{m}{2}(\frac{c}{k})^2$. Then, it is easy to see that $w \in X$ and

$$\begin{aligned} ||w||^2 &= \int_{B(x_0,\tau)\setminus B(x_0,\sigma)} \frac{d^2}{(\tau-\sigma)^2}dx \\ &= \frac{d^2}{(\tau-\sigma)^2}(|B(x_0,\tau)| - |B(x_0,\sigma)|) \\ &= \frac{2d^2}{\tau-\sigma}. \end{aligned}$$

Then, from (E8), we have $\widetilde{K}(||w||^2) > 2r$, i.e., (E3) holds. Moreover, it is trivial to check that $w(x) \in S_d$ for $x \in B(x_0,\tau) \setminus B(x_0,\sigma)$. Thus, by (E9), we have

$$\int_\Omega F(x,w(x))dx \geq \int_{B(x_0,\sigma)} F(x,d)dx.$$

Combining this with (E10), we obtain (E4). Hence, taking into account that

$$\left(\frac{\widetilde{K}\left(\frac{2d^2}{\tau-\sigma}\right)}{2\int_{B(x_0,\sigma)} F(x,d)dx}, \frac{mc^2}{2k^2 \int_\Omega \sup_{|t| \leq c} F(x,t)dx}\right) \subseteq \Lambda_r,$$

the conclusion readily follows from Theorem 5.2.1. $\qquad\square$

Now assume that $f : \mathbb{R} \to \mathbb{R}$ is a continuous function and set

$$F(t) = \int_0^t f(\xi)d\xi, \quad t \in \mathbb{R}.$$

We next present a special case of Corollary 5.2.1.

Corollary 5.2.2. *Let $N = 1$. Assume that (E8) holds and there exist two constants $c > 0$ and $d \neq 0$ such that*

(E11) $F(t) \geq 0$ *for* $t \in S_d$, *where S_d is defined by (5.2.16);*
(E12) $\limsup\limits_{|t| \to \infty} \frac{F(t)}{t^2} \leq 0;$
(E13) $|\Omega| \sup_{|t| \leq c} F(t) < m \left(\frac{c}{k}\right)^2 \frac{2\sigma F(d)}{\tilde{K}\left(\frac{2d^2}{\tau - \sigma}\right)}.$

Then, for each compact interval $[a, b] \subset \left(\dfrac{\tilde{K}\left(\frac{2d^2}{\tau-\sigma}\right)}{4\sigma F(d)}, \dfrac{mc^2}{2k^2|\Omega|\sup_{|t|\leq c} F(t)}\right)$, there exists $\rho > 0$ with the following property: for every $\lambda \in [a, b]$ and every L^1-Carathéodory function g, there exists $\delta > 0$ such that, for each $\mu \in [0, \delta]$, the problem

$$\begin{cases} -K\left(\int_\Omega |\nabla u(x)|^2\right)\Delta u(x) = \lambda f(u) + \mu g(x, u), & \text{in } \Omega, \\ u = 0, & \text{on } \partial\Omega, \end{cases}$$

has at least three distinct weak solutions in X whose norms are less than ρ.

Proof. With $f(x, t) = f(t)$ and $F(x, t) = F(t)$ for every $(x, t) \in \Omega \times \mathbb{R}$, it is easy to see that all the conditions of Corollary 5.2.1 are satisfied and the conclusion follows. $\square$

The following result is a consequence of Corollary 5.2.2.

Corollary 5.2.3. *Let $N = 1$. Assume that (E11) and (E12) hold and there exist four constants $a > 0$, $b > 0$, $c > 0$, and $d \neq 0$ such that*

(E14) $\frac{2ad^2}{\tau - \sigma} + \frac{2bd^4}{(\tau - \sigma)^2} > a(\frac{c}{k})^2;$
(E15) $|\Omega| \sup_{|t| \leq c} F(t) < a\left(\frac{c}{k}\right)^2 \dfrac{2\sigma F(d)}{\frac{2ad^2}{\tau - \sigma} + \frac{2bd^4}{(\tau - \sigma)^2}}$, *where k is defined by* (5.2.8).

Then, for each compact interval $[a, b] \subset \left(\dfrac{\frac{2ad^2}{\tau-\sigma} + \frac{2bd^4}{(\tau-\sigma)^2}}{4\sigma F(d)}, \dfrac{ac^2}{2k^2|\Omega|\sup_{|t|\leq c} F(t)}\right)$, there exists $\rho > 0$ with the following property: for every $\lambda \in [a, b]$ and every

L^1-Carathéodory function g, there exists $\delta > 0$ such that, for each $\mu \in [0, \delta]$, the problem

$$\begin{cases} -(a + b \int_\Omega |\nabla u(x)|^2) \Delta u(x) = \lambda f(u) + \mu g(x, u), & in \ \Omega, \\ u = 0, & on \ \partial\Omega, \end{cases}$$

has at least three distinct weak solutions in X whose norms are less than ρ.

Proof. With $K(t) = a + bt$ and $m = a$, it is easy to check that all the conditions of Corollary 5.2.2 are satisfied. The conclusion then follows from Corollary 5.2.2. □

We end this section with the following remark.

Remark 5.2.2. Applications similar to Corollaries 5.2.1–5.2.3 can also be made to Theorem 5.2.2. We omit the details here and leave them to the interested reader.

In conclusion, we wish to point out that an alternate approach to the one used here is to use Theorem 2.6 in [53] in place of Theorem 1.1.15. In that case, it may even be possible to estimate the value of δ; for example, see [61].

5.3 Biharmonic Systems

Let $\Omega \subset R^N$, $N \geq 1$, be a non-empty bounded open set with a smooth boundary $\partial\Omega$, $n \geq 1$ be an integer, $p_i > 1$, $i = 1, \ldots, n$, and $\lambda > 0$ be a parameter. In this section, we are interested in establishing the existence of a definite interval about λ in which the nonlinear elliptic equation of $(p_1, \ldots, p_n)$-biharmonic type under Navier boundary conditions

$$\begin{cases} \Delta(|\Delta u_i|^{p_i-2} \Delta u_i) = \lambda F_{u_i}(x, u_1, \ldots, u_n), & in \ \Omega, \\ u_i = \Delta u_i = 0, & on \ \partial\Omega, \end{cases} \tag{5.3.1}$$

for $1 \leq i \leq n$, admits at least three weak solutions. Here, $F : \Omega \times \mathbb{R}^n \to \mathbb{R}$ is a function such that the mapping $(t_1, t_2, \ldots, t_n) \to F(x, t_1, t_2, \ldots, t_n)$ is in C^1 in $\mathbb{R}^n$ for all $x \in \Omega$, F_{t_i} is continuous in $\Omega \times \mathbb{R}^n$ for $i = 1, \ldots, n$, where F_{t_i} denotes the partial derivative of F with respect to t_i, $F(x, 0, \ldots, 0) = 0$ for all $x \in \Omega$, and for every $\varrho > 0$,

$$\sup_{\sum_{i=1}^n \frac{|t_i|^{p_i}}{p_i} \leq \varrho} |F(\cdot, t_1, \ldots, t_n)| \in L^1(\Omega).$$

Throughout this section, we let X denote the Cartesian product of n Sobolev spaces $W^{2,p_i}(\Omega) \cap W_0^{1,p_i}(\Omega)$ for $i = 1, \ldots, n$, i.e.,

$$X = (W^{2,p_1}(\Omega) \cap W_0^{1,p_1}(\Omega)) \times \cdots \times (W^{2,p_n}(\Omega) \cap W_0^{1,p_n}(\Omega)),$$

endowed with the norm

$$\| (u_1, \ldots, u_n) \| = \sum_{i=1}^{n} \|u_i\|_{p_i},$$

where

$$\|u_i\|_{p_i} = \left(\int_\Omega |\Delta u_i(x)|^{p_i} dx \right)^{1/p_i}, \quad i = 1, \ldots, n.$$

We say that $u = (u_1, \ldots, u_n)$ is a *weak solution* to the system (5.3.1) if $u = (u_1, \ldots, u_n) \in X$ and

$$\int_\Omega \sum_{i=1}^{n} |\Delta u_i(x)|^{p_i-2} \Delta u_i(x) \Delta v_i(x) dx$$

$$- \lambda \int_\Omega \sum_{i=1}^{n} F_{u_i}(x, u_1(x), \ldots, u_n(x)) v_i(x) dx = 0$$

for every $(v_1, \ldots, v_n) \in X$.

In the literature, a number of papers (for example, see [3, 4, 6, 16, 26, 37, 56, 94, 103–105, 119, 175, 197, 206, 254, 259, 276, 279, 293]) discuss quasilinear elliptic systems that have been used in a great variety of applications. For example, Boccardo and De Figueiredo [37] studied a class of quasilinear elliptic systems involving the p-Laplacian operator. The right hand sides of their systems are closely related to the critical Sobolev exponent, and under some additional assumptions on the nonlinearities, they proved the existence of at least one nontrivial solution. In [64], Bozhkov and Mitidieri used the fibering method introduced by Pokhozhaev [238] to prove the existence of multiple solutions for a Dirichlet problem associated to a quasilinear system involving (p, q)-Laplacian operators. Kristály [197] employed an abstract critical point result of Ricceri and the principle of symmetric criticality to show the existence of an interval $\Lambda \subseteq [0, \infty)$ such that, for each $\lambda \in \Lambda$, the quasilinear elliptic system

$$\begin{cases} -\Delta_p u = \lambda F_u(x, u, v), & \text{in } \Omega, \\ -\Delta_q v = \lambda F_v(x, u, v), & \text{in } \Omega, \\ u = v = 0, & \text{on } \partial\Omega, \end{cases} \quad (5.3.2)$$

where $\Delta_r u = |u|^{r-2}u$ and Ω is a strip-like domain, has at least two distinct nontrivial solutions. In [293], Zhang et al., studying the Nehari manifold for quasilinear elliptic systems with a pair of (p, q)-Laplacian operators and a parameter, proved the existence of a nonnegative nonsemitrivial solution and obtained a global bifurcation result. Li and Tang [206] used Ricceri's three critical points theorem [245] to establish the existence of an interval $\Lambda \subseteq [0, \infty)$ and a positive number ρ such that, for each $\lambda \in \Lambda$, the problem (5.3.2) admits at least three weak solutions whose norms in $W_0^{1,p}(\Omega) \times W_0^{1,q}(\Omega)$ are less than ρ. The results in [206] were generalized by Afrouzi and Heidarkhani in [3].

There has been an increasing interest in studying fourth-order boundary value problems due to their application to deformations of a rigid body and especially the study of traveling waves in suspension bridges (see, for example, Lazer and McKenna [201]). Recently, more general nonlinear fourth-order elliptic boundary value problems have been studied in [6, 25, 31, 43, 67, 76, 156, 159, 207, 208, 215, 216, 228, 277, 294]. In [156], Grossinho et al. used variational methods and a suitable set of assumptions involving the two parameters α and β to prove existence of two nontrivial solutions to the problem

$$\begin{cases} u^{(iv)} + \alpha u'' + \beta u = f(x, u), & x \in (0, 1), \\ u(0) = u(1) = 0, \\ u''(0) = u''(1) = 0, \end{cases} \qquad (5.3.3)$$

where α and β are real constants, $\alpha^2 - 4\beta > 0$, and $f : [0, 1] \times R \to R$ is a continuous function. Han and Zu [159] applied Morse theory to obtain three solutions of (5.3.3) with $\alpha = \beta = 0$. In [216], Liu and Li used the fixed-point index in cones to prove existence of multiple solutions to the problem

$$\begin{cases} u^{(iv)} + \alpha u'' + \beta u = \lambda f(x, u), & x \in (0, 1), \\ u(0) = u(1) = 0, \\ u''(0) = u''(1) = 0, \end{cases} \qquad (5.3.4)$$

where α, β are real constants, $\alpha^2 - 4\beta > 0$, $\lambda > 0$ is a parameter, and $f : [0, 1] \times R \to R$ is a continuous function. In [44], Bonanno and Bella established multiple solutions to the problem (5.3.4) (see [44, Theorem 2.1]) by using a three critical points theorem established by Averno and Bonanno in [24].

Afrouzi et al. [6], using Ricceri's three critical points theorem [248], showed the existence of at least three solutions of the fourth-order boundary value problem

$$\begin{cases} u^{(iv)} + \alpha u'' + \beta u = \lambda f(x, u) + \mu g(x, u), & x \in (0, 1), \\ u(0) = u(1) = 0, \\ u''(0) = u''(1) = 0, \end{cases} \tag{5.3.5}$$

where α, β are real constants, $f, g : [0, 1] \times R \to R$ are Carathéodory functions, and λ, $\mu > 0$ are parameters. Liu and Squassina [215] studied the superlinear p-biharmonic elliptic problem with Navier boundary conditions

$$\begin{cases} \Delta(|\Delta u|^{p-2}\Delta u) = g(x, u), & \text{in } \Omega, \\ u = \Delta u = 0, & \text{on } \partial\Omega, \end{cases} \tag{5.3.6}$$

where Ω is a bounded domain in R^n, $n \geq 2p + 1$, $p > 1$, with a smooth boundary $\partial\Omega$ and $g : \Omega \times R \to R$ is a Carathéodory function. Utilizing Morse theory, they proved the existence of a nontrivial solution to (5.3.6) having a linking structure around the origin. In addition, in both resonance at zero and nonresonance at ∞, they showed the existence of two nontrivial solutions.

In a very interesting paper, Li and Tang [207] employed Ricceri's three critical points theorem [248] to investigate the problem

$$\begin{cases} \Delta(|\Delta u|^{p-2}\Delta u) = \lambda f(x, u) + \mu g(x, u), & \text{in } \Omega, \\ u = \Delta u = 0, & \text{on } \partial\Omega, \end{cases} \tag{5.3.7}$$

where $\Omega \subset R^N$ $(N \geq 1)$ is a non-empty bounded open set with a boundary $\partial\Omega$ of class C^1, $p > \max\{1, \frac{N}{2}\}$, λ, $\mu > 0$, and $f : \Omega \times R \to R$ is a continuous function. They established the existence of an open interval $\Lambda \subseteq [0, \infty)$ and a positive real number q with the property that for every $\lambda \in \Lambda$ and Carathéodory function $g : \Omega \times R \to R$ with

$$\sup_{|t| \leq \zeta} |g(\cdot, t)| \in L^1(\Omega)$$

for all $\zeta > 0$, there is a $\delta > 0$ such that, for each $\mu \in [0, \delta]$ the system (5.3.7) admits at least three weak solutions in $W^{2,p}(\Omega) \cap W_0^{1,p}(\Omega)$ whose norms are less than q. Li and Tang [208] generalized those results to the system

$$\begin{cases} \Delta(|\Delta u|^{p-2}\Delta u) = \lambda F_u(x, u, v) + \mu G_u(x, u, v), & \text{in } \Omega, \\ \Delta(|\Delta v|^{q-2}\Delta v) = \lambda F_v(x, u, v) + \mu G_v(x, u, v), & \text{in } \Omega, \\ u = \Delta u = v = \Delta v = 0, & \text{on } \partial\Omega, \end{cases}$$

where again $\Omega \subset R^N$ ($N \geq 1$) is a non-empty bounded open set with smooth boundary $\partial\Omega$, $p > \max\{1, \frac{N}{2}\}$, $q > \max\{1, \frac{N}{2}\}$, λ, $\mu > 0$, $F :$ $\Omega \times R^2 \to R$ is a function such that $F(\cdot, t_1, t_2)$ is continuous in Ω for all $(t_1, t_2) \in R^2$, $F(x, \cdot, \cdot)$ is C^1 in R^2 for every $x \in \Omega$, $F(x, 0, 0) = 0$ for all $x \in \Omega$, $G : \Omega \times R^2 \to R$ is measurable function with respect to x in Ω for every $(t_1, t_2) \in R^2$, and is C^1 with respect to $(t_1, t_2) \in R^2$ for every x in Ω. (Here, F_t and G_t denote the partial derivative of F and G with respect to t, respectively.)

We begin with the following lemma; its proof is motivated by the proof of [208, Lemma 2].

Lemma 5.3.1. *Let $T : X \to X^*$ be the operator defined by*

$$T(u_1, \ldots, u_n)(h_1, \ldots, h_n) = \int_\Omega \sum_{i=1}^n |\Delta u_i(x)|^{p_i - 2} \Delta u_i(x) \Delta h_i(x) dx$$

for every $(u_1, \ldots, u_n)$, $(h_1, \ldots, h_n) \in X$. Then T admits a continuous inverse on X^*.

Proof. Since

$$T(u_1, \ldots, u_n)(u_1, \ldots, u_n) = \sum_{i=1}^n \int_\Omega |\Delta u_i(x)|^{p_i} dx = \sum_{i=1}^n \|u_i\|_{p_i}^{p_i},$$

T is coercive. For any $p > 1$, by (2.2) in [253], there exists $C_p > 0$ such that

$$\langle |x|^p x - |y|^p y, x - y \rangle \geq \begin{cases} C_p |x - y|^p, & \text{if } p \geq 2, \\ \dfrac{C_p |x - y|^2}{(|x| + |y|)^{2-p}}, & \text{if } 1 < p < 2 \end{cases} \qquad (5.3.8)$$

for any $x, y \in \mathbb{R}^n$. Now, for any $u = (u_1, \ldots, u_n) \in X$ and $v = (v_1, \ldots, v_n) \in X$, we have

$$\langle T(u) - T(v), u - v \rangle$$
$$= \sum_{i=1}^n \int_0^1 (|\Delta u_i(x)|^{p_i - 2} \Delta u_i(x) - |\Delta v_i(x)|^{p_i - 2} \Delta v_i(x))(\Delta u_i(x) - \Delta v_i(x)) dx.$$

Then, from (5.3.8), we see that for $u, v \in X$ with $u \neq v$,

$$\langle T(u) - T(v), u - v \rangle \geq \sum_{i \in I_1} C_{p_i} \int_0^1 |\Delta u_i(x)|$$
$$- \Delta v_i(x)|^{p_i} dx + \sum_{i \in I_2} C_{p_i} \int_0^1 \frac{|\Delta u_i(x) - \Delta v_i(x)|^2}{(|\Delta u_i(x)| + |\Delta v_i(x)|)^{2-p_i}} dx > 0,$$

where $I_1 = \{i \in \{1, \ldots, n\} : p_i \geq 2\}$ and $I_2 = \{i \in \{1, \ldots, n\} : 1 < p_i < 2\}$. Thus, T is strictly monotone. Clearly, T is also demicontinuous, so by [289, Theorem 26.A(d)], the inverse operator T^{-1} of T exists.

The continuity of T^{-1} can be proved essentially by the same way as the latter part of the proof of [208, Lemma 2]. We omit the details. $\qquad\square$

We recall that for any $p \geq 1$ and any positive integer l (see [289, page 1026]), $W_0^{l,p}(\Omega)$ is compactly embedded in

$$\begin{cases} L^q(\Omega) \text{ for } q \in [1, lN/(N - lp)), \ p < N/l, \\ L^q(\Omega) \text{ for } q \geq 1, \ p = N/l, \\ C^0(\overline{\Omega}), \ p > N/l. \end{cases}$$

Thus, for $i = 1, \ldots, n$, if $p_i > N/2$, the embedding $W^{2,p_i}(\Omega) \cap W_0^{1,p_i}(\Omega) \hookrightarrow C_0(\overline{\Omega})$ is compact, and if $p_i \leq N/2$, the embedding $W^{2,p_i}(\Omega) \cap W_0^{1,p_i}(\Omega) \hookrightarrow L^{q_i}(\Omega)$ for all $q_i \in [p_i, \frac{p_i N}{N - 2p_i})$ is compact.

Let

$$k = \max \left\{ \sup_{u_i \in W^{2,p_i}(\Omega) \cap W_0^{1,p_i}(\Omega) \setminus \{0\}} \frac{\max_{x \in \overline{\Omega}} |u_i(x)|^{p_i}}{||u_i||_{p_i}^{p_i}} \ : \ i = 1, \ldots, n \right\}.$$

For the case where $p_i > N/2$, $i = 1, \ldots, n$, since the embedding $W^{2,p_i}(\Omega) \cap W_0^{1,p_i}(\Omega) \hookrightarrow C^0(\overline{\Omega})$ is compact, we have that $k < \infty$. Moreover, for $u = (u_1, \ldots, u_n) \in X$ and $i = 1, \ldots, n$,

$$\sup_{x \in \Omega} |u_i(x)|^{p_i} \leq k ||u_i||_{p_i}^{p_i}. \tag{5.3.9}$$

For any $\gamma > 0$ we define the set $K(\gamma)$ by

$$K(\gamma) = \left\{ (t_1, \ldots, t_n) \in \mathbb{R}^n \ : \ \sum_{i=1}^{n} \frac{|t_i|^{p_i}}{p_i} \leq \gamma \right\}. \tag{5.3.10}$$

We now state our main result, which provides sufficient conditions for the existence of three weak solutions of the system (5.3.1).

Theorem 5.3.1. *Assume that there exist $r > 0$ and a function $w = (w_1, \ldots, w_n) \in X$ such that*

$$\sum_{i=1}^{n} \frac{||w_i||_{p_i}^{p_i}}{p_i} > r. \tag{5.3.11}$$

In addition, assume that:

(E16) *if $p_i > N/2$ for all $i = 1, \ldots, n$, then*

$$0 < \int_\Omega \sup_{(t_1, \ldots, t_n) \in K(kr)} F(x, t_1, \ldots, t_n) dx$$

$$< r \prod_{i=1}^n p_i \frac{\int_\Omega F(x, w_1(x), \ldots, w_n(x)) dx}{2 \sum_{i=1}^n \prod_{j=1, j \neq i}^n p_j \|w_i\|_{p_i}^{p_i}} \qquad (5.3.12)$$

and

$$\limsup_{|t_1| \to \infty, \ldots, |t_n| \to \infty} \frac{F(x, t_1, \ldots, t_n)}{\sum_{i=1}^n \frac{|t_i|^{p_i}}{p_i}}$$

$$< \frac{\int_\Omega \sup_{(t_1, \ldots, t_n) \in K(kr)} F(x, t_1, \ldots, t_n) dx}{krm(\Omega)} \qquad (5.3.13)$$

uniformly with respect to $x \in \Omega$, where $m(\Omega)$ denotes the Lebesgue measure of Ω;

and

(E17) *if $p_i \leq N/2$ for all $i = 1, \ldots, n$, then*

$$\int_\Omega F(x, w_1(x), \ldots, w_n(x)) dx$$

$$- \int_\Omega \sup_{(t_1, \ldots, t_n) \in K(kr)} F(x, t_1, \ldots, t_n) dx > 0 \qquad (5.3.14)$$

and there exist positive constants b_1, θ, and s with $\theta > p_i$ and $1 \leq s < p_i$, $i = 1, \ldots, n$, satisfying

$$|F(x, t_1, \ldots, t_n)| \leq b_1 \left(1 + \sum_{i=1}^n |t_i|^s\right) \qquad \text{for all } t_i \in \mathbb{R}, \qquad (5.3.15)$$

and

$$\limsup_{\sum_{i=1}^n |t_i| \to 0} \frac{|F(x, t_1, \ldots, t_n)|}{\sum_{i=1}^n |t_i|^\theta} < \infty. \qquad (5.3.16)$$

Then, for each $\lambda \in (\underline{\lambda}, \overline{\lambda})$ the system (5.3.1) has at least three weak solutions, where

$$\underline{\lambda} = \frac{\sum_{i=1}^n \frac{\|w_i\|_{p_i}^{p_i}}{p_i}}{\int_\Omega F(x, w_1(x), \ldots, w_n(x)) dx - \int_\Omega \sup_{(t_1, \ldots, t_n) \in K(kr)} F(x, t_1, \ldots, t_n) dx}$$

and

$$\overline{\lambda} = \begin{cases} \dfrac{r}{\int_\Omega \sup_{(t_1, \ldots, t_n) \in K(kr)} F(x, t_1, \ldots, t_n) dx}, & \text{if } p_i > N/2 \text{ for } i = 1, \ldots, n, \\[4mm] \infty, & \text{if } p_i \leq N/2 \text{ for } i = 1, \ldots, n. \end{cases}$$

$$(5.3.17)$$

Proof. For each $u = (u_1, \ldots, u_n) \in X$, let

$$\Phi(u) = \sum_{i=1}^{n} \frac{||u_i||_{p_i}^{p_i}}{p_i}$$

and

$$\Psi(u) = -\int_{\Omega} F(x, u_1(x), \ldots, u_n(x))dx.$$

Then, it is well known that Ψ is a Gâteaux differentiable functional whose Gâteaux derivative at the point $u \in X$ is the functional $\Psi'(u) \in X^*$ given by

$$\Psi'(u)(v) = -\int_{\Omega} \sum_{i=1}^{n} F_{u_i}(x, u_1(x), \ldots, u_n(x))v_i(x)dx$$

for every $v = (v_1, \ldots, v_n) \in X$, and that $\Psi' : X \to X^*$ is a continuous and compact operator. Moreover, Φ is a continuously Gâteaux differentiable and sequentially weakly lower semicontinuous functional whose Gâteaux derivative at the point $u \in X$ is the functional $\Phi'(u) \in X^*$ given by

$$\Phi'(u)(v) = \int_{\Omega} \sum_{i=1}^{n} |\Delta u_i(x)|^{p_i-2} \Delta u_i(x) \Delta v_i(x)dx.$$

In addition, Lemma 5.3.1 implies that $\Phi' : X \to X^*$ admits a continuous inverse on X^*.

In what follows, we discuss two cases.

Case 1. $p_i > N/2$ for all $i = 1, \ldots, n$.

From (5.3.13), there exist two constants γ and $\tau \in \mathbb{R}$ with

$$0 < \gamma < \frac{\int_{\Omega} \sup_{(t_1, \ldots, t_n) \in K(kr)} F(x, t_1, \ldots, t_2)dx}{r} \qquad (5.3.18)$$

such that

$$km(\Omega)F(x, t_1, \ldots, t_n) \leq \gamma \sum_{i=1}^{n} \frac{|t_i|^{p_i}}{p_i} + \tau \quad \text{for all } x \in \overline{\Omega} \text{ and } (t_1, \ldots, t_n) \in \mathbb{R}^n.$$

Then, for any fixed $(u_1, \ldots, u_n) \in X$, we have

$$F(x, u_1(x), \ldots, u_n(x)) \leq \frac{1}{km(\Omega)} \left(\gamma \sum_{i=1}^{n} \frac{|u_i(x)|^{p_i}}{p_i} + \tau \right) \quad \text{for all } x \in \overline{\Omega}.$$
$$(5.3.19)$$

Thus, for any fixed $\lambda \in (\underline{\lambda}, \overline{\lambda})$, from (5.3.9), (5.3.17), and (5.3.19), we have

$$\Phi(u) + \lambda\Psi(u) = \sum_{i=1}^{n} \frac{||u_i||_{p_i}^{p_i}}{p_i} - \lambda \int_{\Omega} F(x, u_1(x), \ldots, u_n(x)) dx$$

$$\geq \sum_{i=1}^{n} \frac{||u_i||_{p_i}^{p_i}}{p_i} - \frac{\lambda\gamma}{km(\Omega)} \left(\sum_{i=1}^{n} \frac{1}{p_i} \int_{\Omega} |u_i(x)|^{p_i} dx \right) - \frac{\lambda\tau}{k}$$

$$\geq \sum_{i=1}^{n} \frac{||u_i||_{p_i}^{p_i}}{p_i} - \frac{\lambda\gamma}{km(\Omega)} \left(km(\Omega) \sum_{i=1}^{n} \frac{||u_i||_{p_i}^{p_i}}{p_i} \right) - \frac{\lambda\tau}{k}$$

$$= \sum_{i=1}^{n} \frac{||u_i||_{p_i}^{p_i}}{p_i} - \lambda\gamma \sum_{i=1}^{n} \frac{||u_i||_{p_i}^{p_i}}{p_i} - \frac{\lambda\tau}{k}$$

$$\geq \left(1 - \gamma \frac{r}{\int_{\Omega} \sup_{(t_1,\ldots,t_n) \in K(kr)} F(x, t_1, \ldots, t_n) dx} \right) \sum_{i=1}^{n} \frac{||u_i||_{p_i}^{p_i}}{p_i}$$

$$- \frac{\lambda\tau}{k}.$$

Hence, from (5.3.18), we see that

$$\lim_{||u|| \to \infty} (\Phi(u) + \lambda\Psi(u)) = \infty,$$

i.e., the functional $\Phi + \lambda\Psi$ is coercive.

From (5.3.12), it follows that

$$\frac{\int_{\Omega} \sup_{(t_1,\ldots,t_n) \in K(kr)} F(x, t_1, \ldots, t_n) dx}{r \prod_{i=1}^{n} p_i}$$

$$< \frac{\int_{\Omega} F(x, w_1(x), \ldots, w_n(x)) dx}{\sum_{i=1}^{n} \prod_{j=1, j \neq i}^{n} p_j ||w_i||_{p_i}^{p_i}} - \frac{\int_{\Omega} \sup_{(t_1,\ldots,t_n) \in K(kr)} F(x, t_1, \ldots, t_n) dx}{r \prod_{i=1}^{n} p_i}.$$

Since (5.3.11) implies that $\sum_{i=1}^{n} \prod_{j=1, j \neq i}^{n} p_j ||w_i||_{p_i}^{p_i} > r \prod_{i=1}^{n} p_i$, we have

$$\frac{\int_{\Omega} \sup_{(t_1,\ldots,t_n) \in K(kr)} F(x, t_1, \ldots, t_n) dx}{r \prod_{i=1}^{n} p_i}$$

$$< \frac{\int_{\Omega} F(x, w_1(x), \ldots, w_n(x)) dx - \int_{\Omega} \sup_{(t_1,\ldots,t_n) \in K(kr)} F(x, t_1, \ldots, t_n) dx}{\sum_{i=1}^{n} \prod_{j=1, j \neq i}^{n} p_j ||w_i||_{p_i}^{p_i}},$$

which, multiplying by $\prod_{i=1}^{n} p_i$, yields

$$\frac{\int_{\Omega} \sup_{(t_1,\ldots,t_n) \in K(kr)} F(x, t_1, \ldots, t_n) dx}{r}$$

$$< \frac{\int_{\Omega} F(x, w_1(x), \ldots, w_n(x)) dx - \int_{\Omega} \sup_{(t_1,\ldots,t_n) \in K(kr)} F(x, t_1, \ldots, t_n) dx}{\sum_{i=1}^{n} \frac{||w_i||_{p_i}^{p_i}}{p_i}}.$$

$$(5.3.20)$$

We claim that

$$\varphi_1(r) \leq \frac{\int_\Omega \sup_{(t_1,\ldots,t_n) \in K(kr)} F(x,t_1,\ldots,t_n)dx}{r} \tag{5.3.21}$$

and

$$\varphi_2(r) \geq \left[\int_\Omega F(x,w_1(x),\ldots,w_n(x))dx \right.$$
$$\left. - \int_\Omega \sup_{(t_1,\ldots,t_n) \in K(kr)} F(x,t_1,\ldots,t_n)dx \right] \times \left[\sum_{i=1}^n \frac{||w_i||^{p_i}_{p_i}}{p_i} \right]^{-1}. \tag{5.3.22}$$

For $u = (u_1,\ldots,u_n) \in X$ and $i = 1,\ldots,n$, from (5.3.9), we have

$$\sup_{x \in \Omega} \sum_{i=1}^n \frac{|u_i(x)|^{p_i}}{p_i} \leq k \sum_{i=1}^n \frac{||u_i||^{p_i}_{p_i}}{p_i}. \tag{5.3.23}$$

Note that $\Phi(0) = \Psi(0) = 0$. Then, from (1.1.9), we obtain that

$$\varphi_1(r) \leq \frac{\sup_{\overline{\Phi^{-1}((-\infty,r))}^w} \int_\Omega F(x,u_1(x),\ldots,u_n(x))dx}{r}$$
$$= \frac{\sup_{\Phi^{-1}((-\infty,r])} \int_\Omega F(x,u_1(x),\ldots,u_n(x))dx}{r}$$
$$\leq \frac{\int_\Omega \sup_{(t_1,\ldots,t_n) \in K(kr)} F(x,t_1,\ldots,t_n)dx}{r},$$

i.e., (5.3.21) holds. By (5.3.11), $\Phi(w) > r$. Then, from (1.1.10),

$$\varphi_2(r) \geq \inf_{u \in \Phi^{-1}((-\infty,r))} \frac{1}{\Phi(w) - \Phi(u)} \left\{ \int_\Omega F(x,w_1(x),\ldots,w_n(x))dx \right.$$
$$\left. - \int_\Omega F(x,u_1(x),\ldots,u_n(x))dx \right\}$$
$$\geq \inf_{u \in \Phi^{-1}((-\infty,r))} \frac{1}{\Phi(w) - \Phi(u)} \left\{ \int_\Omega F(x,w_1(x),\ldots,w_n(x))dx \right.$$
$$\left. - \int_\Omega \sup_{(t_1,\ldots,t_n) \in K(kr)} F(x,t_1,\ldots,t_n)dx \right\}.$$

Now (5.3.12) and (5.3.20) imply that

$$\int_\Omega F(x,w_1(x),\ldots,w_n(x))dx > \int_\Omega \sup_{(t_1,\ldots,t_n) \in K(kr)} F(x,t_1,\ldots,t_n)dx.$$

Then, since $0 < \Phi(w) - \Phi(u) \le \Phi(w)$ for any $u \in X$ with $\Phi(u) < r$, we have

$$\varphi_2(r) \ge \left[\int_\Omega F(x, w_1(x), \dots, w_n(x))dx \right.$$
$$\left. - \int_\Omega \sup_{(t_1,\dots,t_n)\in K(kr)} F(x, t_1, \dots, t_n)dx \right] \times \left[\sum_{i=1}^n \frac{||w_i||_{p_i}^{p_i}}{p_i} \right]^{-1},$$

i.e., (5.3.22) holds, and so our claim holds. By (5.3.21) and (5.3.22), we observe that

$$\frac{1}{\varphi_2(r)} \le \frac{\sum_{i=1}^n \frac{||w_i||_{p_i}^{p_i}}{p_i}}{\int_\Omega F(x, w_1(x), \dots, w_n(x))dx - \int_\Omega \sup_{(t_1,\dots,t_n)\in K(kr)} F(x, t_1, \dots, t_n)dx}$$

and

$$\frac{1}{\varphi_1(r)} \ge \frac{r}{\int_\Omega \sup_{(t_1,\dots,t_n)\in K(kr)} F(x, t_1, \dots, t_n)dx}.$$

Thus, taking into account the fact that the weak solutions of the system (5.3.1) are exactly the solutions of the equation $\Phi'(u) + \lambda \Psi'(u) = 0$, the conclusion follows from Theorem 1.1.16.

Case 2. $p_i \le N/2$ for all $i = 1, \dots, n$.

We first note that

$$\Phi^{-1}((-\infty, r]) = \{u \in X \ : \ \Phi(u) \le r\}$$
$$= \left\{ u = (u_1, \dots, u_n) \in X \ : \ \sum_{i=1}^n \frac{||u_i||_{p_i}^{p_i}}{p_i} \le r \right\}$$
$$\subseteq \{u = (u_1, \dots, u_n) \in X \ : \ ||u_i||_{p_i} \le \sqrt[p_i]{p_i r}, \ 1 \le i \le n\}.$$
$$(5.3.24)$$

From (5.3.16), for some positive constant b_2, there exists $\eta \in (0,1)$ satisfying

$$|F(x, t_1, \dots, t_n)| \le b_2 \sum_{i=1}^n |t_i|^\theta \quad \text{for all } |t_i| \le \eta.$$

In view of (5.3.15), if we let

$$b_3 = \max \left\{ b_2, \ \sup_{|t_i| > \eta} \frac{b_1 \left(1 + \sum_{i=1}^n |t_i|^s \right)}{\sum_{i=1}^n |t_i|^\theta} \right\},$$

then we have $|F(x, t_1, \ldots, t_n)| \leq b_3 \sum_{i=1}^{n} |t_i|^\theta$ for all $t_i \in \mathbb{R}$. Therefore, by (1.1.9), (5.3.9), and (5.3.24), we have (for a suitable constant $b_4 > 0$)

$$
\begin{aligned}
\varphi_1(r) &\leq \frac{\sup_{\overline{\Phi^{-1}((-\infty,r))}^w} \int_\Omega F(x, u_1(x), \ldots, u_n(x)) dx}{r} \\
&= \frac{\sup_{\Phi^{-1}((-\infty,r])} \int_\Omega F(x, u_1(x), \ldots, u_n(x)) dx}{r} \\
&= \frac{\sup_{\sum_{i=1}^{n} \frac{||u_i||_{p_i}^{p_i}}{p_i} \leq r} \int_\Omega |F(x, u_1(x), \ldots, u_n(x))| dx}{r} \\
&\leq \frac{b_3 \sup_{\sum_{i=1}^{n} \frac{||u_i||_{p_i}^{p_i}}{p_i} \leq r} \int_\Omega \sum_{i=1}^{n} |u_i(x)|^\theta dx}{r} \\
&\leq \frac{b_4 \sup_{\sum_{i=1}^{n} \frac{||u_i||_{p_i}^{p_i}}{p_i} \leq r} \sum_{i=1}^{n} ||u_i||_{p_i}^\theta}{r} \\
&\leq \frac{b_4 n (\sqrt[p_i]{p_i r})^\theta}{r} \quad \text{for } i = 1, \ldots, n.
\end{aligned}
$$

Thus,

$$
\lim_{r \to 0} \varphi_1(r) = 0. \tag{5.3.25}
$$

Under (5.3.11) and (5.3.14), as in Case 1, we have that (5.3.22) holds, i.e.,

$$
\varphi_2(r) \geq \frac{\int_\Omega F(x, w_1(x), \ldots, w_n(x)) dx - \int_\Omega \sup_{(t_1, \ldots, t_n) \in K(kr)} F(x, t_1, \ldots, t_n) dx}{\sum_{i=1}^{n} \frac{||w_i||_{p_i}^{p_i}}{p_i}}.
$$

For any fixed $\lambda \in (\underline{\lambda}, \infty)$, from (5.3.15),

$$
\Phi(u) + \lambda \Psi(u) \geq \sum_{i=1}^{n} \frac{||u_i||_{p_i}^{p_i}}{p_i} - \lambda \int_\Omega b_1 \left(1 + \sum_{i=1}^{n} |u_i(x)|^s \right) dx.
$$

Since, for $i = 1, \ldots, n$, the embedding $W^{2,p_i}(\Omega) \cap W_0^{1,p_i}(\Omega) \hookrightarrow L^{p_i}(\Omega)$ is compact and the embedding $L^{p_i}(\Omega) \hookrightarrow L^s(\Omega)$ is continuous, there exists $C > 0$ such that

$$
\int_\Omega |u_i(x)|^s dx \leq C ||u_i||_{p_i}^s, \quad i = 1, \ldots, n.
$$

Thus,

$$
\Phi(u) + \lambda \Psi(u) \geq \sum_{i=1}^{n} \frac{||u_i||_{p_i}^{p_i}}{p_i} - \lambda C \sum_{i=1}^{n} ||u_i||_{p_i}^s - \lambda b_1 m(\Omega).
$$

Since $s < p_i$, we have

$$\lim_{||u|| \to \infty} (\Phi(u) + \lambda \Psi(u)) = \infty.$$

Note that from (5.3.25), $\lim_{r \to 0} 1/\varphi_1(r) = \infty$, and taking into account the fact that the weak solutions of the system (5.3.1) are exactly the solutions of the equation $\Phi'(u) + \lambda \Psi'(u) = 0$, the conclusion follows from Theorem 1.1.16. This completes the proof of the theorem. $\square$

Now, we want to present a verifiable consequence of Theorem 5.3.1 where the test function w is specified.

Fix $x^0 \in \Omega$ and pick r_1, r_2 with $0 < r_1 < r_2$ such that

$$S(x^0, r_1) \subset S(x^0, r_2) \subseteq \Omega,$$

where $S(x^0, r_i)$ denotes the ball with center at x^0 and radius r_i for $i = 1, 2$. For $i = 1, \ldots, n$, let

$$\sigma_i = \sigma_i(N, p_i, r_1, r_2) = \frac{12(N+2)^2(r_1 + r_2)}{(r_2 - r_1)^3} \left(\frac{k\pi^{\frac{N}{2}}(r_2^N - r_1^N)}{\Gamma(1 + \frac{N}{2})} \right)^{1/p_i} \tag{5.3.26}$$

and $\theta_i = \theta_i(N, p_i, r_1, r_2)$

$$= \begin{cases} \frac{3N}{(r_2 - r_1)(r_1 + r_2)} \left(\frac{k\pi^{\frac{N}{2}}((r_1 + r_2)^N - (2r_1)^N)}{2^N \Gamma(1 + \frac{N}{2})} \right)^{1/p_i}, & N < \frac{4r_1}{r_2 - r_1}, \\[3mm] \frac{12r_1}{(r_2 - r_1)^2(r_1 + r_2)} \left(\frac{k\pi^{\frac{N}{2}}((r_1 + r_2)^N - (2r_1)^N)}{2^N \Gamma(1 + \frac{N}{2})} \right)^{1/p_i}, & N \geq \frac{4r_1}{r_2 - r_1}, \end{cases} \tag{5.3.27}$$

where $\Gamma(\cdot)$ is the Gamma function.

Corollary 5.3.1. *Assume there exist two positive constants c and d with* $\sum_{i=1}^{n} \frac{(d\theta_i)^{p_i}}{p_i} > \frac{c}{\prod_{i=1}^{n} p_i}$ *such that*

(E18) $F(x, t_1, \ldots, t_n) \geq 0$ *for each* $(x, t_1, \ldots, t_n) \in (\overline{\Omega} \setminus S(x^0, r_1)) \times [0, d] \times \cdots \times [0, d]$.

Furthermore, assume that

(E19) *if $p_i > N/2$ for all $i = 1, \ldots, n$, then*

$$0 < \int_{\Omega} \sup_{(t_1, \ldots, t_n) \in K(\frac{c}{\prod_{i=1}^{n} p_i})} F(x, t_1, \ldots, t_n) dx$$

$$< \frac{c}{2} \frac{\int_{S(x^0, r_1)} F(x, d, \ldots, d) dx}{\sum_{i=1}^{n} \prod_{j=1, j \neq i}^{n} p_j (d\sigma_i)^{p_i}} \tag{5.3.28}$$

and

$$\limsup_{|t_1|\to\infty,\dots,|t_n|\to\infty} \frac{F(x,t_1,\dots,t_n)}{\sum_{i=1}^n \frac{|t_i|^{p_i}}{p_i}}$$

$$< \frac{\prod_{i=1}^n p_i}{cm(\Omega)} \int_\Omega \sup_{(t_1,\dots,t_n)\in K(\frac{c}{\prod_{i=1}^n p_i})} F(x,t_1,\dots,t_n)dx \quad (5.3.29)$$

uniformly with respect to $x \in \overline{\Omega}$; *and*

(E20) *if* $p_i \le N/2$ *for all* $i = 1,\dots,n$, *then* (5.3.15) *and* (5.3.16) *in Theorem 5.3.1 hold, and*

$$\int_{S(x^0,r_1)} F(x,d,\dots,d)dx$$

$$- \int_\Omega \sup_{(t_1,\dots,t_n)\in K(\frac{c}{\prod_{i=1}^n p_i})} F(x,t_1,\dots,t_n)dx > 0.$$

Then, for each $\lambda \in (\underline{\lambda},\overline{\lambda})$, *the system* (5.3.1) *has at least three weak solutions, where*

$$\underline{\lambda} = \frac{\sum_{i=1}^n \frac{(d\sigma_i)^{p_i}}{kp_i}}{\int_{S(x^0,r_1)} F(x,d,\dots,d)dx - \int_\Omega \sup_{(t_1,\dots,t_n)\in K(\frac{c}{\prod_{i=1}^n p_i})} F(x,t_1,\dots,t_n)dx}$$

and

$$\overline{\lambda} = \begin{cases} \dfrac{k\prod_{i=1}^n p_i}{\int_\Omega \sup_{(t_1,\dots,t_n)\in K(\frac{c}{\prod_{i=1}^n p_i})} F(x,t_1,\dots,t_n)dx}, & \text{if } p_i > N/2, \ i = 1,\dots,n, \\ \\ \infty, & \text{if } p_i \le N/2, \ i = 1,\dots,n. \end{cases}$$

Proof. Let $r = \frac{c}{k\prod_{i=1}^n p_i}$ and $w(x) = (w_1(x),\dots,w_n(x))$ be such that for $i = 1,\dots,n$,

$$w_i(x) = \begin{cases} 0, & \text{if } x \in \overline{\Omega} \setminus S(x^0,r_2), \\ \frac{d\left(3(l^4-r_2^4)-4(r_1+r_2)(l^3-r_2^3)+6r_1r_2(l^2-r_2^2)\right)}{(r_2-r_1)^3(r_1+r_2)}, & \text{if } x \in S(x^0,r_2)\setminus S(x^0,r_1), \\ d, & \text{if } x \in S(x^0,r_1), \end{cases}$$

where $l = \text{dist}(x,x^0) = \sqrt{\sum_{i=1}^N (x_i - x_i^0)^2}$. *By a simple calculation, we have*

$$\frac{\partial w_i(x)}{\partial x_i} = \begin{cases} 0, & x \in \overline{\Omega}\setminus S(x^0,r_2) \cup S(x^0,r_1), \\ \frac{12d\left(l^2(x_i-x_i^0)-(r_1+r_2)l(x_i-x_i^0)+r_1r_2(x_i-x_i^0)\right)}{(r_2-r_1)^3(r_1+r_2)}, & x \in S(x^0,r_2)\setminus S(x^0,r_1), \end{cases}$$

$$\frac{\partial^2 w_i(x)}{\partial^2 x_i} = \begin{cases} 0, & x \in \overline{\Omega} \backslash S(x^0, r_2) \cup S(x^0, r_1), \\ \frac{12d\left(r_1 r_2 + (2l - r_1 - r_2)(x_i - x_i^0)^2/l - (r_2 + r_1 - l)l\right)}{(r_2 - r_1)^3(r_1 + r_2)}, & x \in S(x^0, r_2) \backslash S(x^0, r_1), \end{cases}$$

and

$$\sum_{i=1}^{N} \frac{\partial^2 w_i(x)}{\partial^2 x_i} = \begin{cases} 0, & x \in \overline{\Omega} \backslash S(x^0, r_2) \cup S(x^0, r_1), \\ \frac{12d\left((N+2)l^2 - (N+1)(r_1 + r_2)l + N r_1 r_2\right)}{(r_2 - r_1)^3(r_1 + r_2)}, & x \in S(x^0, r_2) \backslash S(x^0, r_1). \end{cases}$$

It is easy to see that $w = (w_1, \ldots, w_n) \in X$ and, in particular, for $i = 1, \ldots, n$,

$$\|w_i\|_{p_i}^{p_i} = \frac{(12d)^{p_i} 2\pi^{\frac{N}{2}}}{(r_2 - r_1)^{3p_i}(r_1 + r_2)^{p_i} \Gamma(\frac{N}{2})}$$
$$\times \int_{r_1}^{r_2} |(N+2)\xi^2 - (N+1)(r_1 + r_2)\xi + N r_1 r_2|^{p_i} \xi^{N-1} d\xi.$$

$$(5.3.30)$$

Hence, from (5.3.26), (5.3.27), and (5.3.30), we see that

$$\frac{(d\theta_i)^{p_i}}{k} < \|w_i\|_{p_i}^{p_i} < \frac{(d\sigma_i)^{p_i}}{k}, \quad i = 1, \ldots, n. \tag{5.3.31}$$

Now, taking into account the assumption that $\sum_{i=1}^{n} \frac{(d\theta_i)^{p_i}}{p_i} > \frac{c}{\prod_{i=1}^{n} p_i}$, from (5.3.31), we have

$$\sum_{i=1}^{n} \frac{\|w_i\|_{p_i}^{p_i}}{p_i} > r,$$

which is (5.3.11) in Theorem 5.3.1.

Since $0 \le w(x) \le (d, \ldots, d)$ for each $x \in \Omega$, condition (E18) ensures that

$$\int_{\overline{\Omega} \backslash S(x^0, r_2)} F(x, w_1(x), \ldots, w_n(x)) dx$$
$$+ \int_{S(x^0, r_2) \backslash S(x^0, r_1)} F(x, w_1(x), \ldots, w_n(x)) dx \ge 0.$$

Combining this with (5.3.28) and (5.3.31) yields

$$\int_{\Omega} \sup_{(t_1, \ldots, t_n) \in K(\frac{c}{\prod_{i=1}^{n} p_i})} F(x, t_1, \ldots, t_n) dx < \frac{c}{2} \frac{\int_{S(x^0, r_1)} F(x, d, \ldots, d) dx}{\sum_{i=1}^{n} \prod_{j=1, j\ne i}^{n} p_j(d\sigma_i)^{p_i}}$$
$$\le \frac{c}{2} \frac{\int_{\Omega} F(x, w_1(x), \ldots, w_n(x)) dx}{k \sum_{i=1}^{n} \prod_{j=1, j\ne i}^{n} p_j \|w_i\|_{p_i}^{p_i}}$$
$$= \left(r \prod_{i=1}^{n} p_i\right) \frac{\int_{\Omega} F(x, w_1(x), \ldots, w_n(x)) dx}{2 \sum_{i=1}^{n} \prod_{j=1, j\ne i}^{n} p_j \|w_i\|_{p_i}^{p_i}},$$

i.e., (5.3.12) in Theorem 5.3.1 holds.

Moreover, it is easy to check that (5.3.29) implies that (5.3.13) in Theorem 5.3.1 holds. Finally, (E18) and (E20) guarantee that (5.3.14) in Theorem 5.3.1 holds. Thus, all the conditions of Theorem 5.3.1 are satisfied. The conclusion then follows from Theorem 5.3.1. $\qquad\square$

5.4 An Elliptic Problem with a $p(x)$-Biharmonic Operator

Differential equations and variational problems with nonstandard $p(x)$-growth conditions have many applications in mathematical physics such as in the modeling of electrorheological fluids and of other phenomena related to image processing, elasticity, and the flow in porous media ([243, 301]). Such problems have been studied by many authors in the literature. The reader is referred to [27, 28, 111, 115–117, 188, 194, 195, 229, 230] for some recent work on this subject. It is well-known that problems with $p(x)$-growth conditions possess more complicated nonlinearities than the constant case. For instance, it is not homogeneous, and thus many techniques which can be applied when $p(x)$ is a positive constant fail to work in this new setting.

In this section, we are concerned with the existence of weak solutions of the following fourth order nonlinear elliptic equation with a $p(x)$-biharmonic operator

$$\begin{cases} \Delta^2_{p(x)}u + a(x)|u|^{p(x)-2}u = \lambda w(x)f(u) & \text{in } \Omega, \\ u = \Delta u = 0 & \text{on } \partial\Omega, \end{cases} \tag{5.4.1}$$

where $\Omega \subset \mathbb{R}^N$ with $N \geq 1$ is a bounded domain with smooth boundary, $p \in C(\overline{\Omega})$ with $p(x) > 1$ on $\overline{\Omega}$, $\Delta^2_{p(x)}u = \Delta(|\Delta u|^{p(x)-2}\Delta u)$ is the so-called $p(x)$-biharmonic operator, $\lambda > 0$ is a parameter, $a \in C(\overline{\Omega})$ is nonnegative, $f \in C(\mathbb{R})$, and $w \in L^{r(x)}(\Omega)$ for some $r \in C(\overline{\Omega})$.

Several variations of problem (5.4.1) has been studied in the literature. For instance, Ayoujil and El Amrouss [27, 28] studied the problem

$$\begin{cases} \Delta^2_{p(x)}u = \lambda|u|^{q(x)-2}u & \text{in } \Omega, \\ u = \Delta u = 0 & \text{on } \partial\Omega. \end{cases} \tag{5.4.2}$$

In [27], the case $p(x) = q(x)$ was considered. By the Lusternik-Schnirelmann principle on C^1-manifolds, the existence of a sequence of eigenvalues was proved. Let Λ be the set of all nonnegative eigenvalues. It was shown that $\sup \Lambda = \infty$. Sufficient conditions were also found to guarantee that $\inf \Lambda = 0$. We comment that when $p(x) = p > 1$ (a positive constant), we always have inf $\Lambda > 0$. In [28], using the Mountain Pass Theorem and Ekeland's variational principle, several existence criteria for eigenvalues were established for problem (5.4.2) when $p(x) \neq q(x)$. The existence of three weak solutions of the problem

$$\begin{cases} \Delta^2_{p(x)} u + a(x)|u|^{p(x)-2}u = f(x,u) + \lambda g(x,u), & \text{in } \Omega, \\ u = \Delta u = 0, & \text{on } \partial\Omega, \end{cases}$$

was investigated in [111] by using Ricceri's variational principle ([115, 247]). In [194], the existence of weak solutions was obtained for the problem

$$\begin{cases} \Delta^2_{p(x)} u + a(x)|u|^{p(x)-2}u = \lambda \left(b(x)u^{\gamma(x)-1} - c(x)u^{\beta(x)-1} \right), & \text{in } \Omega, \\ u = \Delta u = 0, & \text{on } \partial\Omega, \end{cases}$$

$$(5.4.3)$$

when λ is large by applying variational arguments.

We also want to point out that when $p(x)$ is a positive constant, a number of variations of problem (5.4.2) have been investigated in the literature. See, for example, [34, 129, 139, 212, 215] and the references therein.

In this section, by using simple variational arguments based on Ekeland's variational principle and the theory of the generalized Lebesgue–Sobolev spaces, we study the existence of a continuous family of eigenvalues for problem (5.4.1) in a neighborhood of the origin. More precisely, under some appropriate conditions, we show that there exists $\overline{\lambda} > 0$ such that any $\lambda \in (0, \overline{\lambda})$ is an eigenvalue of problem (5.4.1). For more applications of Ekeland's variational principle to other problems, see, for example, [28, 188, 230]. The result in this section is partly motivated by these nice papers.

Below, we recall some definitions and basic properties of variable spaces $L^{p(x)}(\Omega)$ and $W^{k,p(x)}(\Omega)$, where Ω is given as in problem (5.4.1). The presentation here can be found in, for example, [27, 110, 111, 116, 117, 196, 288].

Let

$$C_+(\overline{\Omega}) = \left\{ h \ : \ h \in C(\overline{\Omega}) \text{ and } h(x) > 1 \text{ on } \overline{\Omega} \right\}.$$

Throughout this section, for any $h \in C(\overline{\Omega})$, we use the notations

$$h^+ := \max_{x \in \overline{\Omega}} h(x) \quad \text{and} \quad h^- := \min_{x \in \overline{\Omega}} h(x).$$

Let $p \in C_+(\overline{\Omega})$ be fixed. We define the variable exponent Lebesgue space

$$L^{p(x)}(\Omega)$$

$$= \left\{ u : u \text{ is a measurable real-valued function with } \int_\Omega |u(x)|^{p(x)} dx < \infty \right\}.$$

Then, equipped with the so-called Luxemburg norm

$$|u|_{p(x)} = \inf\left\{\lambda > 0 \ : \ \int_\Omega \left|\frac{u}{\lambda}\right|^{p(x)} dx \leq 1\right\},$$

$L^{p(x)}(\Omega)$ is a separable and reflexive Banach space. It is clear that, when $p(x) = p > 1$ (a positive constant), the space $L^{p(x)}(\Omega)$ becomes the well-known Lebesgue space $L^p(\Omega)$ and the norm $|u|_{p(x)}$ reduces to the stand norm $|u|_p = \left(\int_\Omega |u|^p\right)^{1/p}$ in $L^p(\Omega)$.

As in the constant exponent case, for any positive integer k, let

$$W^{k,p(x)}(\Omega) = \left\{u \in L^{p(x)}(\Omega) \ : \ D^\alpha u \in L^{p(x)}(\Omega), \ |\alpha| \leq k\right\},$$

where $\alpha = (\alpha_1, \ldots, \alpha_N)$ is a multi-index, $|\alpha| = \sum_{i=1}^N \alpha_i$, and $D^\alpha u = \frac{\partial^{|\alpha|} u}{\partial^{\alpha_1} x_1 \ldots \partial^{\alpha_N} x_N}$. Then, equipped with the norm

$$||u||_{k,p(x)} = \sum_{|\alpha| \leq k} |D^\alpha u|_{p(x)},$$

$W^{k,p(x)}(\Omega)$ is also a separable and reflexive Banach space. We denote by $W_0^{k,p(x)}(\Omega)$ the closure of $C_0^\infty(\Omega)$ in $W^{k,p(x)}(\Omega)$.

In the sequel, we let

$$X = W_0^{1,p(x)}(\Omega) \cap W^{2,p(x)}(\Omega).$$

Define a norm $|| \cdot ||_X$ of X by

$$||u||_X = ||u||_{1,p(x)} + ||u||_{2,p(x)}.$$

Then, endowed with $|| \cdot ||_X$, X is a separable and reflexive Banach space. Moreover, by [288, Theorem 4.4], $||u||$ and $|\Delta u|_{p(x)}$ are two equivalent norms of X.

Let

$$||u||_a = \inf\left\{\lambda > 0 \ : \ \int_\Omega \left(\left|\frac{\Delta u}{\lambda}\right|^{p(x)} + a(x)\left|\frac{u}{\lambda}\right|^{p(x)}\right) dx \leq 1\right\} \quad \text{for } u \in X.$$

In view of $a^- \geq 0$, it is easy to see that $||u||_a$ is equivalent to the norms $||u||$ and $|\Delta u|_{p(x)}$ in X. In this section, for the convenience of discussion, we use the norm $||u||_a$ for X.

Proposition 5.4.1. ([111, Proposition 2.3]) *Let*

$$\rho_a(u) = \int_\Omega \left(|\Delta u|^{p(x)} + a(x)|u|^{p(x)}\right) dx \quad for \quad u \in X.$$

Then, we have

(a) *if* $||u||_a \geq 1$, *then* $||u||_a^{p^-} \leq \rho_a(u) \leq ||u||_a^{p^+}$;

(b) *if* $||u||_a \leq 1$, *then* $||u||_a^{p^+} \leq \rho_a(u) \leq ||u||_a^{p^-}$.

Proposition 5.4.2. ([116, Propositions 2.4 and 2.5] or [196, Theorem 2.1 and Corollary 2.7]) *The conjugate space of* $L^{p(x)}(\Omega)$ *is* $L^{q(x)}(\Omega)$, *where* $1/p(x) + 1/q(x) = 1$. *Moreover, for* $u \in L^{p(x)}(\Omega)$ *and* $v \in L^{q(x)}(\Omega)$, *we have the following inequality of Hölder type*

$$\left| \int_\Omega uv dx \right| \leq \left(\frac{1}{p^-} + \frac{1}{q^-} \right) |u|_{p(x)} |v|_{q(x)} \leq 2|u|_{p(x)} |v|_{q(x)}.$$

Moreover, if $h_i \in C_+(\overline{\Omega})$ *with* $1/h_1(x) + 1/h_2(x) + 1/h_3(x) = 1$, *then, for* $u \in L^{h_1(x)}(\Omega)$, $v \in L^{h_2(x)}(\Omega)$, *and* $w \in L^{h_3(x)}(\Omega)$, *we have*

$$\left| \int_\Omega uvw dx \right| \leq \left(\frac{1}{h_1^-} + \frac{1}{h_2^-} + \frac{1}{h_3^-} \right) |u|_{h_1(x)} |v|_{h_2(x)} |w|_{h_3(x)}$$

$$\leq 3|u|_{h_1(x)} |v|_{h_2(x)} |w|_{h_3(x)}.$$

Proposition 5.4.3. ([110, Lemma 2.1]) *Let* $q \in L^\infty(\Omega)$ *be such that* $1 \leq p(x)q(x) \leq \infty$ *for a.e.* $x \in \Omega$. *Let* $u \in L^{p(x)}(\Omega)$, $u \neq 0$. *Then, we have*

(a) *if* $|u|_{p(x)q(x)} \leq 1$, *then* $|u|_{p(x)q(x)}^{q^+} \leq ||u|^{q(x)}|_{p(x)} \leq |u|_{p(x)q(x)}^{q^-}$;

(b) *if* $|u|_{p(x)q(x)} \geq 1$, *then* $|u|_{p(x)q(x)}^{q^-} \leq ||u|^{q(x)}|_{p(x)} \leq |u|_{p(x)q(x)}^{q^+}$.

For any $x \in \overline{\Omega}$, let

$$p^*(x) = \begin{cases} \frac{Np(x)}{N - 2p(x)} & \text{if } p(x) < \frac{N}{2}, \\ \infty & \text{if } p(x) \geq \frac{N}{2}. \end{cases}$$

Proposition 5.4.4. ([27, Theorem 3.2]) *Assume that* $q \in C_+(\overline{\Omega})$ *satisfy* $q(x) < p^*(x)$ *on* $\overline{\Omega}$. *Then, there exists a continuous and compact embedding* $X \hookrightarrow L^{q(x)}(\Omega)$.

In this section, we need the following assumptions.

(E21) There exist $k_1 \geq k_2 > 0$, $0 < \delta < 1$ and $q_1, q_2, r \in C_+(\overline{\Omega})$ such that

$$1 < q_1(x) \leq q_2(x) < p(x) \leq \frac{N}{2} < r(x) \quad \text{on } \overline{\Omega} \tag{5.4.4}$$

$$0 \leq tf(t) \leq k_1 |t|^{q_1(x)} \quad \text{for } t \in \mathbb{R}, \tag{5.4.5}$$

and

$$tf(t) \geq k_2 |t|^{q_2(x)} \quad \text{for } t \in [-\delta, \delta]. \tag{5.4.6}$$

(E22) $w \in L^{r(x)}(\Omega)$ and there exists a subset $\Omega_1 \subset \Omega$ with meas $(\Omega_1) > 0$ such that $w(x) > 0$ for $x \in \Omega_1$, where meas $(\cdot)$ denotes the Lebesgue measure of a set.

Remark 5.4.1. Regarding the condition (E21), we make the following comments.

(a) Let

$$r'(x) = \frac{r(x)}{r(x) - 1} \quad \text{and} \quad s_i(x) = \frac{r(x)q_i(x)}{r(x) - q_i(x)}, \quad i = 1, 2.$$

Then, $1/r(x) + 1/r'(x) = 1$ and it is easy to check that (2.1) implies

$$r'(x)q_i(x) < p^*(x) \quad \text{and} \quad s_i(x) < p^*(x) \quad \text{for } i = 1, 2 \text{ and } x \in \overline{\Omega}.$$

Thus, by Proposition 5.4.4, the embeddings $X \hookrightarrow L^{r'(x)q_i(x)}(\Omega)$ and $X \hookrightarrow L^{s_i(x)}(\Omega)$, $i = 1, 2$, are continuous and compact.

(b) There are many functions f satisfying both (5.4.5) and (5.4.6). For instance, it is easy to check that the following are several simple examples of such functions $f(t)$:

$$f(t) = |t|^{q(x)-2}t, \quad t \in \mathbb{R},$$

or

$$f(t) = t|\sin t^{q(x)-2}|, \quad t \in \mathbb{R},$$

or

$$f(t) = |t|^{\gamma(x)-2}t - |t|^{\beta(x)-2}t, \quad t \in \mathbb{R},$$

where $q, \beta, \gamma \in C_+(\overline{\Omega})$ satisfy $q(x) < p(x)$ and $\gamma(x) < \beta(x) < p(x)$ on $\overline{\Omega}$.

We say that $\lambda \in \mathbb{R}$ is an *eigenvalue* of problem (5.4.1) if there exists $u \in X \setminus \{0\}$ such that

$$\int_{\Omega} |\Delta u|^{p(x)-2} \Delta u \Delta v \, dx + \int_{\Omega} a(x)|u|^{p(x)-2}uv \, dx - \lambda \int_{\Omega} w(x)f(u)v \, dx = 0$$

for all $v \in X$. When λ is an eigenvalue of problem (5.4.1), the corresponding function $u \in X \setminus \{0\}$ is a *weak solution* of problem (5.4.1).

In the rest of this section, we assume that (E21) and (E22) hold. Define functionals $\Phi, \Psi, I_\lambda : X \to R$ by

$$\Phi(u) = \int_{\Omega} \frac{1}{p(x)} \left(|\Delta u|^{p(x)} + a(x)|u|^{p(x)} \right) dx,$$

$$\Psi(u) = \int_\Omega w(x)F(u)dx,$$

and

$$I_\lambda(u) = \Phi(u) - \lambda\Psi(u),$$

where $F(t) = \int_0^t f(s)ds$.

Lemma 5.4.1. *We have the following:*

    (a) Φ *is weakly lower semicontinuous,* $\Phi \in C^1(X,\mathbb{R})$, *and*

$$\langle\Phi'(u), v\rangle = \int_\Omega |\Delta u|^{p(x)-2}\Delta u\Delta v dx + \int_\Omega a(x)|u|^{p(x)-2}uvdx$$

    for all $u, v \in X$;

    (b) $\Phi'(u) : X \to X^*$ *is of type* (S_+), *i.e., if* $u_n \rightharpoonup u$ *and* $\liminf_{n\to\infty}\langle\Phi'(u_n), u_n - u\rangle \leq 0$, *then* $u_n \to u$, *where* X^* *is the dual space of* X, $\to$ *and* $\rightharpoonup$ *denote the strong and weak convergence, respectively.*

    (c) Ψ *is weakly lower semicontinuous,* $\Psi \in C^1(X,\mathbb{R})$, *and*

$$\langle\Psi'(u), v\rangle = \int_\Omega w(x)f(u)vdx$$

    for all $u, v \in X$.

Proof. Parts (a) and (b) follows from [111, Proposition 2.5]. Part (c) can be proved by a standard argument, and hence the details are omitted. $\square$

Remark 5.4.2. By Lemma 5.4.1 (a) and (c), I_λ is (weakly) lower semicontinuous, $I_\lambda \in C^1(X,\mathbb{R})$, and

$$\langle I_\lambda'(u), v\rangle = \int_\Omega |\Delta u|^{p(x)-2}\Delta u\Delta v dx$$
$$+ \int_\Omega a(x)|u|^{p(x)-2}uvdx - \lambda\int_\Omega w(x)f(u)vdx$$

for all $v \in X$. Thus, u is a critical point of I_λ if and only if u is weak solution of problem (5.4.1).

Note from Remark 5.4.1 (a) that the embedding $X \hookrightarrow L^{r'(x)q_1(x)}(\Omega)$ is continuous. Then, there exists a constant $C > 1$ such that

$$|u|_{r'(x)q_1(x)} \leq C\|u\|_a \quad \text{for any } u \in X. \tag{5.4.7}$$

Lemma 5.4.2. *For any* $\varrho \in (0, 1/C)$, *there exists* $\overline{\lambda} > 0$ *and* $\kappa > 0$ *such that* $I_\lambda(u) \geq \kappa$ *for any* $\lambda \in (0, \overline{\lambda})$ *and* $u \in X$ *with* $\|u\|_a = \varrho$,

Proof. Let $\varrho \in (0, 1/C)$ be fixed. Then, $\varrho < 1$, and from (5.4.7), it is clear that

$$|u|_{r'(x)q_1(x)} < 1 \quad \text{for any } u \in X \text{ and } ||u||_a = \varrho.$$

By (5.4.5), we see that

$$0 \le F(t) \le \frac{1}{q_1(x)} k_1 |t|^{q_1(x)} \quad \text{for } t \in \mathbb{R}.$$

Thus, for $u \in X$ with $||u||_a = \varrho$, from Propositions 5.4.1–5.4.3 and (5.4.7), it follows that

$$
\begin{aligned}
I_\lambda(u) &\ge \frac{1}{p^+} \int_\Omega \left(|\Delta u|^{p(x)} + a(x)|u|^{p(x)} \right) dx - \lambda \int_\Omega |w(x)| F(u) dx \\
&\ge \frac{1}{p^+} ||u||_a^{p^+} - \frac{1}{q_1^-} \lambda k_1 \int_\Omega |w(x)| \, |u|^{q_1(x)} dx \\
&\ge \frac{1}{p^+} ||u||_a^{p^+} - \frac{2}{q_1^-} \lambda k_1 |w|_{r(x)} \left| |u|^{q_1(x)} \right|_{r'(x)} \\
&\ge \frac{1}{p^+} ||u||_a^{p^+} - \frac{2}{q_1^-} \lambda k_1 |w|_{r(x)} |u|^{q_1^-}_{r'(x)q(x)} \\
&\ge \frac{1}{p^+} ||u||_a^{p^+} - \frac{2}{q_1^-} \lambda k_1 C^{q_1^-} |w|_{r(x)} ||u||_a^{q_1^-} \\
&= \frac{1}{p^+} \varrho^{p^+} - \frac{2}{q_1^-} \lambda k_1 C^{q_1^-} |w|_{r(x)} \varrho^{q_1^-} \\
&= \varrho^{q_1^-} \left(\frac{1}{p^+} \varrho^{p^+ - q_1^-} - \frac{2}{q_1^-} \lambda k_1 C^{q_1^-} |w|_{r(x)} \right). \quad (5.4.8)
\end{aligned}
$$

Hence, if we let

$$\overline{\lambda} = \frac{\varrho^{p^+ - q_1^-}}{2\varrho^+} \frac{q_1^-}{2k_1 C^{q_1^-} |w|_{r(x)}}, \quad (5.4.9)$$

then, for any $\lambda \in (0, \overline{\lambda})$ and $u \in X$ with $||u||_a = \varrho$, there exists $\kappa = \frac{\varrho^{p^+}}{2p^+} > 0$ such that $I_\lambda(u) \ge \kappa$. This completes the proof of the lemma. $\square$

Lemma 5.4.3. *There exists $\phi \in X$ such that $\phi \ge 0$, $\phi \ne 0$, and $I_\lambda(t\phi) < 0$ for $t > 0$ small enough.*

Proof. Let $\Omega_1 \subset \Omega$ be given in (E22). Then, (ch5.4-2.1) implies that $q_2(x) < p(x)$ on $\overline{\Omega}_1$. If we let $\hat{q} = \min_{x \in \overline{\Omega}_1} q_2(x)$ and $\hat{p} = \min_{x \in \overline{\Omega}_1} p(x)$, then there exists $\epsilon_0 > 0$ such that $\hat{q} + \epsilon_0 < \hat{p}$. Moreover, since $q_2 \in C(\overline{\Omega}_1)$, there exists an open set $\Omega_2 \subset \Omega_1$ such that $\text{meas}(\Omega_2) > 0$ and $|q_2(x) - \hat{q}| < \epsilon_0$ for $x \in \Omega_2$. Thus, $q_2(x) < \hat{q} + \epsilon_0 < \hat{p}$ in Ω_2.

Let $\phi \in C_0^\infty(\Omega)$ be nontrivial such that supp $(\phi) \subset \Omega_2 \subset \Omega_1$, $\phi \geq 0$, and $\phi \neq 0$ in Ω_2. Note from (5.4.6) that

$$F(t) \geq \frac{1}{q_2(x)} k_2 |t|^{q_2(x)} \quad \text{for } t \in [-\delta, \delta].$$

Then, for $0 < t < \min \left\{ 1, \frac{\delta}{\max_{x \in \Omega_2} \phi(x)} \right\}$, we have

$$
\begin{aligned}
I_\lambda(t\phi) &= \int_\Omega \frac{1}{p(x)} \left(|\Delta(t\phi)|^{p(x)} + a(x)|t\phi|^{p(x)} \right) dx - \lambda \int_\Omega w(x) F(t\phi) dx \\
&= \int_{\Omega_2} \frac{t^{p(x)}}{p(x)} \left(|\Delta\phi|^{p(x)} + a(x)|\phi|^{p(x)} \right) dx - \lambda \int_{\Omega_2} w(x) F(t\phi) dx \\
&\leq \frac{t^{\hat{p}}}{p^-} \int_{\Omega_2} \left(|\Delta\phi|^{p(x)} + a(x)|\phi|^{p(x)} \right) dx - \lambda k_2 \int_{\Omega_2} \frac{t^{q_2(x)}}{q_2(x)} w(x)|\phi|^{q_2(x)} dx \\
&\leq \frac{t^{\hat{p}}}{p^-} \int_{\Omega_2} \left(|\Delta\phi|^{p(x)} + a(x)|\phi|^{p(x)} \right) dx - \frac{\lambda k_2 t^{\hat{q}+\epsilon_0}}{q_2^+} \int_{\Omega_2} w(x)|\phi|^{q_2(x)} dx.
\end{aligned}
$$

Hence, $I_\lambda(t\phi) < 0$ for $0 < t < \eta^{1/(\hat{p}-\hat{q}-\epsilon_0)}$, where

$$0 < \eta < \min \left\{ 1, \frac{\delta}{\max_{x \in \Omega_2} \phi(x)}, \frac{\lambda k_2 p^- \int_{\Omega_2} w(x)|\phi|^{q_2(x)} dx}{q_2^+ \int_{\Omega_2} \left(|\Delta\phi|^{p(x)} + a(x)|\phi|^{p(x)} \right) dx} \right\}.$$

Here, we point out that

$$\int_{\Omega_2} \left(|\Delta\phi|^{p(x)} + a(x)|\phi|^{p(x)} \right) dx > 0.$$

In fact, if this is not true, then

$$\int_\Omega \left(|\Delta\phi|^{p(x)} + a(x)|\phi|^{p(x)} \right) dx = 0.$$

By Proposition 5.4.1, we have $\|\phi\|_a = 0$, and so $\phi \equiv 0$ in Ω, which is a contradiction. This completes the proof of the lemma. $\square$

We now state the main theorem in this section.

Theorem 5.4.1. *Assume that (E21) and (E22) hold. Then, there exists $\overline{\lambda} > 0$ such that any $\lambda \in (0, \overline{\lambda})$ is an eigenvalue of problem (5.4.1).*

Remark 5.4.3. We make the following comments.

(a) By Remark 5.4.1 (b), we see that Theorem 5.4.1 can be applied to problems (5.4.2) and (5.4.3) with $b(x) = c(x)$ in Ω as well as to some other problems.

(b) In [194], it was proved that if $p^+ < p^*(x)$ for all $x \in \overline{\Omega}$, $b(x) = c(x)$ in Ω, and the following condition holds:

(E23) $a^- \geq 0$, $b^- > 0$, and $1 < \beta^- \leq \beta^+ < \gamma^- \leq \gamma^+ < p^-$.

Then, there exists $\lambda^* > 0$ such that, for all $\lambda > \lambda^*$, problem (5.4.3) has at least one nontrivial weak solution $u(x)$.

Clearly, this result complements the above Theorem 5.4.1.

Proof. Let $\overline{\lambda}$ be defined by (5.4.9) and $\lambda \in (0, \overline{\lambda})$. By Lemma 5.4.2, we have

$$\inf_{\partial B_\varrho(0)} I_\lambda > 0, \tag{5.4.10}$$

where $B_\varrho(0)$ is the ball in X centered at the origin and of radius ϱ, and $\partial B_\varrho(0)$ is the boundary of $B_\varrho(0)$. For any $u \in B_\varrho(0)$, by an argument similar to those used in obtaining (5.4.8), we can derive that

$$I_\lambda(u) \geq \frac{1}{p^+} ||u||_a^{p^+} - \frac{2}{q_1^-} \lambda k_1 C^{q_1^-} |w|_{r(x)} ||u||_a^{q_1^-}. \tag{5.4.11}$$

Note from Lemma 5.4.3 that there exists $\phi \in X$ such that $I_\lambda(t\phi) < 0$ for $t > 0$ small enough. Then, from (5.4.10) and (5.4.11), it follows that

$$-\infty < \underline{c}_\lambda := \inf_{B_\varrho(0)} I_\lambda < 0.$$

Let

$$0 < \epsilon < \inf_{\partial B_\varrho(0)} I_\lambda - \underline{c}_\lambda.$$

Applying Theorem 1.1.17 to the functional $I_\lambda : \overline{B_\varrho(0)} \to \mathbb{R}$, we see that there exists $u_\epsilon \in \overline{B_\varrho(0)}$ such that

$$\underline{c}_\lambda \leq I_\lambda(u_\epsilon) \leq \underline{c}_\lambda + \epsilon \tag{5.4.12}$$

and

$$I_\lambda(u_\epsilon) < I_\lambda(u) + \epsilon ||u - u_\epsilon||_a \quad \text{for } u \neq u_\epsilon. \tag{5.4.13}$$

Since $I_\lambda(u_\epsilon) \leq \underline{c}_\lambda + \epsilon < \inf_{\partial B_\varrho(0)} I_\lambda$, we have $u_\epsilon \in B_\varrho(0)$.

Now, define a functional $J_\lambda : \overline{B_\varrho(0)} \to \mathbb{R}$ by

$$J_\lambda(u) = I_\lambda(u) + \epsilon ||u - u_\epsilon||_a.$$

Obviously, by (5.4.13), u_ϵ is a minimum point of J_λ, and so

$$\frac{J_\lambda(u_\epsilon + tv) - J_\lambda(u_\epsilon)}{t} \geq 0$$

for $t > 0$ small enough and all $v \in B_\varrho(0)$. Then,
$$\frac{I_\lambda(u_\epsilon + tv) - I_\lambda(u_\epsilon)}{t} + \epsilon\|v\| \geq 0$$
for $t > 0$ small enough and all $v \in B_\varrho(0)$. Letting $t \to 0$, we obtain
$$\langle I_\lambda'(u_\epsilon), v \rangle + \epsilon\|v\| \leq 0 \quad \text{for all } v \in B_\varrho(0).$$
Hence, $\|I_\lambda'(u_\epsilon)\|_{X^*} \leq \epsilon$. This, together with (5.4.12), implies that there exists a sequence $\{u_n\} \subset B_\varrho(0)$ such that
$$I_\lambda(u_n) \to \underline{c}_\lambda \quad \text{and} \quad I_\lambda'(u_n) \to 0. \tag{5.4.14}$$
Obviously, $\{u_n\}$ is bounded in X. Then, by the reflexivity of X, there exists $u_0 \in X$ such that, up to a subsequence, $u_n \rightharpoonup u_0$ in X. Note that
$$|\langle I_\lambda'(u_n), u_n - u_0 \rangle| \leq |\langle I_\lambda'(u_n), u_n \rangle| + |\langle I_\lambda'(u_n), u_0 \rangle|$$
$$\leq \|\langle I_\lambda'(u_n) \rangle\| \, \|u_n\| + \|\langle I_\lambda'(u_n) \rangle\| \, \|u_0\|.$$
Then, from (5.4.14) and the fact that $\{u_n\}$ is bounded in X, it follows that
$$\lim_{n \to \infty} \langle I_\lambda'(u_n), u_n - u_0 \rangle = 0. \tag{5.4.15}$$
Now, we claim that
$$\lim_{n \to \infty} \langle \Psi'(u_n), u_n - u_0 \rangle = 0. \tag{5.4.16}$$
In fact, from (5.4.5) and Propositions 5.4.2 and 5.4.3, we have
$$|\langle \Psi'(u_n), u_n - u_0 \rangle| = \left| \int_\Omega w(x) f(u_n)(u_n - u_0) dx \right|$$
$$\leq \int_\Omega |w(x)| \, |f(u_n)| \, |(u_n - u_0)| dx$$
$$\leq k_1 \int_\Omega |w(x)| \, |u_n|^{q_1(x)-1} \, |(u_n - u_0)| dx$$
$$\leq 3k_1 |w|_{r(x)} \left| |u_n|^{q_1(x)-1} \right|_{\frac{q_1(x)}{q_1(x)-1}} |u_n - u_0|_{s_1(x)}$$
$$\leq 3k_1 |w|_{r(x)} \max \left\{ |u_n|_{q_1(x)}^{q_1^+ - 1}, \; |u_n|_{q_1(x)}^{q_1^- - 1} \right\} |u_n - u_0|_{s_1(x)},$$
where $s_1(x)$ is defined in Remark 5.4.1 (a). Then, by the continuous and compact embedding of $X \hookrightarrow L^{q_1(x)}(\Omega)$ and $X \hookrightarrow L^{s_1(x)}(\Omega)$, we see that (5.4.16) holds. Now, from (5.4.15) and (5.4.16), we conclude that
$$\lim_{n \to \infty} \langle \Phi'(u_n), u_n - u_0 \rangle = 0.$$
Thus, by Lemma 5.4.1 (b), we have $u_n \to u_0$ in X. Hence, from (5.4.14), we see that
$$I_\lambda(u_0) = \underline{c}_\lambda < 0 \quad \text{and} \quad I_\lambda'(u_0) = 0.$$
Therefore, u_0 is a nontrivial weak solution of problem (5.4.1), and so any $\lambda \in (0, \overline{\lambda})$ is an eigenvalue of problem (5.4.1). This completes the proof of the theorem. $\qquad \square$

5.5 Elliptic Systems of $(p_1, \ldots, p_n)$-Kirchhoff Type

Throughout this section, we let $\Omega \subset \mathbb{R}^N$ ($N \geq 1$) be a non-empty bounded open set with a smooth boundary $\partial\Omega$, $K_i : [0, \infty) \to \mathbb{R}$, for $1 \leq i \leq n$, be n continuous functions such that there exist $2n$ positive numbers m_i and M_i, with $m_i \leq K_i(t) \leq M_i$, for all $t \geq 0$ and for $1 \leq i \leq n$, $a_i \in L^\infty(\Omega)$ with essinf$_\Omega a_i(x) \geq 0$, and $p_i > N$, for $1 \leq i \leq n$.

Consider the following nonlocal elliptic Kirchhoff type system

$$\begin{cases} -\Big[K_i(\int_\Omega(|\nabla u_i(x)|^{p_i} + a_i(x)|u_i(x)|^{p_i})dx)\Big]^{p_i-1} \\ \quad \times \Big(\text{div}(|\nabla u_i|^{p_i-2}\nabla u_i) + a_i(x)|u_i|^{p_i-2}u\Big) = \lambda F_{u_i}(x, u_1, \ldots, u_n), \text{ in } \Omega, \\ u_i = 0, \qquad\qquad\qquad\qquad\qquad\qquad\qquad\qquad\qquad\qquad\quad \text{on } \partial\Omega, \end{cases}$$
$$(5.5.1)$$

for $1 \leq i \leq n$, where λ is a positive parameter and $F : \Omega \times \mathbb{R}^n \to \mathbb{R}$ is a function such that the mapping $(t_1, t_2, \ldots, t_n) \to F(x, t_1, t_2, \ldots, t_n)$ is in C^1 in $\mathbb{R}^n$ for all $x \in \Omega$, F_{t_i} is continuous in $\Omega \times \mathbb{R}^n$, for $i = 1, \ldots, n$, and $F(x, 0, \ldots, 0) = 0$ for all $x \in \Omega$. Here, F_{t_i} denotes the partial derivative of F with respect to t_i.

The system (5.5.1) is related to a model given by the equation of elastic strings

$$\rho \frac{\partial^2 u}{\partial t^2} - \left(\frac{P_0}{h} + \frac{E}{2L}\int_0^L |\frac{\partial u}{\partial x}|^2 dx\right) \frac{\partial^2 u}{\partial^2 x} = 0, \qquad (5.5.2)$$

where ρ is the mass density, P_0 is the initial tension, h is the area of the cross-section, E is the Young modulus of the material, and L is the length of the string, was proposed by Kirchhoff [190] as a extension of the classical D'Alembert's wave equation for free vibrations of elastic strings. Kirchhoff's model takes into account the changes in length of the string produced by transverse vibrations. Similar nonlocal problems also model several physical and biological systems where u describes a process that depends on the average of itself, for example, the population density. Later, the equation (5.5.2) was extended to the equation

$$\frac{\partial^2 u}{\partial t^2} - K(\int_\Omega |\nabla u(x)|^2 dx)\Delta u = g(x, u) \quad \text{in } \Omega,$$

where $\Omega \subset \mathbb{R}^N$ ($N \geq 1$) is a non-empty bounded open set with a given $\partial\Omega$ and $K : [0, \infty) \to \mathbb{R}$ is a continuous function.

Denote by $W_0^{1,p_i}(\Omega)$ the closure of $C_0^\infty(\Omega)$ with respect to the norm

$$||u_i||_{p_i} = \left(\int_\Omega |\nabla u_i(x)|^{p_i} dx\right)^{1/p_i} \quad \text{for } 1 \leq i \leq n.$$

Define

$$c_i := \max \left\{ \sup_{u_i \in W_0^{1,p_i}(\Omega) \setminus \{0\}} \frac{\max_{x \in \overline{\Omega}} |u_i(x)|}{\|u_i\|_{p_i}} \quad 1 \le i \le n \right\}.$$

Since $p_i > N$ for $1 \le i \le n$, one has $c_i < \infty$. Moreover, from [259, formula (6b)], we have

$$\sup_{u \in W_0^{1,p_i}(\Omega) \setminus \{0\}} \frac{\max_{x \in \overline{\Omega}} |u_i(x)|}{\|u_i\|_{p_i}}$$

$$\le \frac{N^{-1/p_i}}{\sqrt{\pi}} \left[\Gamma\left(1 + \frac{N}{2}\right) \right]^{1/N} \left(\frac{p_i - 1}{p_i - N} \right)^{1 - 1/p_i} |\Omega|^{1/N - 1/p_i}$$

for $1 \le i \le n$, where $|\Omega|$ is the Lebesgue measure of the set Ω, and equality occurs when Ω is a ball.

Let X be the Cartesian product of the n Sobolev spaces $W_0^{1,p_1}(\Omega)$, ..., $W_0^{1,p_n}(\Omega)$, i.e., $X = \prod_{i=1}^n W_0^{1,p_i}(\Omega)$ equipped with the norm

$$\| (u_1, u_2, \ldots, u_n) \| = \sum_{i=1}^n \|u_i\|_*,$$

where

$$\|u_i\|_* = \left(\int_\Omega (|\nabla u_i(x)|^{p_i} + a_i(x)|u_i(x)|^{p_i}) dx \right)^{1/p_i}$$

is a norm in $W_0^{1,p_i}(\Omega)$ that is equivalent to the usual one. Let

$$C := \max \left\{ \sup_{u_i \in W_0^{1,p_i}(\Omega) \setminus \{0\}} \frac{\max_{x \in \overline{\Omega}} |u_i(x)|^{p_i}}{\|u_i\|_*^{p_i}}; \quad 1 \le i \le n \right\}. \qquad (5.5.3)$$

In the sequel, let $\underline{p} := \min\{p_i; \ 1 \le i \le n\}$, $\overline{p} := \max\{p_i; \ 1 \le i \le n\}$, and $\underline{m} := \min\{m_i; \ 1 \le i \le n\}$.

Following the construction given in [58], define

$$\sigma(p_i, N) := \inf_{\mu \in (0,1)} \frac{1 - \mu^N}{\mu^N (1 - \mu)^{p_i}},$$

and consider $\overline{\mu}_i \in (0,1)$ such that $\sigma(p_i, N) := \frac{1 - \overline{\mu}_i^N}{\overline{\mu}_i^N (1 - \overline{\mu}_i)^{p_i}}$. Let

$$\overline{\mu} := \max \overline{\mu}_i, \quad \underline{\mu} := \min \overline{\mu}_i,$$

and

$$\tau := \sup \operatorname{dist}(x, \partial\Omega).$$

Simple calculations show that there is an $x_0 \in \Omega$ such that $B(x_0, \tau) \subseteq \Omega$, where $B(x_0, s)$ denotes the ball with center at x_0 and radius of s. Further, put

$$g_{\overline{\mu}_i}(p_i, N) := \overline{\mu}_i^N + \frac{1}{(1 - \overline{\mu}_i)^{p_i}} N B_{(\overline{\mu}_i, 1)}(N, p_i + 1),$$

where $B_{(\overline{\mu}_i, 1)}(N, p_i + 1)$ denotes the generalized incomplete beta function defined as follows:

$$B_{(\overline{\mu}_i, 1)}(N, p_i + 1) := \int_{\overline{\mu}_i}^1 t^{N-1}(1 - t)^{(p_i+1)-1} dt.$$

We also denote by $\omega_\tau := \tau^N \frac{\pi^{\frac{N}{2}}}{\Gamma(1+\frac{N}{2})}$ the measure of the N-dimensional ball of radius τ.

Let

$$\upsilon := \max_{1 \le i \le n} \left\{ \frac{\sigma(p_i, N)}{\tau^{p_i}} + \|a_i\|_\infty \frac{g_{\mu_i}(p_i, N)}{\overline{\mu}_i^N} \right\}.$$

Corresponding to K_i we introduce the functions $\tilde{K}_i : [0, \infty) \to \mathbb{R}$ as follows

$$\tilde{K}_i(t) = \int_0^t K_i(s) ds \quad \text{for all } t \ge 0$$

for $1 \le i \le n$.

For all $\gamma > 0$ we denote by $Q(\gamma)$ the set

$$\left\{ (t_1, \ldots, t_n) \in \mathbb{R}^n : \sum_{i=1}^n |t_i| \le \gamma \right\}. \tag{5.5.4}$$

By a weak solution of system (5.5.1), we mean a vector $u = (u_1, \ldots, u_n) \in X$ such that

$$\sum_{i=1}^n \left(\left[K_i \left(\int_\Omega (|\nabla u_i(x)|^{p_i} + a_i(x)|u_i(x)|^{p_i}) dx \right) \right]^{p_i-1} \right.$$
$$\times \int_\Omega \left(|\nabla u_i(x)|^{p_i-2} \nabla u_i(x) \nabla v_i(x) + |u_i(x)|^{p_i-2} u_i(x) v_i(x) \right) dx \right)$$
$$- \lambda \int_\Omega \sum_{i=1}^n F_{u_i}(x, u_1(x), \ldots, u_n(x)) v_i(x) dx = 0$$

for every $v = (v_1, \ldots, v_n) \in X$.

Theorem 5.5.1. *Assume that*

(E23) $F(x, t_1, \ldots, t_n) \geq 0$, for each $(x, t_1, \ldots, t_n) \in \Omega \times \mathbb{R}_+^n$, where

$$\mathbb{R}_+^n = \{(t_1, \ldots, t_n) \in \mathbb{R}^n : t_i \geq 0, \ i = 1, \ldots, n\};$$

(E24) *the following inequality holds:*

$$\liminf_{\xi \to \infty} \frac{\int_\Omega \sup_{(t_1, \ldots, t_n) \in Q(\xi)} F(x, t_1, \ldots, t_n) dx}{\xi^{\underline{p}}}$$

$$< \frac{1}{\left(\sum_{i=1}^n (p_i \frac{C}{m})^{\frac{1}{p_i}} \right)^{\underline{p}}}$$

$$\times \limsup_{(t_1, \ldots, t_n) \to (\infty, \ldots, \infty)_{(t_1, \ldots, t_n) \in \mathbb{R}_+^n}} \frac{\int_{B(x_0, \mu\tau)} F(x, t_1, \ldots, t_n) dx}{\sum_{i=1}^n \frac{\tilde{K}_i(\overline{\mu}^N \omega_\tau v |t_i|^{p_i})}{p_i}}.$$

Then, for each $\lambda \in \Lambda$, *the system (5.5.1) has an unbounded sequence of weak solutions in* X, *where*

$$\Lambda = \left(\frac{1}{\limsup_{(t_1, \ldots, t_n) \to (\infty, \ldots, \infty)} \frac{\int_{B(x_0, \mu\tau)} F(x, t_1, \ldots, t_n) dx}{\sum_{i=1}^n \frac{\tilde{K}_i(\overline{\mu}^N \omega_\tau v |t_i|^{p_i})}{p_i}}}, \right.$$

$$\left. \frac{\frac{1}{\left(\sum_{i=1}^n (p_i \frac{C}{m})^{\frac{1}{p_i}} \right)^{\overline{p}}}}{\liminf_{\xi \to \infty} \frac{\int_\Omega \sup_{(t_1, \ldots, t_n) \in Q(\xi)} F(x, t_1, \ldots, t_n) dx}{\xi^{\underline{p}}}} \right).$$

Proof. In order to apply Theorem 1.1.5 to our problem, we introduce the functionals $\Phi, \Psi : X \to \mathbb{R}$, for each $u = (u_1, \ldots, u_n) \in X$, defined as follows

$$\Phi(u) = \sum_{i=1}^n \frac{\tilde{K}_i(\|u_i\|_*^{p_i})}{p_i}$$

and

$$\Psi(u) = \int_\Omega F(x, u_1(x), \ldots, u_n(x)) dx.$$

Let us prove that the functionals Φ and Ψ satisfy the required conditions. It is well known that Ψ is a differentiable functional whose differential at the point $u \in X$ is

$$\Psi'(u)(v) = \int_\Omega \sum_{i=1}^n F_{u_i}(x, u_1(x), \ldots, u_n(x)) v_i(x) dx,$$

for every $v = (v_1, \ldots, v_n) \in X$, as well as is sequentially weakly upper semicontinuous. Furthermore, $\Psi' : X \to X^*$ is a compact operator. Indeed, it is enough to show that Ψ' is strongly continuous on X. For this, for fixed $(u_1, \ldots, u_n) \in X$, let $(u_{1k}, \ldots, u_{nk}) \to (u_1, \ldots, u_n)$ weakly in X as $k \to \infty$. Then, we have $(u_{1k}, \ldots, u_{nk})$ converges uniformly to $(u_1, \ldots, u_n)$ on Ω as $k \to \infty$ (see [289]). Since $F(x, \cdot, \ldots, \cdot)$ is C^1 in $\mathbb{R}^n$ for every $x \in \Omega$, the derivatives of F are continuous in $\mathbb{R}^n$ for every $x \in \Omega$, so for $1 \leq i \leq n$, $F_{u_i}(x, u_{1k}, \ldots, u_{nk}) \to F_{u_i}(x, u_1, \ldots, u_n)$ strongly as $k \to \infty$, from which it follows $\Psi'(u_{1k}, \ldots, u_{nk}) \to \Psi'(u_1, \ldots, u_n)$ strongly as $k \to \infty$. Thus, we have that Ψ' is strongly continuous on X, which implies that Ψ' is a compact operator by [289, Proposition 26.2]. Moreover, bearing in mind the conditions $0 < m_i \leq K_i(t) \leq M_i$ for all $t \geq 0$ for $1 \leq i \leq n$, we see that Φ is continuously differentiable and whose differential at the point $u \in X$ is

$$\Phi'(u)(v) = \sum_{i=1}^{n} \left(\left[K_i \left(\int_{\Omega} (|\nabla u_i(x)|^{p_i} + a_i(x)|u_i(x)|^{p_i}) dx \right) \right]^{p_i - 1} \right. $$
$$\left. \times \int_{\Omega} \left(|\nabla u_i(x)|^{p_i - 2} \nabla u_i(x) \nabla v_i(x) + |u_i(x)|^{p_i - 2} u_i(x) v_i(x) \right) dx \right) $$

for every $v \in X$, and Φ' admits a continuous inverse on X^*. Furthermore, Φ is sequentially weakly lower semicontinuous. Indeed, for any $(u_{1k}, \ldots, u_{nk}) \in X$ with $(u_{1k}, \ldots, u_{nk}) \to (u_1, \ldots, u_n)$ weakly in X, then $u_{ik} \to u_i$ in $W_0^{1, p_i}(\Omega)$ for $1 \leq i \leq n$. Therefore, taking the norm of weakly lower semicontinuity, we have

$$\liminf_{k \to \infty} ||u_{ik}||_* \geq ||u_i||_*, \quad i = 1, \ldots, n.$$

Hence, since $\tilde{K}_i$ is continuous and monotone for $1 \leq i \leq n$, we obtain

$$\tilde{K}_i(||u_i||_*^{p_i}) \leq \tilde{K}_i(\liminf_{k \to \infty} ||u_{ik}||_*^{p_i}) \leq \liminf_{k \to \infty} \tilde{K}_i(||u_{ik}||_*^{p_i})$$

for $1 \leq i \leq n$, from which it follows that Φ is sequentially weakly lower semicontinuous. Put $I_\lambda := \Phi - \lambda \Psi$. Clearly, the weak solutions of the system (5.5.1) are exactly the solutions of the equation $I'_\lambda(u_1, \ldots, u_n) = 0$. Moreover, since for $1 \leq i \leq n$, $m_i \leq K_i(s)$ for all $s \in [0, \infty)$, from the definition of Φ, we have

$$\Phi(u) \geq \sum_{i=1}^{n} \frac{m_i ||u_i||_*^{p_i}}{p_i} \geq m \sum_{i=1}^{n} \frac{||u_i||_*^{p_i}}{p_i} \quad \text{for all } u = (u_1, \ldots, u_n) \in X.$$

$$(5.5.5)$$

Now, let us verify that

$$\gamma < \infty.$$

Let $\{\xi_k\}$ be a real sequence such that $\xi_k \to \infty$ as $k \to \infty$ and

$$\lim_{k \to \infty} \frac{\int_\Omega \sup_{(t_1,\ldots,t_n) \in Q(\xi_k)} F(x, t_1, \ldots, t_n) dx}{\xi_k^p}$$

$$= \liminf_{\xi \to \infty} \frac{\int_\Omega \sup_{(t_1,\ldots,t_n) \in Q(\xi)} F(x, t_1, \ldots, t_n) dx}{\xi^p}. \qquad (5.5.6)$$

Set $r_k = \dfrac{\xi_k^p}{\left(\sum_{i=1}^n (p_i \frac{C}{m})^{\frac{1}{p_i}} \right)^p}$ for all $k \in \mathbb{N}$. Since

$$\sup_{x \in \overline{\Omega}} |u_i(x)|^{p_i} \le C \|u_i\|_{p_i}^{p_i} \quad \text{for all } u_i \in W_0^{1,p_i}(\Omega)$$

for $1 \le i \le n$, we have

$$\sup_{x \in \overline{\Omega}} \sum_{i=1}^n \frac{|u_i(x)|^{p_i}}{p_i} \le C \sum_{i=1}^n \frac{\|u_i\|_*^{p_i}}{p_i} \qquad (5.5.7)$$

for each $u = (u_1, \ldots, u_n) \in X$. So, from (5.5.5) and (5.5.7), we have

$$\Phi^{-1}(-\infty, r_k] = \{u = (u_1, u_2, \ldots, u_n) \in X; \ \Phi(u) \le r_k\}$$

$$\subseteq \left\{ u \in X; \ \underline{m} \sum_{i=1}^n \frac{\|u_i\|_*^{p_i}}{p_i} \le r_k \right\}$$

$$\subseteq \left\{ u \in X; \ \sum_{i=1}^n \frac{|u_i(x)|^{p_i}}{p_i} \le \frac{Cr_k}{\underline{m}} \quad \text{for each } x \in \Omega \right\}$$

$$\subseteq \left\{ u \in X; \ \sum_{i=1}^n |u_i(x)| \le \xi_k \quad \text{for each } x \in \Omega \right\}.$$

Hence, taking into account that $\Phi(0, \ldots, 0) = \Psi(0, \ldots, 0) = 0$, we have for every k large enough,

$$\varphi(r_k) = \inf_{u \in \Phi^{-1}(-\infty, r_k)} \frac{\sup_{v \in \Phi^{-1}(-\infty, r_k]} \Psi(v) - \Psi(u)}{r_k - \Phi(u)}$$

$$\le \frac{\sup_{v \in \Phi^{-1}(-\infty, r_k]} \Psi(v)}{r_k}$$

$$\le \left(\sum_{i=1}^n \left(p_i \frac{C}{\underline{m}} \right)^{\frac{1}{p_i}} \right)^p \frac{\int_\Omega \sup_{(t_1,\ldots,t_n) \in Q(\xi_k)} F(x, t_1, \ldots, t_n) dx}{\xi_k^p}.$$

Moreover, from Assumption (E24), we also have

$$\lim_{k \to \infty} \frac{\int_\Omega \sup_{(t_1,\ldots,t_n) \in Q(\xi_k)} F(x, t_1, \ldots, t_n) dx}{\xi_k^p} < \infty.$$

Therefore,

$$\gamma \leq \liminf_{k \to \infty} \varphi(r_k)$$

$$\leq \left(\sum_{i=1}^{n} \left(p_i \frac{C}{m} \right)^{\frac{1}{p_i}} \right)^p \lim_{k \to \infty} \frac{\int_{\Omega} \sup_{(t_1,\ldots,t_n) \in Q(\xi_k)} F(x, t_1, \ldots, t_n) dx}{\xi_k^{\frac{p}{p}}}$$

$$< \infty. \tag{5.5.8}$$

Assumption (E24), in conjunction with (5.5.8), implies that

$$\Lambda \subseteq \left(0, \frac{1}{\gamma} \right).$$

Fix $\lambda \in \Lambda$. The inequality (5.5.8) yields that the condition (b) of Theorem 1.1.5 can be applied, and either I_λ has a global minimum or there exists a sequence $\{u_k = (u_{1k}, \ldots, u_{nk})\}$ of weak solutions of the system (5.5.1) such that $\lim_{k \to \infty} \|(u_{1k}, \ldots, u_{nk})\| = \infty$.

The other step is to show that the functional I_λ has no global minimum. For the fixed λ, let us verify that the functional I_λ is unbounded from below. Arguing as in [58], consider n positive real sequences $\{d_{i,k}\}_{i=1}^n$ such that $\sqrt{\sum_{i=1}^n d_{i,k}^2} \to \infty$ as $k \to \infty$ and

$$\lim_{k \to \infty} \frac{\int_{\Omega} F(x, d_{1,k}, \ldots, d_{n,k}) dx}{\sum_{i=1}^{n} \frac{\tilde{K}_i(\overline{\mu}^N \omega_\tau \upsilon |d_{i,k}|^{p_i})}{p_i}}$$

$$= \limsup_{(t_1,\ldots,t_n) \to (\infty,\ldots,\infty)} \frac{\int_{B(x_0, \underline{\mu}\tau)} F(x, t_1, \ldots, t_n) dx}{\sum_{i=1}^{n} \frac{\tilde{K}_i(\overline{\mu}^N \omega_\tau \upsilon |t_i|^{p_i})}{p_i}}. \tag{5.5.9}$$

Let $\{w_k = (w_{1k}, \ldots, w_{nk})\}$ be a sequence in X defined by

$$w_{ik}(x) = \begin{cases} 0 & \text{if } x \in \Omega \setminus B(x_0, \tau), \\ \frac{d_{i,k}}{\tau(1-\overline{\mu}_i)}(\tau - |x - x_0|) & \text{if } x \in B(x_0, \tau) \setminus B(x_0, \overline{\mu}_i \tau), \\ d_{i,k} & \text{if } x \in B(x_0, \overline{\mu}_i \tau), \end{cases} \tag{5.5.10}$$

for $1 \leq i \leq n$. For any fixed $k \in \mathbb{N}$, it is easy to see that $w_k \in X$, and in particular, one has

$$\|w_{ik}\|_*^{p_i} = \int_{\Omega} (|\nabla w_{ik}(x)|^{p_i} + a_i(x)|w_{ik}(x)|^{p_i}) dx$$

$$\leq |d_{i,k}|^{p_i} \omega_\tau \left[\frac{1 - \overline{\mu}_i^N}{\tau^{p_i}(1 - \overline{\mu}_i)^{p_i}} + \|a_i\|_\infty g_{\overline{\mu}_i}(p_i, N) \right]$$

$$\leq \overline{\mu}^N \omega_\tau \upsilon |d_{i,k}|^{p_i}$$

for $1 \leq i \leq n$. Taking into account $\inf_{t \geq 0} K(t) > 0$, it follows that

$$\Phi(w_k) = \sum_{i=1}^{n} \frac{\tilde{K}_i(\|w_{ik}\|_*^{p_i})}{p_i} \leq \sum_{i=1}^{n} \frac{\tilde{K}_i(\overline{\mu}^N \omega_\tau \upsilon |d_{i,k}|^{p_i})}{p_i}. \qquad (5.5.11)$$

On the other hand, bearing in mind Assumption (E23), from the definition of Ψ, we infer that

$$\Psi(w_k) \geq \int_{B(x_0, \underline{\mu}\tau)} F(x, d_{1,k}, \ldots, d_{n,k}) dx. \qquad (5.5.12)$$

So, according to (5.5.11) and (5.5.12), we obtain

$$I_\lambda(w_k) \leq \sum_{i=1}^{n} \frac{\tilde{K}_i(\overline{\mu}^N \omega_\tau \upsilon |d_{i,k}|^{p_i})}{p_i} - \lambda \int_{B(x_0, \underline{\mu}\tau)} F(x, d_{1,k}, \ldots, d_{n,k}) dx$$

for every $k \in \mathbb{N}$. Now, if

$$\limsup_{(t_1, \ldots, t_n) \to (\infty, \ldots, \infty)} \frac{\int_{B(x_0, \underline{\mu}\tau)} F(x, t_1, \ldots, t_n) dx}{\sum_{i=1}^{n} \frac{\tilde{K}_i(\overline{\mu}^N \omega_\tau \upsilon |t_i|^{p_i})}{p_i}} < \infty,$$

we fix

$$\epsilon \in \left(\frac{1}{\limsup_{(t_1, \ldots, t_n) \to (\infty, \ldots, \infty)} \frac{\int_{B(x_0, \underline{\mu}\tau)} F(x, t_1, \ldots, t_n) dx}{\sum_{i=1}^{n} \frac{\tilde{K}_i(\overline{\mu}^N \omega_\tau \upsilon |t_i|^{p_i})}{p_i}}}, 1 \right).$$

From (5.5.9), there exists ϑ_ϵ such that

$$\int_{B(x_0, \underline{\mu}\tau)} F(x, d_{1,k}, \ldots, d_{n,k}) dx$$

$$> \epsilon \left(\limsup_{(t_1, \ldots, t_n) \to (\infty, \ldots, \infty)} \frac{\int_{B(x_0, \underline{\mu}\tau)} F(x, t_1, \ldots, t_n) dx}{\sum_{i=1}^{n} \frac{\tilde{K}_i(\overline{\mu}^N \omega_\tau \upsilon |t_i|^{p_i})}{p_i}} \right)$$

$$\times \sum_{i=1}^{n} \frac{\tilde{K}_i(\overline{\mu}^N \omega_\tau \upsilon |d_{i,k}|^{p_i})}{p_i}$$

for $k > \vartheta_\epsilon$. Therefore,

$$I_\lambda(w_k) \leq \left(1 - \lambda \epsilon \limsup_{(t_1, \ldots, t_n) \to (\infty, \ldots, \infty)} \frac{\int_{B(x_0, \underline{\mu}\tau)} F(x, t_1, \ldots, t_n) dx}{\sum_{i=1}^{n} \frac{\tilde{K}_i(\overline{\mu}^N \omega_\tau \upsilon |t_i|^{p_i})}{p_i}} \right)$$

$$\times \sum_{i=1}^{n} \frac{\tilde{K}_i(\overline{\mu}^N \omega_\tau \upsilon |d_{i,k}|^{p_i})}{p_i}$$

for any $k > \vartheta_\epsilon$, and by the choice of ϵ, one then has

$$\lim_{m \to \infty} [\Phi(w_k) - \lambda \Psi(w_k)] = -\infty.$$

If

$$\limsup_{(t_1,\dots,t_n) \to (\infty,\dots,\infty)} \frac{\int_{B(x_0,\mu\tau)} F(x, t_1, \dots, t_n) dx}{\sum_{i=1}^n \frac{\tilde{K}_i(\overline{\mu}^N \omega_\tau \upsilon |t_i|^{p_i})}{p_i}} = \infty,$$

let us consider $M > \frac{1}{\lambda}$. From (5.5.9), there exists ϑ_M such that

$$\int_{B(x_0,\mu\tau)} F(x, d_{1,k}, \dots, d_{n,k}) dx > M \sum_{i=1}^n \frac{\tilde{K}_i(\overline{\mu}^N \omega_\tau \upsilon |d_{i,k}|^{p_i})}{p_i} \quad \text{for } k > \vartheta_M,$$

and therefore

$$I_\lambda(w_k) \le (1 - \lambda M) \sum_{i=1}^n \frac{\tilde{K}_i(\overline{\mu}^N \omega_\tau \upsilon |d_{i,k}|^{p_i})}{p_i} \quad \text{for } k > \vartheta_M,$$

and by the choice of M, one then has

$$\lim_{k \to \infty} [\Phi(w_k) - \lambda \Psi(w_k)] = -\infty.$$

Hence, our claim is proved. Since all assumptions of Theorem 1.1.5 are satisfied, the functional I_λ admits a sequence $\{u_k = (u_{1k}, \dots, u_{nk})\} \subset X$ of critical points such that

$$\lim_{k \to \infty} \|(u_{1k}, \dots, u_{nk})\| = \infty,$$

and we have the desired conclusion. $\qquad\square$

Remark 5.5.1. We point out that if $K_i(t) = 1$ for each $t \ge 0$ for $1 \le i \le n$, Theorem 5.5.1 reduces to [58, Theorem 3.1].

Now, we want to point out the following existence result, in which instead of Assumption (E24) in Theorem 5.5.1 a more general condition is assumed.

Theorem 5.5.2. *Assume that Assumption (A1) in Theorem 5.5.1 holds. Suppose that*

(E25) *there exist a sequence $\{a_k\}$ and n positive real sequence $\{b_{i,k}\}$ with*

$$\frac{a_k^p}{\left(\sum_{i=1}^n (p_i \frac{C}{m})^{\frac{1}{p_i}}\right)^p} > \sum_{i=1}^n \frac{\tilde{K}_i(\overline{\mu}^N \omega_\tau \upsilon |b_{ik}|^{p_i})}{p_i} \quad \text{and} \quad \lim_{k \to \infty} a_k = \infty$$

such that

$$\lim_{k\to\infty}\left[\int_{\Omega}\sup_{(t_1,\dots,t_n)\in Q(a_k)} F(x,t_1,\dots,t_n)dx - \int_{B(x_0,\mu\tau)} F(x,b_{1k},\dots,b_{nk})dx\right]$$

$$\times\left[\frac{a_k^{\frac{p}{}}}{\left(\sum_{i=1}^n (p_i\frac{C}{\underline{m}})^{\frac{1}{p_i}}\right)^{\frac{p}{}}} - \sum_{i=1}^n \frac{\tilde{K}_i(\overline{\mu}^N\omega_\tau\upsilon|b_{ik}|^{p_i})}{p_i}\right]^{-1}$$

$$< \limsup_{(t_1,\dots,t_n)\to(\infty,\dots,\infty)} \frac{\int_{B(x_0,\underline{\mu}\tau)} F(x,t_1,\dots,t_n)dx}{\sum_{i=1}^n \frac{\tilde{K}_i(\overline{\mu}^N\omega_\tau\upsilon|t_i|^{p_i})}{p_i}}.$$

Then, for each $\lambda \in \Lambda'$, the system (5.5.1) has an unbounded sequence of weak solutions in X, where

$$\Lambda' = \left(\frac{1}{\limsup_{(t_1,\dots,t_n)\to(\infty,\dots,\infty)} \frac{\int_{B(x_0,\underline{\mu}\tau)} F(x,t_1,\dots,t_n)dx}{\sum_{i=1}^n \frac{\tilde{K}_i(\overline{\mu}^N\omega_\tau\upsilon|t_i|^{p_i})}{p_i}}},\right.$$

$$\left.\frac{\frac{a_k^{\frac{p}{}}}{\left(\sum_{i=1}^n (p_i\frac{C}{\underline{m}})^{\frac{1}{p_i}}\right)^{\frac{p}{}}}}{\lim_{k\to\infty} \frac{\int_{\Omega}\sup_{(t_1,\dots,t_n)\in Q(a_k)} F(x,t_1,\dots,t_n)dx - \int_{B(x_0,\underline{\mu}\tau)} F(x,b_{1k},\dots,b_{nk})dx}{\frac{a_k^{\frac{p}{}}}{\left(\sum_{i=1}^n (p_i\frac{C}{\underline{m}})^{\frac{1}{p_i}}\right)^{\frac{p}{}}} - \sum_{i=1}^n \frac{\tilde{K}_i(\overline{\mu}^N\omega_\tau\upsilon|b_{ik}|^{p_i})}{p_i}}}\right).$$

Proof. Clearly, from (E25), we obtain (E24) by choosing $b_{i,k} = 0$ for all $k \in \mathbb{N}$ and for $1 \leq i \leq n$. Moreover, if we assume (E25) instead of (E24) and set $r_k = \dfrac{a_k^{\frac{p}{}}}{\left(\sum_{i=1}^n (p_i\frac{C}{\underline{m}})^{\frac{1}{p_i}}\right)^{\frac{p}{}}}$ for all $k \in \mathbb{N}$, by the same argument as in Theorem 5.5.1, we obtain that

$$\varphi(r_k) = \inf_{u\in\Phi^{-1}(-\infty,r_k)} \frac{\sup_{v\in\Phi^{-1}(-\infty,r_k]}\Psi(v)-\Psi(u)}{r_k-\Phi(u)}$$

$$\leq \frac{\sup_{v\in\Phi^{-1}(-\infty,r_k]}\Psi(v) - \int_a^b F(x,w_{1k}(x),\dots,w_{nk}(x))dx}{r_k-\sum_{i=1}^n \frac{\tilde{K}_i(\|w_{ik}\|_*^{p_i})}{p_i}}$$

$$\leq \frac{\int_{\Omega}\sup_{(t_1,\dots,t_n)\in Q(a_k)} F(x,t_1,\dots,t_n)dx - \int_{B(x_0,\underline{\mu}\tau)} F(x,b_{1k},\dots,b_{nk})dx}{\frac{a_k^{\frac{p}{}}}{\left(\sum_{i=1}^n (p_i\frac{C}{\underline{m}})^{\frac{1}{p_i}}\right)^{\frac{p}{}}} - \sum_{i=1}^n \frac{\tilde{K}_i(\overline{\mu}^N\omega_\tau\upsilon|b_{ik}|^{p_i})}{p_i}}$$

where $w_k = (w_{1k},\dots,w_{nk})$, with w_{ik} for $1 \leq i \leq n$, as given in (5.5.10) with $b_{i,k}$ instead of $d_{i,k}$. Hence, we have the desired conclusion. $\qquad\square$

Here, we point out a consequence of Theorem 5.5.1.

Corollary 5.5.1. *Suppose that condition (E23) in Theorem 5.5.1 holds. Assume that*

(E26) $\displaystyle \liminf_{\xi \to \infty} \frac{\int_\Omega \sup_{(t_1,\ldots,t_n)\in Q(\xi)} F(x,t_1,\ldots,t_n)dx}{\xi^p} < \frac{1}{\left(\sum_{i=1}^n (p_i \frac{C}{m})^{\frac{1}{p_i}}\right)^p};$

(E27) $\displaystyle \limsup_{(t_1,\ldots,t_n)\to(\infty,\ldots,\infty)} \frac{\int_{B(x_0,\mu\tau)} F(x,t_1,\ldots,t_n)dx}{\sum_{i=1}^n \frac{\tilde{K}_i(\overline{\mu}^N \omega_\tau \upsilon |t_i|^{p_i})}{p_i}} > 1.$

Then, the system

$$\begin{cases} -\Big[K_i(\int_\Omega (|\nabla u_i(x)|^{p_i} + a_i(x)|u_i(x)|^{p_i})dx)\Big]^{p_i-1} \\ \qquad \times \Big(div(|\nabla u_i|^{p_i-2}\nabla u_i) + a_i(x)|u_i|^{p_i-2}u \Big) = F_{u_i}(x,u_1,\ldots,u_n) \; \text{in } \Omega, \\ u_i = 0 \qquad\qquad\qquad\qquad\qquad\qquad\qquad\qquad\qquad\qquad\quad \text{on } \partial\Omega, \end{cases}$$

for $1 \le i \le n$, has an unbounded sequence of weak solutions in X.

As an example, we state here a special case of our main result.

Theorem 5.5.3. *Let $\Omega \subset \mathbb{R}^2$ be a non-empty bounded open set with a smooth boundary $\partial\Omega$. Let $f,g : \mathbb{R}^2 \to \mathbb{R}$ be two positive $C^0(\mathbb{R}^2)$-functions such that the differential 1-form $w := f(\xi,\eta)d\xi + g(\xi,\eta)d\eta$ is integrable and let F be a primitive of w such that $F(0,0) = 0$. Fix $p,q > 2$, with $p \le q$, and assume that*

$$\liminf_{\xi \to \infty} \frac{F(\xi,\xi)}{\xi^p} = 0$$

and

$$\limsup_{\xi \to \infty} \frac{F(\xi,\xi)}{\frac{\tilde{K}_1(\overline{\mu}^2\tau^2\pi\upsilon|t_1|^p)}{p} + \frac{\tilde{K}_2(\overline{\mu}^2\tau^2\pi\upsilon|t_2|^q)}{q}} = \infty,$$

where

$$\upsilon := \max\left\{ \frac{\sigma(p,2)}{\tau^p} + \|a_1\|_\infty \frac{g_{\mu_1}(p,2)}{\overline{\mu_1}^2}, \; \frac{\sigma(q,2)}{\tau^q} + \|a_2\|_\infty \frac{g_{\mu_2}(q,2)}{\overline{\mu_2}^2} \right\}.$$

Then, the system

$$\begin{cases} -\Big[K_1(\int_\Omega (|\nabla u(x)|^p + a_1(x)|u(x)|^p)dx)\Big]^{p-1} \\ \qquad\qquad \times \Big(div(|\nabla u|^{p-2}\nabla u) + a_1(x)|u|^{p-2}u \Big) = f(u,v), \; \text{in } \Omega, \\[2mm] -\Big[K_2(\int_\Omega (|\nabla v(x)|^q + a_2(x)|v(x)|^q)dx)\Big]^{q-1} \\ \qquad\qquad \times \Big(div(|\nabla v|^{q-2}\nabla v) + a_2(x)|v|^{q-2}v \Big) = g(u,v), \; \text{in } \Omega, \\[2mm] u = v = 0, \qquad\qquad\qquad\qquad\qquad\qquad\qquad\qquad\qquad\quad \text{on } \partial\Omega, \end{cases}$$

admits a sequence of pairwise distinct positive weak solutions in $W_0^{1,p}(\Omega) \times W_0^{1,q}(\Omega)$.

Proof. Take $n = 2$ and set $F(x, t_1, t_2) = F(t_1, t_2)$ for all $x \in \Omega$ and $t_1, t_2 \in \mathbb{R}$. From the conditions

$$\liminf_{\xi \to \infty} \frac{F(\xi, \xi)}{\xi^p} = 0$$

and

$$\limsup_{\xi \to \infty} \frac{F(\xi, \xi)}{\frac{\tilde{K}_1(\bar{\mu}^2 \tau^2 \pi v |t_1|^p)}{p} + \frac{\tilde{K}_2(\bar{\mu}^2 \tau^2 \pi v |t_2|^q)}{q}} = \infty,$$

we see that the assumptions (E26) and (E27), respectively, are satisfied. Then, taking into account that $F_{t_1}(t_1, t_2) = f(t_1, t_2)$, $F_{t_2}(t_1, t_2) = g(t_1, t_2)$ for all $(t_1, t_2) \in \mathbb{R}^2$, and $f, g : \mathbb{R}^2 \to \mathbb{R}$ are positive, the conclusion follows from Corollary 5.5.1. $\qquad \square$

Remark 5.5.2. We observe that, in Theorem 5.5.1, if we replace $\xi \to \infty$ with $\xi \to 0^+$, then by the same way as in the proof of Theorem 5.5.1 but using conclusion (c) of Theorem 1.1.5 instead of (b), we can show that the system (5.5.1) has a sequence of weak solutions, which strongly converges to 0 in X.

Now, we point out a particular situation of Theorem 5.5.1.

Corollary 5.5.2. *Fix α_i, $\beta_i > 0$ for $1 \le i \le n$, and denote $\underline{\alpha} = \min\{\alpha_i; 1 \le i \le n\}$. In addition to condition (E23) in Theorem 5.5.1, assume*

(E28) *the inequality*

$$\liminf_{\xi \to \infty} \frac{\int_\Omega \sup_{(t_1,\ldots,t_n) \in Q(\xi)} F(x, t_1, \ldots, t_n) dx}{\xi^{\underline{p}}}$$

$$< \frac{1}{\left(\sum_{i=1}^n \left(p_i \frac{C}{\underline{\alpha}} \right)^{\frac{1}{p_i}} \right)^{\underline{p}}}$$

$$\times \limsup_{(t_1,\ldots,t_n) \to (\infty,\ldots,\infty)} \frac{\int_{B(x_0,\mu\tau)} F(x, t_1, \ldots, t_n) dx}{\sum_{i=1}^n \frac{\alpha_i \bar{\mu}^N \omega_\tau v |t_i|^{p_i} + \frac{\beta_i}{2}(\bar{\mu}^N \omega_\tau v |t_i|^{p_i})^2}{p_i}}$$

holds. Then, for each

$$\lambda \in \left(\frac{1}{\limsup_{(t_1,\ldots,t_n) \to (\infty,\ldots,\infty)} \frac{\int_{B(x_0,\mu\tau)} F(x,t_1,\ldots,t_n) dx}{\sum_{i=1}^n \frac{\alpha_i \bar{\mu}^N \omega_\tau v |t_i|^{p_i} + \frac{\beta_i}{2}(\bar{\mu}^N \omega_\tau v |t_i|^{p_i})^2}{p_i}}} \right.$$

$$, \ \frac{\left(\sum_{i=1}^n (p_i \frac{C}{\alpha})^{\frac{1}{p_i}}\right)^{\frac{1}{p}}}{\liminf_{\xi \to \infty} \frac{\int_\Omega \sup_{(t_1,\dots,t_n) \in Q(\xi)} F(x,t_1,\dots,t_n)dx}{\xi^{\underline{p}}}}\right),$$

the system

$$\begin{cases} -\left[\alpha_i + \beta_i \int_\Omega (|\nabla u_i(x)|^{p_i} + a_i(x)|u_i(x)|^{p_i})dx\right]^{p_i-1} \\ \quad \times \left(div(|\nabla u_i|^{p_i-2}\nabla u_i) + a_i(x)|u_i|^{p_i-2}u\right) = \lambda F_{u_i}(x, u_1, \dots, u_n), \text{ in } \Omega, \\ u_i = 0, \hspace{6cm} on \ \partial\Omega, \end{cases}$$

has an unbounded sequence of weak solutions in X.

Proof. For fixed α_i, $\beta_i > 0$ for $1 \le i \le n$, let $K_i(t) = \alpha_i + \beta_i t$ for all $t \ge 0$. Bearing in mind that $m_i = \alpha_i$ for $1 \le i \le n$, the conclusion follows immediately from Theorem 5.5.1. □

We illustrate the result by giving the following example whose construction is motivated by [58, Example 3.1].

Example 5.5.1. Let $\Omega \subset \mathbb{R}^2$ be a non-empty open set with a smooth boundary $\partial\Omega$ and consider the increasing sequence of positive real numbers given by

$$a_1 = 2, \quad a_{n+1} = n!(a_n)^{\frac{7}{3}} + 2$$

for every $n \ge 1$. Define the function $F : \Omega \times \mathbb{R}^2 \to \mathbb{R}$ by

$$F(x,y,t_1,t_2) = \begin{cases} (a_{n+1})^7 e^{x^2+y^2 - \frac{1}{1-(t_1-a_{n+1})^2-(t_2-a_{n+1})^2}+1}, \\ \qquad \text{if } (x,y,t_1,t_2) \in \Omega \times \bigcup_{n \ge 1} S((a_{n+1},a_{n+1}),1), \\ 0, \hspace{5cm} \text{otherwise}, \end{cases}$$

where $S((a_{n+1},a_{n+1}),1)$ denotes the open unit ball with center at (a_{n+1},a_{n+1}). It is clear that $F : \Omega \times \mathbb{R}^2 \to \mathbb{R}$ is a non-negative function such that the mapping $(t_1,t_2) \to F(x,t_1,t_2)$ is in C^1 in $\mathbb{R}^2$ for all $x \in \Omega$, F_{t_i} is continuous in $\Omega \times \mathbb{R}^2$, for $i = 1,2$, and $F(x,y,0,0) = 0$ for all $(x,y) \in \Omega$. Now, for every $n \in \mathbb{N}$,

$$\int_{B(x_0,\underline{\mu}\tau)} \sup_{(t_1,t_2) \in S((a_{n+1},a_{n+1}),1)} F(x,y,t_1,t_2)dxdy$$

$$= \int_{B(x_0,\underline{\mu}\tau)} F(x,y,a_{n+1},a_{n+1})dxdy$$

$$= (a_{n+1})^7 \int_{B(x_0,\underline{\mu}\tau)} e^{x^2+y^2} dxdy.$$

We will denote by f and g the partial derivative of F respect to t_1 and t_2, respectively. Since

$$\lim_{n\to\infty} \frac{\int_{B(x_0,\underline{\mu}\tau)} F(x,y,a_{n+1},a_{n+1})dxdy}{\sum_{i=1}^{2} \frac{\bar{\mu}^2\tau^2\pi va_{n+1}^3 + \frac{1}{2}(\bar{\mu}^2\tau^2\pi va_{n+1}^3)^2}{3}} = \infty,$$

where

$$v := \max\left\{ \frac{\sigma(3,2)}{\tau^3} + \|a_1\|_\infty \frac{g_{\mu_1}(3,2)}{\bar{\mu}_1^2}, \ \frac{\sigma(3,2)}{\tau^3} + \|a_2\|_\infty \frac{g_{\mu_2}(3,2)}{\bar{\mu}_2^2} \right\},$$

we see that

$$\limsup_{\substack{(t_1,t_2)\to(\infty,\infty) \\ (t_1,t_2)\in\mathbb{R}^2_+}} \frac{\int_{B(x_0,\underline{\mu}\tau)} F(x,y,t_1,t_2)dx}{\sum_{i=1}^{2} \frac{\bar{\mu}^2\tau^2\pi v|t_i|^3 + \frac{1}{2}(\bar{\mu}^2\tau^2\pi v|t_i|^3)^2}{3}} = \infty.$$

Moreover, by choosing $\xi_n = a_{n+1} - 1$ for every $n \in \mathbb{N}$, we obtain

$$\int_{B(x_0,\underline{\mu}\tau)} \sup_{(t_1,t_2)\in K(\xi)} F(x,y,t_1,t_2)dxdy = (a_n)^7 \int_{B(x_0,\underline{\mu}\tau)} e^{x^2+y^2}dxdy$$

for $n \in \mathbb{N}$. Then,

$$\lim_{n\to\infty} \frac{\int_{B(x_0,\underline{\mu}\tau)} \sup_{(t_1,t_2)\in K(\xi)} F(x,y,t_1,t_2)dxdy}{(a_{n+1}-1)^3} = 0,$$

and so

$$\liminf_{\xi\to\infty} \frac{\int_{B(x_0,\underline{\mu}\tau)} \sup_{(t_1,t_2)\in K(\xi)} F(x,y,t_1,t_2)dxdy}{\xi^3} = 0.$$

Therefore,

$$0 = \liminf_{\xi\to\infty} \frac{\int_{B(x_0,\underline{\mu}\tau)} \sup_{(t_1,t_2)\in K(\xi)} F(x,y,t_1,t_2)dxdy}{\xi^3}$$

$$< \frac{1}{24C} \limsup_{\substack{(t_1,t_2)\to(\infty,\infty) \\ (t_1,t_2)\in\mathbb{R}^2_+}} \frac{\int_{B(x_0,\underline{\mu}\tau)} F(x,y,t_1,t_2)dx}{\sum_{i=1}^{2} \frac{\bar{\mu}^2\tau^2\pi v|t_i|^3 + \frac{1}{2}(\bar{\mu}^2\tau^2\pi v|t_i|^3)^2}{3}} = \infty.$$

Hence, all the assumptions of Corollary 5.5.2 are satisfied. Then, Corollary 5.5.2 is applicable to the system

$$\begin{cases} -\left[1 + \int_\Omega(|\nabla u(x)|^3 + a_1(x)|u(x)|^3)dx\right]^2 \\ \qquad \times \left(\mathrm{div}(|\nabla u|\nabla u) + a_1(x)|u|u\right) = \lambda f(x,y,u,v), \text{ in } \Omega, \\[2mm] -\left[1 + \int_\Omega(|\nabla v(x)|^3 + a_2(x)|v(x)|^3)dx\right]^2 \\ \qquad \left(\mathrm{div}(|\nabla v|\nabla v) + a_2(x)|v|v\right) = \lambda g(x,y,u,v), \quad \text{ in } \Omega, \\[2mm] u = v = 0, \qquad\qquad\qquad\qquad\qquad\qquad\qquad\quad \text{ on } \partial\Omega, \end{cases}$$

for every $\lambda \in (0, \infty)$.

Finally, as an application of Theorem 1.1.5, we consider the problem

$$
\begin{cases}
-\Big[\alpha + \beta \int_{\Omega}(|\nabla u(x)|^p + a(x)|u(x)|^p)dx\Big]^{p-1} \\
\quad \times \Big(\text{div}(|\nabla u|^{p-2}\nabla u) + a(x)|u|^{p-2}u\Big) = \lambda f(x,u), \text{ in } \Omega, \\
u = 0, \qquad\qquad\qquad\qquad\qquad\qquad\qquad\qquad \text{on } \partial\Omega,
\end{cases}
$$
$$(5.5.13)$$

where $p > N$, $\lambda > 0$, $\alpha, \beta > 0$, $f : \Omega \times \mathbb{R} \to \mathbb{R}$ is an L^1-Caratéodory function and $a \in L^{\infty}(\Omega)$ with $\text{essinf}_{\Omega}a(x) \geq 0$. Let

$$
F(x,t) = \int_0^t f(x,\xi)d\xi \quad \text{for all } (x,t) \in \Omega \times \mathbb{R}.
$$

The following existence result is an immediate consequence of Theorem 5.5.1.

Theorem 5.5.4. *Assume that*

(E29) $F(x,t) \geq 0$ *for each* $(x,t) \in \Omega \times \mathbb{R}_+$;

(E30) $\liminf_{\xi \to \infty} \dfrac{\int_{\Omega} \sup_{|t| \leq \xi} F(x,t)dx}{\xi^p}$

$$
< \frac{\alpha}{C^p} \limsup_{t \to \infty} \frac{\int_{B(x_0, \mu\tau)} F(x,t)dx}{\alpha \overline{\mu}^N \omega_\tau v|t|^p + \frac{\beta}{2}(\overline{\mu}^N \omega_\tau v|t|^p)^2},
$$

where

$$
C := \sup_{u \in W_0^{1,p}(\Omega)\setminus\{0\}} \frac{\max_{x \in \overline{\Omega}} |u(x)|}{\left(\int_{\Omega} |\nabla u(x)|^p dx\right)^{1/p}}.
$$

Then, for each

$$
\lambda \in \left(\frac{\frac{1}{p}}{\limsup_{t \to \infty} \frac{\int_{B(x_0, \mu\tau)} F(x,t)dx}{\alpha \overline{\mu}^N \omega_\tau v|t|^p + \frac{\beta}{2}(\overline{\mu}^N \omega_\tau v|t|^p)^2}}, \frac{\frac{\alpha}{pC^p}}{\liminf_{\xi \to \infty} \frac{\int_{\Omega} \sup_{|t| \leq \xi} F(x,t)dx}{\xi^p}} \right),
$$

the problem (5.5.13) has an unbounded sequence of weak solutions in $W_0^{1,p}(\Omega)$.

Chapter 6

Difference Equations

6.1 Introduction

The theory of nonlinear discrete dynamical systems has been widely used to examine discrete models appearing in such areas as statistics, computing, electrical circuit analysis, dynamical systems, economics, biology, etc [9, 189]. Variational methods are a powerful tool in the investigation of existence theory for nonlinear problems of this type. In recent years, there has been an increasing interest in the literature on the use of variational methods to study the existence and multiplicity of solutions for discrete systems, and we refer the reader to [72, 81, 83, 127, 152, 183, 220, 231, 261, 264, 285, 286, 298] and the references therein for recent work of this kind. In this chapter, we will examine some applications of the variational method to several discrete problems. In Sections 2 and 3 we look at fourth order problems with one and two parameters, and in Section 4 we consider multi-point problems for fourth order equations. In Section 5, we consider homoclinic solutions, and in Section 6, we study anti-periodic solutions of a higher order equation.

In recent years, many researchers have devoted a lot of attention to discrete BVPs with various BCs. For example, see [18, 71, 161, 163, 186, 223, 290] for work on fourth order discrete problems with separated BCs and [17, 21, 68, 69] for work on periodic problems. Cabada and Dimitrov [68] studied the periodic BVP

$$\begin{cases} u(t+4) + Mu(t) = \lambda g(t)f(u(t)) + h(t), \ t \in [0, N-1]_{\mathbb{Z}}, \\ u(i) = u(N+i), \ i = 0, 1, 2, 3, \end{cases} \tag{6.1.1}$$

where f, g, and h are functions satisfying some appropriate conditions, $\lambda > 0$ is a parameter, and M is a real parameter for which the associated linear problem

$$\begin{cases} u(t+4) + Mu(t) = 0, \ t \in [0, N-1]_{\mathbb{Z}}, \\ u(i) = u(N+i), \ i = 0, 1, 2, 3 \end{cases}$$

has only the trivial solution and the related Green's function has constant sign. In [68], the Krasnosel'skii fixed point theorem was employed to investigate the existence of positive solutions of BVP (6.1.1). The properties of the associated Green's function were carefully analyzed and played an important role in the proofs of their main results.

Throughout this chapter, for any integers c and d with $c < d$, let $[c, d]_{\mathbb{Z}}$ denote the discrete interval $\{c, c+1, \ldots, d\}$, let Δ be the forward difference operator defined by $\Delta u(t) = u(t+1) - u(t)$, $\Delta^0 u(t) = u(t)$, $\Delta^i u(t) = \Delta^{i-1}(\Delta u(t))$ for $i \in \mathbb{N}$.

6.2   Periodic Problems with One Parameter

It is well known that variational methods and critical point theory are important tools used in the study of the existence of solutions to various BVPs. In this section, we use them to obtain conditions for the existence of at least one and possibly multiple solutions. The eigenvalues of a symmetric matrix associated with the problem are used in the statements of our theorems.

Consider the discrete nonlinear fourth order periodic BVP consisting of the equation

$$\Delta^4 u(t-2) - \Delta\big(p(t-1)\Delta u(t-1)\big) + q(t)u(t) = \lambda f(t, u(t)), \quad t \in [1, N]_{\mathbb{Z}},$$
$$(6.2.1)$$

and the BC

$$\Delta^i u(-1) = \Delta^i u(N-1), \quad i = 0, 1, 2, 3, \tag{6.2.2}$$

where $\lambda > 0$ is a parameter, $N \geq 1$ is an integer, $p \in C([0, N]_{\mathbb{Z}}, \mathbb{R})$ with $p(0) = p(N)$, $q \in C([1, N]_{\mathbb{Z}}, \mathbb{R})$, and $f \in C([1, N]_{\mathbb{Z}} \times \mathbb{R}, \mathbb{R})$. By a *solution* of BVP (6.2.1), (6.2.2), we mean a function $u : [-1, N+2]_{\mathbb{Z}} \to \mathbb{R}$ that satisfies both (6.2.1) and (6.2.2).

BVP (6.2.1), (6.2.2) can be regarded as a discrete analogue of the fourth order BVP

$$u^{(4)}(t) - \big(p(t)u'(t)\big)' + q(t)u(t) = \lambda f(t, u(t)), \quad t \in (0, 1), \tag{6.2.3}$$

$$u^{(i)}(0) = u^{(i)}(1), \quad i = 0, 1, 2, 3. \tag{6.2.4}$$

Various special cases of BVP (6.2.3), (6.2.4) have been studied by many researchers using a variety of methods. See, for example, [29,36,90,204,205] and the references therein.

Let X be given by

$$X = \left\{ u : [-1, N+2]_{\mathbb{Z}} \to \mathbb{R} \; : \; \Delta^i u(-1) = \Delta^i u(N-1), \; i = 0, 1, 2, 3 \right\},$$
$$(6.2.5)$$

and for any $u \in X$, define

$$\|u\| = \left(\sum_{t=1}^{N} |u(t)|^2 \right)^{1/2}$$

and

$$\|u\|_\infty = \max_{t \in [1, N]_{\mathbb{Z}}} |u(t)|.$$

Remark 6.2.1. It is easy to see that, for any $u \in X$, we have

$$u(-1) = u(N-1), \; u(0) = u(N), \; u(1) = u(N+1), \; \text{and} \; u(2) = u(N+2).$$
$$(6.2.6)$$

Then, equipped with $\| \cdot \|$, X is an N dimensional reflexive Banach space. In fact, X is isomorphic to $\mathbb{R}^N$. When we write the vector $u = (u(1), \dots, u(N)) \in \mathbb{R}^N$, we always imply that u can be extended as a vector in X so that (6.2.6) holds, i.e., u can be extended to the vector

$$\big(u(N-1), u(N), u(1), \dots, u(N), u(1), u(2) \big) \in X.$$

Moreover, when we write $X = \mathbb{R}^N$, we mean the elements in $\mathbb{R}^N$ have been extended in the above sense.

For $u \in X$ and $\lambda > 0$, let the functional I_λ be defined by

$$I_\lambda(u) = \frac{1}{2} \sum_{t=1}^{N} \left(|\Delta^2 u(t-2)|^2 + p(t-1)|\Delta u(t-1)|^2 + q(t)|u(t)|^2 \right)$$
$$- \lambda \sum_{t=1}^{N} F(t, u(t)), \tag{6.2.7}$$

where $F(t, x) = \int_0^x f(t, s)ds$ for $x \in \mathbb{R}$. Then, it is easy to see that I_λ is continuously differentiable and its derivative $I_\lambda'(u)$ at $u \in X$ is given by

$$I_\lambda'(u)(v) = \sum_{t=1}^{N} \big(\Delta^2 u(t-2)\Delta^2 v(t-2) + p(t-1)\Delta u(t-1)\Delta v(t-1)$$
$$+ q(t)u(t)v(t) - \lambda f(t, u(t))v(t) \big) \tag{6.2.8}$$

for any $v \in X$.

Lemma 6.2.1. *For any $u, v \in X$, we have*

$$\sum_{t=1}^{N} \Delta^2 u(t-2)\Delta^2 v(t-2) = \sum_{t=1}^{N} \Delta^4 u(t-2)v(t) \qquad (6.2.9)$$

and

$$\sum_{t=1}^{N} p(t-1)\Delta u(t-1)\Delta v(t-1) = -\sum_{t=1}^{N} \Delta\big(p(t-1)\Delta u(t-1)\big)v(t). \quad (6.2.10)$$

Proof. We first prove (6.2.9). For any $u, v \in X$, from the summation by parts formula and Remark 6.2.1, it follows that

$$\sum_{t=1}^{N} \Delta^2 u(t-2)\Delta^2 v(t-2) = \Delta^2 u(N-1)\Delta v(N-1) - \Delta^2 u(-1)\Delta v(-1)$$

$$- \sum_{t=1}^{N} \Delta^3 u(t-2)\Delta v(t-1)$$

$$= - \sum_{t=1}^{N} \Delta^3 u(t-2)\Delta v(t-1)$$

$$= -\Delta^3 u(N-1)v(N) + \Delta^3 u(-1)v(0)$$

$$+ \sum_{t=1}^{N} \Delta^4 u(t-2)v(t)$$

$$= \sum_{t=1}^{N} \Delta^4 u(t-2)v(t),$$

i.e., (6.2.9) holds.

Now, we show (6.2.10). Again, by the summation by parts formula, Remark 6.2.1, and the fact that $p(0) = p(N)$, we have

$$\sum_{t=1}^{N} p(t-1)\Delta u(t-1)\Delta v(t-1)$$

$$= p(N)\Delta u(N)v(N) - p(0)\Delta u(0)v(0) - \sum_{t=1}^{N} \Delta\big(p(t-1)\Delta u(t-1)\big)v(t)$$

$$= -\sum_{t=1}^{N} \Delta\big(p(t-1)\Delta u(t-1)\big)v(t),$$

i.e., (6.2.10) holds. This completes the proof of the lemma. $\qquad\square$

Lemma 6.2.2. *A function $u \in X$ is a critical point of I_λ if and only if $u(t)$ is a solution of BVP* (6.2.1), (6.2.2).

Proof. By (6.2.8) and Lemma 6.2.1, we see that $u \in X$ is a critical point of I_λ if and only if

$$I'_\lambda(u)(v) = \sum_{t=1}^{N} \left(\Delta^4 u(t-2) - \Delta\big(p(t-1)\Delta u(t-1)\big) \right.$$

$$\left. + q(t)u(t) - \lambda f(t, u(t)) \right) v(t) = 0$$

for any $v \in X$, which is equivalent to

$$\Delta^4 u(t-2) - \Delta\big(p(t-1)\Delta u(t-1)\big) + q(t)u(t) = \lambda f(t, u(t)) \quad \text{for } t \in [1, N]_{\mathbb{Z}}.$$

Note that (6.2.2) clearly holds (since $u \in X$), and this completes the proof of the lemma. $\qquad \square$

Lemma 6.2.3. *For $u \in X$, we have*

$$\sum_{t=1}^{N} |\Delta u(t-1)|^2 \leq 4||u||^2 \tag{6.2.11}$$

and

$$\sum_{t=1}^{N} |\Delta^2 u(t-2)|^2 \leq 16||u||^2. \tag{6.2.12}$$

Proof. For $u \in X$, by Hölder's inequality and Remark 6.2.1, we see that

$$\sum_{t=1}^{N} |\Delta u(t-1)|^2 = \sum_{t=1}^{N} \left(|u(t)|^2 + |u(t-1)|^2 - 2u(t)u(t-1) \right)$$

$$\leq \sum_{t=1}^{N} \left(|u(t)|^2 + |u(t-1)|^2 + 2|u(t)| \, |u(t-1)| \right)$$

$$\leq \sum_{t=1}^{N} |u(t)|^2 + \sum_{t=1}^{N} |u(t-1)|^2 + 2 \left(\sum_{t=1}^{N} |u(t)|^2 \right)^{1/2} \left(\sum_{t=1}^{N} |u(t-1)|^2 \right)^{1/2}$$

$$= \sum_{t=1}^{N} |u(t)|^2 + \sum_{t=1}^{N} |u(t)|^2 + 2 \left(\sum_{t=1}^{N} |u(t)|^2 \right)^{1/2} \left(\sum_{t=1}^{N} |u(t)|^2 \right)^{1/2}$$

$$= ||u||^2 + ||u||^2 + 2||u|| \, ||u|| = 4||u||^2$$

and

$$\sum_{t=1}^{N} |\Delta^2 u(t-2)|^2 = \sum_{t=1}^{N} \left(u(t) - 2u(t-1) + u(t-2) \right)^2$$

$$= \sum_{t=1}^{N} \left(|u(t)|^2 + 4|u(t-1)|^2 + |u(t-2)|^2 - 4u(t)u(t-1) \right.$$

$$\left. + 2u(t)u(t-2) - 4u(t-1)u(t-2) \right)$$

$$\leq \sum_{t=1}^{N} |u(t)|^2 + 4 \sum_{t=1}^{N} |u(t-1)|^2 + \sum_{t=1}^{N} |u(t-2)|^2$$

$$+ 4 \left(\sum_{t=1}^{N} |u(t)|^2 \right)^{1/2} \left(\sum_{t=1}^{N} |u(t-1)|^2 \right)^{1/2}$$

$$+ 2 \left(\sum_{t=1}^{N} |u(t)|^2 \right)^{1/2} \left(\sum_{t=1}^{N} |u(t-2)|^2 \right)^{1/2}$$

$$+ 4 \left(\sum_{t=1}^{N} |u(t-1)|^2 \right)^{1/2} \left(\sum_{t=1}^{N} |u(t-2)|^2 \right)^{1/2}$$

$$= \sum_{t=1}^{N} |u(t)|^2 + 4 \sum_{t=1}^{N} |u(t)|^2 + \sum_{t=1}^{N} |u(t)|^2$$

$$+ 4 \left(\sum_{t=1}^{N} |u(t)|^2 \right)^{1/2} \left(\sum_{t=1}^{N} |u(t)|^2 \right)^{1/2}$$

$$+ 2 \left(\sum_{t=1}^{N} |u(t)|^2 \right)^{1/2} \left(\sum_{t=1}^{N} |u(t)|^2 \right)^{1/2}$$

$$+ 4 \left(\sum_{t=1}^{N} |u(t)|^2 \right)^{1/2} \left(\sum_{t=1}^{N} |u(t)|^2 \right)^{1/2}$$

$$= ||u||^2 + 4||u||^2 + ||u||^2 + 4||u||^2 + 2||u||^2 + 4||u||^2$$

$$= 16||u||^2.$$

Thus, (6.2.11) and (6.2.12) hold. This completes the proof.   □

We wish to give an equivalent form of the functional I_λ. Define the $N \times N$ matrices A, B, and C as follows:

If $N \geq 5$, let

$$A = \begin{pmatrix}
6 & -4 & 1 & 0 & 0 & \cdots & 0 & 0 & 1 & -4 \\
-4 & 6 & -4 & 1 & 0 & \cdots & 0 & 0 & 0 & 1 \\
1 & -4 & 6 & -4 & 1 & \cdots & 0 & 0 & 0 & 0 \\
0 & 1 & -4 & 6 & -4 & \cdots & 0 & 0 & 0 & 0 \\
0 & 0 & 1 & -4 & 6 & \cdots & 0 & 0 & 0 & 0 \\
\vdots & \vdots & \vdots & \vdots & \vdots & \ddots & \vdots & \vdots & \vdots & \vdots \\
0 & 0 & 0 & 0 & 0 & \cdots & 6 & -4 & 1 & 0 \\
0 & 0 & 0 & 0 & 0 & \cdots & -4 & 6 & -4 & 1 \\
1 & 0 & 0 & 0 & 0 & \cdots & 1 & -4 & 6 & -4 \\
-4 & 1 & 0 & 0 & 0 & \cdots & 0 & 1 & -4 & 6
\end{pmatrix}, \tag{6.2.13}$$

and if $N = 1, 2, 3, 4$, let A be respectively given by

$$(0), \quad \begin{pmatrix} 8 & -8 \\ -8 & 8 \end{pmatrix}, \quad \begin{pmatrix} 6 & -3 & -3 \\ -3 & 6 & -3 \\ -3 & -3 & 6 \end{pmatrix}, \quad \text{and} \quad \begin{pmatrix} 6 & -4 & 2 & -4 \\ -4 & 6 & -4 & 2 \\ 2 & -4 & 6 & -4 \\ -4 & 2 & -4 & 6 \end{pmatrix}.$$

If $N \geq 3$, let

$$B = \begin{pmatrix}
p(0)+p(1) & -p(1) & 0 & \cdots & -p(0) \\
-p(1) & p(1)+p(2) & -p(2) & \cdots & 0 \\
0 & -p(2) & p(2)+p(3) & \cdots & 0 \\
\vdots & \vdots & \vdots & \ddots & \vdots \\
0 & 0 & 0 & \cdots & -p(N-1) \\
-p(0) & 0 & 0 & \cdots & p(N-1)+p(0)
\end{pmatrix}, \tag{6.2.14}$$

and if $N = 1, 2$, let B be respectively given by

$$(0) \quad \text{and} \quad \begin{pmatrix} p(0)+p(1) & -p(0)-p(1) \\ -p(0)-p(1) & p(0)+p(1) \end{pmatrix}.$$

Finally, for $N \geq 1$, let

$$C = \begin{pmatrix}
q(1) & 0 & 0 & \cdots & 0 & 0 \\
0 & q(2) & 0 & \cdots & 0 & 0 \\
0 & 0 & q(3) & \cdots & 0 & 0 \\
\vdots & \vdots & \vdots & \ddots & \vdots & \vdots \\
0 & 0 & 0 & \cdots & q(N-1) & 0 \\
0 & 0 & 0 & \cdots & 0 & q(N)
\end{pmatrix}. \tag{6.2.15}$$

Clearly, A, B, and C are symmetric. Let

$$u = (u(1), u(2), \ldots, u(N))^T \in \mathbb{R}^N.$$

Then, for $u \in X$, it is easy to verify that

$$\sum_{t=1}^{N} |\Delta^2 u(t-2)|^2 = u^T A u,$$

$$\sum_{t=1}^{N} p(t-1)|\Delta u(t-1)|^2 = u^T B u,$$

and

$$\sum_{t=1}^{N} q(t)|u(t)|^2 = u^T C u.$$

Thus, from (6.2.7), I_λ can be rewritten as

$$I_\lambda(u) = \frac{1}{2} u^T (A + B + C) u - \lambda \sum_{t=1}^{N} F(t, u(t))$$

$$= \Phi(u) - \lambda \sum_{t=1}^{N} F(t, u(t)), \qquad (6.2.16)$$

where

$$\Phi(u) = \frac{1}{2} u^T (A + B + C) u$$

$$= \frac{1}{2} \sum_{t=1}^{N} \left(|\Delta^2 u(t-2)|^2 + p(t-1)|\Delta u(t-1)|^2 \right.$$

$$\left. + q(t)|u(t)|^2 \right). \qquad (6.2.17)$$

Remark 6.2.2. The matrices A, B, and C satisfy the following properties:

(a) A is positive semidefinite. In fact, it is clear that 0 is an eigenvalue of A with an eigenvector $(1, 1, \ldots, 1)^T$. Moreover, it can be shown that the $(N-1)$th leading principal submatrix of A is positive definite. Thus, A is positive semidefinite.

(b) If $p(t) > 0$ for $t \in [0, N-1]_{\mathbb{Z}}$, then B is positive semidefinite. In fact, it is clear that 0 is an eigenvalue of B with an eigenvector $(1, 1, \ldots, 1)^T$. Moreover, it can be shown that the $(N-1)$th leading principal submatrix of B is positive definite. Thus, B is positive semidefinite.

(c) If $q(t) > 0$ for $t \in [1, N]_\mathbb{Z}$, then C is positive definite.

In what follows, for $i = 1, \ldots, N$, let λ_i be the eigenvalues of $A + B + C$ satisfying

$$\lambda_1 \leq \lambda_2 \leq \cdots \leq \lambda_N,$$

and let ξ_i be an eigenvector of $A + B + C$ associated with λ_i such that

$$(\xi_i, \xi_j) = \begin{cases} 0, & i \neq j, \\ 1, & i = j. \end{cases}$$

Then, for any $u = (u(1), u(2), \ldots, u(N))^T \in \mathbb{R}^N$, it is easy to see that

$$\frac{1}{2}\lambda_1 \|u\|^2 \leq \Phi(u) = \frac{1}{2}u^T(A + B + C)u \leq \frac{1}{2}\lambda_N \|u\|^2. \tag{6.2.18}$$

Moreover, by Remark 6.2.2, we see that if $p(t) > 0$ for $t \in [0, N - 1]_\mathbb{Z}$ and $q(t) > 0$ for $t \in [1, N]_\mathbb{Z}$, then $A + B + C$ is positive definite, and so

$$0 < \lambda_1 \leq \lambda_2 \leq \cdots \leq \lambda_N. \tag{6.2.19}$$

For convenience, we introduce the following notations.

$$f_0 = \liminf_{x \to 0} \min_{t \in [1,N]_\mathbb{Z}} \frac{f(t, x)}{x}, \quad f^0 = \limsup_{x \to 0} \max_{t \in [1,N]_\mathbb{Z}} \frac{f(t, x)}{x},$$

$$f_\infty = \liminf_{|x| \to \infty} \min_{t \in [1,N]_\mathbb{Z}} \frac{f(t, x)}{x}, \quad f^\infty = \limsup_{|x| \to \infty} \max_{t \in [1,N]_\mathbb{Z}} \frac{f(t, x)}{x}.$$

Theorem 6.2.1. *Assume that*

(F1) *there exists δ with $\delta > 16 + 4\overline{p} + \overline{q}$ such that $\lambda \geq \delta/f_\infty$, where*

$$\overline{p} = \max_{t \in [0,N-1]_\mathbb{Z}} |p(t)| \quad and \quad \overline{q} = \max_{t \in [1,N]_\mathbb{Z}} |q(t)|.$$

Then, BVP (6.2.1), (6.2.2) has at least one solution.

If, in addition to (F1), we further assume that

(F2) *there exists $k \in \{1, \ldots, N - 1\}$ such that*

$$\frac{\lambda_k}{f_0} < \lambda < \frac{\lambda_{k+1}}{f^0}. \tag{6.2.20}$$

Then, BVP (6.2.1), (6.2.2) has at least two nontrivial solutions.

Proof. Let $\lambda > 0$ be fixed. Since $f_\infty \geq \delta/\lambda$ by (F1), there exists $C > 0$ such that

$$\frac{f(t, x)}{x} \geq \frac{1}{\lambda}(\delta - \epsilon) \quad \text{for } (t, |x|) \in [1, N]_{\mathbb{Z}} \times (C, \infty), \tag{6.2.21}$$

where $\epsilon > 0$ satisfies

$$\delta - \epsilon > 16 + 4\bar{p} + \bar{q}. \tag{6.2.22}$$

Let $t \in [1, N]_{\mathbb{Z}}$. If $x > C$, then (6.2.21) implies that $f(t, x) \geq (\delta - \epsilon)x/\lambda$, and so

$$\begin{aligned}
F(t, x) &= \int_0^x f(t, s)ds = \int_0^C f(t, s)ds + \int_C^x f(t, s)ds \\
&\geq \int_0^C f(t, s)ds + \frac{1}{\lambda}(\delta - \epsilon)\int_C^x s\,ds \\
&= \int_0^C f(t, s)ds + \frac{\delta - \epsilon}{2\lambda}(x^2 - C^2) \\
&\geq \frac{\delta - \epsilon}{2\lambda}x^2 + \min_{t \in [1, N]_{\mathbb{Z}}}\int_0^C f(t, s)ds - \frac{\delta - \epsilon}{2\lambda}C^2. \tag{6.2.23}
\end{aligned}$$

If $x < -C$, then (6.2.21) implies that $f(t, x) \leq (\delta - \epsilon)x/\lambda$, and so

$$\begin{aligned}
F(t, x) &= -\int_x^0 f(t, s)ds = -\int_x^{-C} f(t, s)ds - \int_{-C}^0 f(t, s)ds \\
&\geq -\frac{1}{\lambda}(\delta - \epsilon)\int_x^{-C} s\,ds - \int_{-C}^0 f(t, s)ds \\
&= \frac{\delta - \epsilon}{2\lambda}(x^2 - C^2) - \int_{-C}^0 f(t, s)ds \\
&\geq \frac{\delta - \epsilon}{2\lambda}x^2 - \max_{t \in [1, N]_{\mathbb{Z}}}\int_{-C}^0 f(t, s)ds - \frac{\delta - \epsilon}{2\lambda}C^2. \tag{6.2.24}
\end{aligned}$$

If $|x| \leq C$, then

$$\begin{aligned}
F(t, x) &= \frac{\delta - \epsilon}{2\lambda}x^2 + F(t, x) - \frac{\delta - \epsilon}{2\lambda}x^2 \\
&\geq \frac{\delta - \epsilon}{2\lambda}x^2 + \min_{\substack{t \in [1, N]_{\mathbb{Z}} \\ x \in [-C, C]}}\int_0^x f(t, s)ds - \frac{\delta - \epsilon}{2\lambda}C^2. \tag{6.2.25}
\end{aligned}$$

From (6.2.23)–(6.2.25), we see that there exists a constant $C_1 \in \mathbb{R}$ such that

$$F(t, x) \geq \frac{\delta - \epsilon}{2\lambda}x^2 + C_1 \quad \text{for } (t, x) \in [1, N]_{\mathbb{Z}} \times \mathbb{R}. \tag{6.2.26}$$

Let I_λ be defined by (6.2.7). For $u \in X$, by (6.2.11), (6.2.12), and (6.2.26), we have

$$I_\lambda(u) = \frac{1}{2} \sum_{t=1}^{N} \left(|\Delta^2 u(t-2)|^2 + p(t-1)|\Delta u(t-1)|^2 + q(t)|u(t)|^2 \right)$$

$$- \lambda \sum_{t=1}^{N} F(t, u(t))$$

$$\leq \frac{1}{2} \left(16 + 4\overline{p} + \overline{q} \right) ||u||^2 - \frac{\delta - \epsilon}{2} \sum_{t=1}^{N} |u(t)|^2 - \lambda C_1 N$$

$$= \frac{1}{2} \left(16 + 4\overline{p} + \overline{q} - (\delta - \epsilon) \right) ||u|| - \lambda C_1 N. \tag{6.2.27}$$

Then, in view of (6.2.22), $I_\lambda(u) \to -\infty$ as $||u||_X \to \infty$. Thus, by Theorem 1.1.9, there exists $u_0 \in X$ such that $I'_\lambda(u_0) = 0$. From Lemma 6.2.2, $u_0(t)$ is a solution of BVP (6.2.1), (6.2.2). This shows the first part of the theorem.

To show the second part of the theorem, let $H(u) = -I_\lambda(u)$. Then, $H(0) = 0$, and for $u \in X$, (6.2.27) implies that

$$H(u) \geq \frac{1}{2} \left(\delta - \epsilon - (16 + 4\overline{p} + \overline{q}) \right) ||u|| + \lambda C_1 N.$$

Then, from (6.2.22), H is bounded below and satisfies the (PS) condition. In fact, any (PS) sequence $\{u_n\} \subset X$ must be bounded, and since the dimension X is finite, $\{u_n\}$ has a convergent subsequence.

Let

$$X_1 = \text{span}\{\xi_1, \ldots, \xi_k\} \quad \text{and} \quad X_2 = \text{span}\{\xi_{k+1}, \ldots, \xi_N\}.$$

Then, $X = \mathbb{R}^N = X_1 \oplus X_2$. From (6.2.20), there exists $\sigma > 0$ such that

$$\lambda_k < \frac{\lambda f(t, x)}{x} < \lambda_{k+1} \quad \text{for } (t, |x|) \in [1, N]_{\mathbb{Z}} \times [0, \sigma]. \tag{6.2.28}$$

Since X is finite dimensional, there is $R > 0$ such that $||u||_\infty < \sigma$ if $u \in X$ and $||u|| \leq R$. Thus, from (6.2.16) and (6.2.28), we have

$$H(u) = -I_\lambda(u) = -\frac{1}{2} u^T (A + B + C) u + \lambda \sum_{t=1}^{N} F(t, u(t))$$

$$\geq -\frac{1}{2} \lambda_k ||u||^2 + \lambda \sum_{t=1}^{N} \int_0^{u(t)} f(t, s) ds$$

$$\geq -\frac{1}{2} \lambda_k ||u||^2 + \frac{1}{2} \lambda_k ||u||^2 = 0$$

for $u \in X_1$ with $||u|| \leq R$, and

$$H(u) = -I_\lambda(u) = -\frac{1}{2}u^T(A + B + C)u + \lambda \sum_{t=1}^{N} F(t, u(t))$$

$$\leq -\frac{1}{2}\lambda_{k+1}||u||^2 + \lambda \sum_{t=1}^{N} \int_{0}^{u(t)} f(t, s)ds$$

$$\leq -\frac{1}{2}\lambda_{k+1}||u||^2 + \frac{1}{2}\lambda_{k+1}||u||^2 = 0$$

for $u \in X_2$ with $||u|| \leq R$. Thus, H has a local linking at 0.

If $\inf_{u \in X} H(u) < 0$, then by Theorem 1.1.10, H has at least two nontrivial critical points, and so does J. Hence, Lemma 6.2.2 implies that BVP (6.2.1), (6.2.2) has at least two nontrivial solutions. If $\inf_{u \in X} H(u) \geq 0$, then $H(u) = \inf_{u \in X_2} H(u) = 0$ for all $u \in X_2$ with $||u|| \leq R$. This shows that H has infinitely many critical points, and so does J. Again by Lemma 6.2.2, BVP (6.2.1), (6.2.2) has infinitely many nontrivial solutions. This completes the proof of the theorem.   □

Theorem 6.2.2. *Assume that*

(F3) $p(t) \geq 0$ and $q(t) \geq 0$ for $t \in [1, N]_{\mathbb{Z}}$ and there exists η with $\eta < \underline{q}$ such that $\lambda \leq \eta/f^\infty$, where $\underline{q} = \min_{t \in [1, N]_{\mathbb{Z}}} q(t)$.

Then, BVP (6.2.1), (6.2.2) has at least one solution.

If, in addition to (F3), we further assume that (F2) holds. Then, BVP (6.2.1), (6.2.2) has at least two nontrivial solutions.

Proof. By (F3), $f^\infty \leq \eta/\lambda$. Then, by a similar argument as obtaining (6.2.26), it follows that there exists a constant $C_2 \in \mathbb{R}$ such that

$$F(t, x) \leq \frac{\eta + \epsilon}{2\lambda}x^2 + C_2 \quad \text{for } (t, x) \in [1, T]_{\mathbb{Z}} \times \mathbb{R},$$

where $\epsilon > 0$ satisfies $\eta + \epsilon < \underline{q}$. Then, for $u \in X$, by (6.2.7), we have

$$I_\lambda(u) = \frac{1}{2}\sum_{t=1}^{N}\left(|\Delta^2 u(t-2)|^2 + p(t-1)|\Delta u(t-1)|^2 + q(t)|u(t)|^2\right)$$

$$-\lambda \sum_{t=1}^{N} F(t, u(t))$$

$$\geq \frac{1}{2}\underline{q}||u||^2 - \frac{\eta + \epsilon}{2}\sum_{t=1}^{N}|u(t)|^2 - \lambda C_2 N$$

$$= \frac{1}{2}(\underline{q} - (\eta + \epsilon))||u|| - \lambda C_2 N.$$

Then, $I_\lambda(u) \to \infty$ as $||u||_X \to \infty$, and I_λ is bounded from below and satisfies the (PS) condition. Applying Theorem 1.1.9 and Lemma 6.2.2 completes the proof of the first part of the theorem.

For the proof of the second part of the theorem, we apply Theorem 1.1.10 to J directly with X_1 and X_2 now defined as follows

$$X_1 = \text{span}\{\xi_{k+1}, \dots, \xi_N\} \quad \text{and} \quad X_2 = \text{span}\{\xi_1, \dots, \xi_k\}.$$

The rest of the proof is very similar to that of Theorem 6.2.1, and hence is omitted. This completes the proof of the theorem. $\square$

Theorem 6.2.3. *In addition to (F1), assume that the following conditions hold:*

(F4) $f(t, x)$ *is odd in* x, *i.e.,* $f(t, -x) = -f(t, x)$ *for* $(t, x) \in [1, N]_\mathbb{Z} \times \mathbb{R}$.

(F5) *there exists* $m \in \{1, \dots, N\}$ *such that* $\lambda < \lambda_m / f^0$.

Then, BVP (6.2.1), (6.2.2) *has at least* $2(N - m + 1)$ *nontrivial solutions.*

Proof. Let $H(u) = -I_\lambda(u)$. Then, $H(0) = 0$ and H is even by (F4). Since (F1) holds, from the proof of Theorem 6.2.1, we know that H is bounded below and satisfies the (PS) condition. Assume first that $f^0 > -\infty$. In view of (F5), $\lambda_m/\lambda - f^0 > 0$. Let $0 < \tau < \lambda_m/\lambda - f^0$. Then, there exists $r > 0$ such that

$$\frac{f(t, x)}{x} \leq \frac{1}{\lambda}(f^0 + \tau) \quad \text{for } (t, |x|) \in [1, T]_\mathbb{Z} \times [0, r].$$

Thus,

$$F(t, x) = \int_0^x f(t, s)\,ds \leq \frac{1}{2}(f^0 + \tau)x^2 \quad \text{for } (t, |x|) \in [1, T]_\mathbb{Z} \times [0, r]. \quad (6.2.29)$$

Define

$$K = \left\{ u = \sum_{i=m}^N c_i \xi_i : \sum_{i=m}^N c_i^2 = r^2 \right\}.$$

Let S^{N-m} be the unit sphere in $\mathbb{R}^{N-m+1}$, and define $T : K \to S^{N-m}$ by

$$T(u) = \frac{1}{r}(c_m, c_{m+1}, \dots, c_N).$$

Then, T is an odd homeomorphism between K and S^{N-m}. For $u \in K$, from (6.2.16) and (6.2.29), we have

$$H(u) = -I_\lambda(u) = -\frac{1}{2}u^T(A + B + C)u + \lambda \sum_{t=1}^N F(t, u(t))$$

$$\leq -\frac{1}{2}\lambda_m ||u||^2 + \frac{\lambda}{2}(f^0 + \tau)\sum_{t=1}^N |u(t)|^2$$

$$= \frac{1}{2}(f^0 + \tau - \lambda_m)r^2 < 0.$$

Thus, $\sup_{u \in K} H(u) < 0$. If $f^0 = -\infty$, it can be similarly shown that this inequality also holds. Hence, all the conditions of Theorem 1.1.11 are satisfied, so H has at least $2(N - m + 1)$ nontrivial critical points, which are nontrivial solutions of BVP (6.2.1), (6.2.2) by Lemma 6.2.2. This completes the proof of the theorem.   □

Theorem 6.2.4. *Assume that (F3), (F4), and the following conditions hold:*

(F6) *there exists $m \in \{1, \ldots, N\}$ such that $\lambda > \lambda_m / f_0$.*

Then, BVP (6.2.1), (6.2.2) has at least $2m$ nontrivial solutions.

Proof. Clearly, $I_\lambda(0) = 0$ and I_λ is even by (F4). Since (F3) holds, from the proof of Theorem 6.2.2, we know that I_λ is bounded below and satisfies the (PS) condition. Assume first that $f_0 < \infty$. By (F6), $f_0 - \lambda_m / \lambda > 0$. Let $0 < \tau < f_0 - \lambda_m / \lambda$. Then, there exists $\rho > 0$ such that

$$\frac{f(t, x)}{x} \geq f_0 - \tau \quad \text{for } (t, |x|) \in [1, T]_{\mathbb{Z}} \times [0, \rho].$$

Thus,

$$F(t, x) = \int_0^x f(t, s)\,ds \geq \frac{1}{2}(f_0 - \tau)x^2 \quad \text{for } (t, |x|) \in [1, T]_{\mathbb{Z}} \times [0, \rho]. \quad (6.2.30)$$

Define

$$K = \left\{ u = \sum_{i=1}^m c_i \xi_i \; : \; \sum_{i=1}^m c_i^2 = \rho^2 \right\}.$$

Let S^{m-1} be the unit sphere in $\mathbb{R}^m$, and define $T : K \to S^{m-1}$ by

$$T(u) = \frac{1}{r}(c_1, c_2, \ldots, c_m).$$

Then, T is an odd homeomorphism between K and S^{m-1}. For $u \in K$, from (6.2.16) and (6.2.30), we have

$$I_\lambda(u) = \frac{1}{2}u^T(A + B + C)u - \lambda \sum_{t=1}^N F(t, u(t))$$

$$\leq \frac{1}{2}\lambda_m \|u\|^2 - \frac{\lambda}{2}(f_0 - \tau)\sum_{t=1}^N |u(t)|^2$$

$$= \frac{1}{2}(\lambda_m - f^0 + \tau)r^2 < 0.$$

Thus, $\sup_{u \in K} I_\lambda(u) < 0$. When $f_0 = \infty$, we can show that this inequality also holds. Hence, all the conditions of Theorem 1.1.11 are satisfied. Now, an application of Theorem 1.1.11 and Lemma 6.2.2 completes the proof of the theorem.   □

To illustrate our results, we present two examples.

Example 6.2.1. In BVP (6.2.1), (6.2.2), let $\lambda = 1$, $N = 8$, $p(0) = 64$, $p(t) = t^2$, $q(t) = -t^2$, and $f(t, x) = t^{1/2}x^3$ for $t \in [1, 8]_{\mathbb{Z}}$. Then, we claim that BVP (6.2.1), (6.2.2) has at least 10 nontrivial solutions.

In fact, it is clear that $f^0 = 0$, $f_\infty = \infty$, and $f(t, x)$ is odd in x. Then, (F1) and (F4) hold.

With the above N, p, and q, let the matrices A, B, C be defined by (6.2.13)–(6.2.15). Then, using MATLAB, we find that the eigenvalues of $A + B + C$ are given by

$$\lambda_1 \approx -38.7900, \quad \lambda_2 \approx -12.7897, \quad \lambda_3 \approx -1.8889, \quad \lambda_4 \approx 10.3216,$$

$$\lambda_5 \approx 21.7289, \quad \lambda_6 \approx 44.7501, \quad \lambda_7 \approx 81.2402, \quad \lambda_8 \approx 147.4279.$$

Thus, $f^0 < \lambda_4$, i.e., (F5) holds with $m = 4$. The claim now follows from Theorem 6.2.3.

Example 6.2.2. In BVP (6.2.1), (6.2.2), let $\lambda = 1$, $N = 6$, $p(0) = 1 + \sin 6$, $p(t) = 1 + \sin t$ and $q(t) = t^2 + 2$ for $t \in [1, 6]_{\mathbb{Z}}$, and

$$f(t, x) = t^{-1/4} \begin{cases} x + 24, & x \in (1, \infty), \\ 25x, & x \in [-1, 1], \\ x - 24, & x \in (-\infty, -1). \end{cases}$$

Then, we claim that BVP (6.2.1), (6.2.2) has at least 6 nontrivial solutions. In fact, it is easy to see that $f_0 \approx 15.9736$, $f^\infty = 1$, and $f(t, x)$ is odd in x. Note that $\underline{q} = \min_{[1,6]_{\mathbb{Z}}} q(t) = 3$. Then, (F3) and (F5) hold.

With the above N, p, and q, let the matrices A, B, C be defined by (6.2.13)–(6.2.15). Then, using MATLAB, we find that the eigenvalues of $A + B + C$ are given by

$$\lambda_1 \approx 6.5981, \quad \lambda_2 \approx 13.7406, \quad \lambda_3 \approx 21.1240,$$

$$\lambda_4 \approx 27.7973, \quad \lambda_5 \approx 34.4156, \quad \lambda_6 \approx 47.1180.$$

Thus, $f_0 < \lambda_3$, i.e., (F5) holds with $m = 3$. The claim now follows from Theorem 6.2.4.

Theorem 6.2.5. *Assume that*

(F7) $p(t) > 0$ *for* $t \in [0, N - 1]_{\mathbb{Z}}$ *and* $q(t) > 0$ *for* $t \in [1, N]_{\mathbb{Z}}$;
(F8) $\lim_{|x| \to 0} \frac{|F(t,x)|}{|x|^2} = 0$ *for* $t \in [1, N]_{\mathbb{Z}}$;
(F9) $\limsup_{|x| \to \infty} \frac{F(t,x)}{|x|^2} \leq 0$ *for* $t \in [1, N]_{\mathbb{Z}}$;

(F10) *there exists $w \in X$ such that $\sum_{t=1}^{N} F(t, w(t)) > 0$.*

Then, for each $\lambda > \underline{\lambda}$, BVP (6.2.1), (6.2.2) has at least two nontrivial solutions, where $\underline{\lambda} = \inf_{u \in S} \underline{\lambda}(u)$ with

$$\underline{\lambda}(u) = \frac{\lambda_N \|u\|^2}{2 \sum_{t=1}^{N} F(t, u(t))} \tag{6.2.31}$$

and

$$S = \left\{ u \in X \; : \; \sum_{t=1}^{N} F(t, u(t)) > 0 \right\}.$$

In order to prove this theorem we need the following lemma which is useful in its own right.

Lemma 6.2.4. *If (F9) holds, then for any $\lambda > 0$, the functional I_λ is coercive and satisfies the (PS) condition.*

Proof. Let $\lambda > 0$ be fixed. We first show that I_λ is coercive, i.e.,

$$\lim_{\|u\| \to \infty} I_\lambda(u) = \infty \quad \text{for any } u \in X. \tag{6.2.32}$$

By (F9), there exists $K > 0$ such that

$$F(t, x) \leq \epsilon |x|^2 \quad \text{for } (t, x) \in [1, N]_{\mathbb{Z}} \times \mathbb{R}^N \text{ with } |x| > K, \tag{6.2.33}$$

where

$$0 < \epsilon < \frac{\lambda_1}{2\lambda}. \tag{6.2.34}$$

On the other hand, by the continuity of f, there exists $c : [1, N]_{\mathbb{Z}} \to \mathbb{R}^+$ such that

$$|F(t, x)| \leq c(t) \quad \text{for } (t, x) \in [1, N]_{\mathbb{Z}} \times \mathbb{R}^N \text{ with } |x| \leq K. \tag{6.2.35}$$

For any $u \in X$, let $S_1 = \{t \in [1, N]_{\mathbb{Z}} : |u(t)| \leq K\}$ and $S_2 = \{t \in [1, N]_{\mathbb{Z}} : |u(t)| > K\}$. Then, from (6.2.18), (6.2.33), and (6.2.35), we have

$$I_\lambda(u) \geq \frac{1}{2}\lambda_1 \|u\|^2 - \lambda \sum_{t \in S_1} F(t, u(t)) - \lambda \sum_{t \in S_2} F(t, u(t))$$

$$\geq \frac{1}{2}\lambda_1 \|u\|^2 - \lambda \sum_{t=1}^{N} c(t) - \lambda \epsilon \sum_{t=1}^{N} |u(t)|^2$$

$$= \left(\frac{1}{2}\lambda_1 - \lambda \epsilon \right) \|u\|^2 - \lambda \sum_{t=1}^{N} c(t).$$

Then, from (6.2.34), we see that $I_\lambda(u) \to \infty$ as $\|u\| \to \infty$, i.e., (6.2.32) holds.

Now, assume that $\{u_n\} \subset X$ is a (PS) sequence of I_λ. From (6.2.32), $\{u_n\}$ is bounded in X. Since the dimension of X is finite, $\{u_n\}$ has a convergent subsequence, i.e., I_λ satisfies the (PS) condition. This completes the proof of the lemma. $\qquad\square$

We can now prove Theorem 6.2.5.

Proof of Theorem 6.2.5. We first show that, for each $\lambda > 0$, 0 is a strict local minimizer of I_λ. Obviously,

$$I_\lambda(0) = \Phi(0) - \lambda\Psi(0) = 0.$$

For ϵ satisfying (6.2.34), by (F8), there exists $\kappa > 0$ such that

$$|F(t,x)| \leq \epsilon|x|^2 \quad \text{for } (t,x) \in [1,N]_{\mathbb{Z}} \times \mathbb{R}^N \text{ with } |x| \leq \kappa.$$

Note that $u(t) \leq \|u\|$ for $t \in [1,N]_{\mathbb{Z}}$. Then, for any $u \in B_\kappa(0) \setminus \{0\}$, from (6.2.18) and (6.2.34), we have

$$I_\lambda(u) \geq \frac{1}{2}\lambda_1\|u\|^2 - \lambda\epsilon \sum_{t=1}^N |u(t)|^2$$

$$= \left(\frac{1}{2}\lambda_1 - \lambda\epsilon\right)\|u\|^2 > 0.$$

Thus, for each $\lambda > 0$, 0 is a strict local minimizer of I_λ.

For $\lambda > \underline{\lambda}$, by the definition of $\underline{\lambda}$ and (6.2.31), there exists $v \in X$ such that

$$\lambda > \frac{\lambda_N\|v\|^2}{2\sum_{t=1}^N F(t,v(t))}.$$

Thus, from (6.2.18), it follows that

$$I_\lambda(v) \leq \frac{1}{2}\lambda_N\|v\|^2 - \lambda \sum_{t=1}^N F(t,v(t)) < 0 \quad \text{if } \lambda > \underline{\lambda}.$$

Thus, 0 is not a global minimizer of I_λ if $\lambda > \underline{\lambda}$.

Below, for any λ satisfying $\lambda > \underline{\lambda}$, we show that I_λ has a global minimizer. Choose $\xi \in \mathbb{R}$ such that

$$I_\lambda(w) < \xi < 0.$$

Let

$$Y = \{u \in X \ : \ I_\lambda(u) \leq \xi\}.$$

Then, $Y \neq \emptyset$ and is bounded since I_λ is coercive by Lemma 6.2.4. We claim that I_λ is bounded below on Y. Assume, to the contrary, that there exists a sequence $\{u_n\} \subset Y$ such that

$$\lim_{n \to \infty} I_\lambda(u_n) = -\infty. \tag{6.2.36}$$

Note that $\{u_n\}$ is bounded. Then, passing to a subsequence, if necessary, we may assume that $u_n \to u$ in X. Since I_λ is continuous in X, we have

$$\lim_{n \to \infty} I_\lambda(u_n) = \lim_{n \to \infty} (\Phi(u_n) - \lambda \Psi(u_n)) = \Phi(u) - \lambda \Psi(u).$$

This contradicts (6.2.36). Thus, we have

$$0 > \eta := \inf_{u \in Y} I_\lambda(u) = \inf_{u \in X} I_\lambda(u) > -\infty.$$

Let $\{u_n\} \subset Y$ be a sequence such that

$$\lim_{n \to \infty} I_\lambda(u_n) = \eta.$$

Arguing as above, we see that there exits $u_1 \in X$ such that, up to a subsequence, $u_n \to u_1$ in X. Hence, we have

$$I_\lambda(u_1) = \eta < 0, \tag{6.2.37}$$

and so $u_1 \not\equiv 0$. Clearly, u_1 is a critical point of I_λ. Then, by Lemma 6.2.2, u_1 is a nontrivial solution of BVP (6.2.1), (6.2.2).

In the following, we apply Theorem 1.1.8 to find a second critical point of I_λ when $\lambda > \underline{\lambda}$. By Lemma 6.2.4, I_λ satisfies the (PS) condition. Since 0 is a strict local minimizer of I_λ, there exists $0 < \rho < \|u_1\|$ such that

$$r := \inf_{u \in \partial B_\rho(u_0)} I_\lambda(u) > 0.$$

Then, in view of the fact that $I_\lambda(0) = 0$ and (6.2.27) holds, we see that all the conditions of Theorem 1.1.8 are satisfied with $u_0 = 0$ and the above u_1. Thus, by Theorem 1.1.8, there exists a critical point u_2 of I_λ such that

$$I_\lambda(u_2) \geq r > 0. \tag{6.2.38}$$

By (6.2.37) and (6.2.38), we see that $u_1 \neq u_2$ and $u_2 \not\equiv 0$. Hence, Lemma 6.2.2 implies that u_2 is a second nontrivial solution of BVP (6.2.1), (6.2.2). This completes the proof of the theorem. $\qquad \square$

Remark 6.2.3. The following observations should come as no surprise.

(a) In view of (F10), $S \neq \emptyset$. The conclusion of Theorem 6.2.5 still holds if $\lambda > \underline{\lambda}(w)$.

(b) Under condition (F7), (6.2.19) clearly holds.

We conclude this section with an example to illustrate the applicability of Theorem 6.2.5.

Example 6.2.3. In BVP (6.2.1), (6.2.2), let $N = 6$, $p(0) = 1 + \sin 6$, $p(t) = 1 + \sin t$ and $q(t) = t^2 + 2$ for $t \in [1, 6]_{\mathbb{Z}}$, and

$$f(t, x) = \begin{cases} (\eta + 1)|x|^{\eta-1}x, & |x| \geq 1, \\ (\zeta + 1)|x|^{\zeta-1}x, & |x| < 1, \end{cases}$$

where $0 < \eta < 1 < \zeta < \infty$. Then, we claim that, for each $\lambda > 23.5590$, BVP (6.2.1), (6.2.2) has at least two nontrivial solutions.

Clearly, (F7) holds, and for f defined above, we have

$$F(t, x) = \begin{cases} |x|^{\eta+1}, & |x| \geq 1, \\ |x|^{\zeta+1}, & |x| < 1, \end{cases}$$

so (F8) and (F9) hold as well. Moreover, (F10) also holds with $w(t) \equiv 1 \in X$.

With the above N, p, and q, let the matrices A, B, C be defined by (6.2.13)–(6.2.15). Then, using MATLAB, we find that the eigenvalues of $A + B + C$ are given by

$$\lambda_1 \approx 6.5981, \quad \lambda_2 \approx 13.7406, \quad \lambda_3 \approx 21.1240,$$

$$\lambda_4 \approx 27.7973, \quad \lambda_5 \approx 34.4156, \quad \lambda_6 \approx 47.1180.$$

In view of (6.2.21), we have $\underline{\lambda}(w) = \nu_6/2 \approx 23.5590$. The claim now follows from Theorem 6.2.5 and Remark 6.2.3.

Remark 6.2.4. In Example 6.2.1, it is interesting to observe that

$$\lim_{u \in S, \, \|u\| \to 0^+} \lambda(u) = \infty \quad \text{and} \quad \lim_{u \in S, \, \|u\| \to \infty} \lambda(u) = \infty.$$

6.3 Periodic Problems with Two Parameters

In this section, we consider the discrete nonlinear fourth order periodic BVP with two parameters

$$\begin{cases} (Lu)(t) = \lambda f(t, u(t)) - \mu g(t, u(t)), & t \in [1, N]_{\mathbb{Z}}, \\ \Delta^i u(-1) = \Delta^i u(N - 1), & i = 0, 1, 2, 3, \end{cases} \tag{6.3.1}$$

where

$$(Lu)(t) = \Delta^4 u(t - 2) - \Delta\big(p(t - 1)\Delta u(t - 1)\big) + q(t)u(t),$$

$N \geq 5$ is an integer, $p \in C([0, N]_{\mathbb{Z}}, \mathbb{R})$ with $p(0) = p(N)$, $q \in C([1, N]_{\mathbb{Z}}, \mathbb{R})$, λ and μ are two parameters with $\lambda \in (0, \infty)$ and $\mu \in [0, \infty)$, and f, $g \in C([1, N]_{\mathbb{Z}} \times \mathbb{R}, \mathbb{R})$. By a *solution* of BVP (6.3.1), we mean a function $u \in C([-1, N + 2]_{\mathbb{Z}}, \mathbb{R})$ such that u satisfies (6.3.1).

The problem (6.3.1) can be regarded as a discrete analogue of the fourth order BVP

$$\begin{cases} u^{(4)}(t) - \big(p(t)u'(t)\big)' + q(t)u(t) = \lambda f(t, u(t)) - \mu g(t, u(t)), & t \in (0, 1), \\ u^{(i)}(0) = u^{(i)}(1), & i = 0, 1, 2, 3, \end{cases}$$

(6.3.2)

which has been studied by many researchers using a variety of methods; see, for example, [29, 36, 70, 90, 204, 205] and the references therein.

Let the space X and the functional Φ be defined by (6.2.5) and (6.2.17), respectively. In addition, for $u \in X$, we define the functional Ψ by

$$\Psi(u) = \sum_{t=1}^{N} F(t, u(t)) - \frac{\mu}{\lambda} \sum_{t=1}^{N} G(t, u(t)), \tag{6.3.3}$$

where

$$F(t, x) = \int_0^x f(t, s) ds \tag{6.3.4}$$

and

$$G(t, x) = \int_0^x g(t, s) ds \tag{6.3.5}$$

for $(t, x) \in [1, N]_{\mathbb{Z}} \times \mathbb{R}$, and set

$$I_\lambda(u) = \Phi(u) - \lambda \Psi(u). \tag{6.3.6}$$

Then, Φ and Ψ are continuously Gâteaux differentiable and their Gâteaux derivatives at $u \in X$ are the functional $\Phi'(u)$ and $\Psi'(u)$ given by

$$\Phi'(u)(v) = \sum_{t=1}^{N} \big[\Delta^2 u(t-2)\Delta^2 v(t-2) + p(t-1)\Delta u(t-1)\Delta v(t-1)$$
$$+ q(t)u(t)v(t)\big]$$

and

$$\Psi'(u)(v) = \sum_{t=1}^{N} f(t, u(t))v(t) - \frac{\mu}{\lambda} \sum_{t=1}^{N} g(t, u(t))v(t)$$

for any $v \in X$.

By essentially the same arguments used to prove Lemma 6.2.2, we can prove the following result.

Lemma 6.3.1. *A function $u \in X$ is a critical point of the functional I_λ if and only if $u(t)$ is a solution of BVP* (6.3.1).

For convenience, we introduce the notations:

$$\mathcal{A} = \liminf_{\xi \to \infty} \frac{\sum_{t=1}^{N} \max_{|x| \le \xi} F(t, x)}{\xi^2}, \quad \mathcal{B} = \limsup_{\xi \to \infty} \frac{\sum_{t=1}^{N} F(t, \xi)}{\xi^2}, \quad (6.3.7)$$

$$\mathcal{C} = \liminf_{\xi \to 0^+} \frac{\sum_{t=1}^{N} \max_{|x| \le \xi} F(t, x)}{\xi^2}, \quad \mathcal{D} = \limsup_{\xi \to 0^+} \frac{\sum_{t=1}^{N} F(t, \xi)}{\xi^2}. \quad (6.3.8)$$

In the remainder of this section, we assume that:

(F11) $p(t) > 0$ for $t \in [0, N-1]_{\mathbb{Z}}$ and $q(t) > 0$ for $t \in [1, N]_{\mathbb{Z}}$;
(F12) $\mathcal{A}, \mathcal{B}, \mathcal{C}, \mathcal{D} \ge 0$,

and use the convention that $1/a = \infty$ if $a = 0$.

Let the matrices A, B, and C be defined as in (6.2.13)–(6.2.15), and as before, let λ_i, $i = 1, \ldots, N$, be the eigenvalues of $A + B + C$ satisfying

$$\lambda_1 \le \lambda_2 \le \cdots \le \lambda_N.$$

Then, in view of condition (F11), we see that (6.2.18) and (6.2.19) hold. The following notations are also needed in the statements of our results.

$$\zeta_1 = \frac{N\lambda_N}{2\mathcal{B}}, \quad \zeta_2 = \frac{\lambda_1}{2\mathcal{A}}, \quad (6.3.9)$$

$$\zeta_3 = \frac{N\lambda_N}{2\mathcal{D}}, \quad \zeta_4 = \frac{\lambda_1}{2\mathcal{C}}.$$

Our first main result in this section is given in the following theorem.

Theorem 6.3.1. *Assume that*

$$\mathcal{A} < \frac{\lambda_1}{N\lambda_N}\mathcal{B}. \quad (6.3.10)$$

Then, for each $\lambda \in (\zeta_1, \zeta_2)$, each function $g \in C([1, N]_{\mathbb{Z}} \times \mathbb{R}, \mathbb{R})$ with

$$g(t, x) \le 0 \text{ on } [1, N]_{\mathbb{Z}} \times \mathbb{R} \quad and \quad G_\infty = \liminf_{\xi \to \infty} \frac{\sum_{t=1}^{N} G(t, \xi)}{\xi^2} > -\infty,$$

$$(6.3.11)$$

and each $\mu \in [0, \overline{\mu}_1)$ with

$$\overline{\mu}_1 = \frac{\lambda_1 - 2\lambda\mathcal{A}}{-2G_\infty}, \quad (6.3.12)$$

BVP (6.3.1) *has a sequence of solutions that is unbounded in X.*

Proof. To prove this theorem we will apply Theorem 1.1.5 (b), so let the functionals Φ, Ψ, $I_\lambda : X \to \mathbb{R}$ be defined by (6.2.17), (6.3.3), and (6.3.6), respectively. Then, it is clear that Φ and Ψ satisfy all the regularity assumptions needed in Theorem 1.1.5.

From the definition of $\mathcal{A}$ in (6.3.7), there exists a sequence $\{x_n\}$ of positive numbers such that $\lim_{n\to\infty} x_n = \infty$ and

$$\mathcal{A} = \lim_{n\to\infty} \frac{\sum_{t=1}^N \max_{|x|\le x_n} F(t,x)}{x_n^2}. \tag{6.3.13}$$

Let $r_n = \frac{\lambda_1 x_n^2}{2}$. Then, for any $u \in X$ with $\Phi(u) < r_n$, from (6.2.18), we have

$$\max_{t\in[1,N]_{\mathbb{Z}}} |u(t)| \le ||u|| < \left(\frac{2\Phi(u)}{\lambda_1}\right)^{1/2} < \left(\frac{2r_n}{\lambda_1}\right)^{1/2} = x_n. \tag{6.3.14}$$

Note that $\Phi(0) = \Psi(0) = 0$ by (6.2.18), so, from (1.1.4), it follows that

$$\varphi(r_n) = \inf_{u\in\Phi^{-1}(-\infty,r_n)} \frac{\left(\sup_{v\in\Phi^{-1}(-\infty,r_n)} \Psi(v)\right) - \Psi(u)}{r_n - \Phi(u)}$$

$$\le \frac{\sup_{v\in\Phi^{-1}(-\infty,r_n)} \Psi(v)}{r_n}$$

$$\le \frac{\sum_{t=1}^N \max_{|x|\le x_n} F(t,x) - \frac{\mu}{\lambda} \sum_{t=1}^N \min_{|x|\le x_n} G(t,x)}{r_n}$$

$$= \frac{2 \sum_{t=1}^N \max_{|x|\le x_n} F(t,x) - \frac{\mu}{\lambda} \sum_{t=1}^N G(t,x_n)}{\lambda_1 \, x_n^2}.$$

Hence, for γ defined in Theorem 1.1.5, (6.3.11) and (6.3.13) imply that

$$\gamma \le \liminf_{n\to\infty} \varphi(r_n) \le \frac{2}{\lambda_1}\left(\mathcal{A} - \frac{\mu}{\lambda}G_\infty\right) < \infty. \tag{6.3.15}$$

We claim that

$$\text{if } \lambda \in (\zeta_1, \zeta_2) \text{ and } \mu \in [0, \overline{\mu}_1), \text{ then } \lambda \in (0, 1/\gamma). \tag{6.3.16}$$

In fact, for $\lambda \in (\zeta_1, \zeta_2)$ and $\mu \in [0, \overline{\mu}_1)$, we have $\lambda > 0$. Then, from (6.3.9), (6.3.12), and (6.3.15), we see that

$$\gamma < \frac{2}{\lambda_1}\left(\mathcal{A} - \frac{\overline{\mu}_1}{\lambda}G_\infty\right) = \frac{2}{\lambda_1}\left(\mathcal{A} + \frac{\lambda_1 - 2\lambda\mathcal{A}}{2\lambda}\right) = \frac{1}{\lambda} \quad \text{if } G_\infty \ne 0,$$

and

$$\gamma \le \frac{2\mathcal{A}}{\lambda_1} = \frac{1}{\zeta_2} < \frac{1}{\lambda} \quad \text{if } G_\infty = 0.$$

Thus, we always have $\lambda < 1/\gamma$. This shows that (6.3.16) holds.

Let $\lambda \in (\zeta_1, \zeta_2)$ and $\mu \in [0, \overline{\mu}_1)$ be fixed. Then, in view of (6.3.15) and (6.3.16), from Theorem 1.1.5 (b), we see that one of the following alternatives holds

(b_1) either $I_\lambda := \Phi - \lambda\Psi$ has a global minimum, or

(b_2) there exists a sequence $\{u_n\}$ of critical points of I_λ such that $\lim_{n\to\infty}\|u_n\|_X = \infty$.

In the following, we show that alternative (b_1) does not hold. From the definition of $\mathcal{B}$ in (6.3.7), there exists a sequence $\{\eta_n\}$ of positive numbers such that $\lim_{n\to\infty}\eta_n = \infty$ and

$$\mathcal{B} = \lim_{n\to\infty}\frac{\sum_{t=1}^{N}F(t,\eta_n)}{\eta_n^2}. \tag{6.3.17}$$

For each $n \in \mathbb{N}$, define

$$w_n = (\eta_n, \eta_n, \ldots, \eta_n)^T \in X = \mathbb{R}^N. \tag{6.3.18}$$

Then, from (6.3.3) and (6.2.18), we have

$$\frac{1}{2}N\lambda_1\eta_n^2 \le \Phi(w_n) \le \frac{1}{2}N\lambda_N\eta_n^2$$

and

$$\Psi(w_n) = \sum_{t=1}^{N}F(t,\eta_n) - \frac{\mu}{\lambda}\sum_{t=1}^{N}G(t,\eta_n).$$

Since $G(t,\eta_n) \le 0$ for $t \in [1,N]_{\mathbb{Z}}$ by (6.3.11), we have

$$I_\lambda(w_n) = \Phi(w_n) - \lambda\Psi(w_n)$$
$$\le \frac{1}{2}N\lambda_N\eta_n^2 - \lambda\sum_{t=1}^{N}F(t,\eta_n) + \mu\sum_{t=1}^{N}G(t,\eta_n)$$
$$\le \frac{1}{2}N\lambda_N\eta_n^2 - \lambda\sum_{t=1}^{N}F(t,\eta_n). \tag{6.3.19}$$

Now, we consider two cases.

Case 1: $\mathcal{B} < \infty$. From $\lambda > \zeta_1$ and the definition of ζ_1 in (6.3.9), we have $\mathcal{B} - \frac{N\lambda_N}{2\lambda} > 0$. Let

$$\epsilon \in \left(0, \mathcal{B} - \frac{N\lambda_N}{2\lambda}\right). \tag{6.3.20}$$

From (6.3.17), there exists $N_1 \in \mathbb{N}$ such that

$$\sum_{t=1}^{N}F(t,\eta_n) > (\mathcal{B} - \epsilon)\eta_n^2 \quad \text{for } n \ge N_1.$$

Then, from (6.3.19),

$$I_\lambda(w_n) \le \left(\frac{1}{2}N\lambda_N - \lambda(\beta - \epsilon)\right)\eta_n^2.$$

Thus, by (6.3.20) and the fact that $\lim_{n \to \infty} \eta_n = \infty$, it follows that $\lim_{n \to \infty} I_\lambda(w_n) = -\infty$.

Case 2: $\mathcal{B} = \infty$. Choose

$$M > \frac{N\lambda_N}{2\lambda}. \tag{6.3.21}$$

From (6.3.17), there exists $N_2 \in \mathbb{N}$ such that

$$\sum_{t=1}^{N} F(t, \eta_n) > M\eta_n^2 \quad \text{for } n \geq N_2.$$

Then, from (6.3.19), we have

$$I_\lambda(w_n) \leq \left(\frac{1}{2} N\lambda_N - \lambda M \right) \eta_n^2.$$

Hence, from (6.3.21) and the fact that $\lim_{n \to \infty} \eta_n = \infty$, we have $\lim_{n \to \infty} I_\lambda(w_n) = -\infty$.

Combining the above two cases, we see that the functional I_λ is always unbounded from below. Thus, the alternative (b_1) does not hold. Therefore, there exists a sequence $\{u_n\}$ of critical points of I_λ such that $\lim_{n \to \infty} \|u_n\| = \infty$. Finally, an application of Lemma 6.3.1 completes the proof of the theorem. $\qquad\square$

In our next result, the sequence of solutions, whose existence we prove, converges to zero.

Theorem 6.3.2. *Assume that*

$$\mathcal{C} < \frac{\lambda_1}{N\lambda_N} \mathcal{D}. \tag{6.3.22}$$

Then, for each $\lambda \in (\zeta_3, \zeta_4)$, each function $g \in C(\mathbb{R}, \mathbb{R})$ satisfying (6.3.11), and each $\mu \in [0, \overline{\mu}_2)$ with

$$\overline{\mu}_2 = \frac{\lambda_1 - 2\lambda\mathcal{C}}{-2G_\infty},$$

BVP (6.3.1) has a sequence of solutions converging uniformly to zero in X.

Proof. The proof of this result uses part (c) of Theorem 1.1.5 but the argument is similar to that used in the proof of Theorem 6.3.1, so we only sketch the details. Again let the functionals Φ, Ψ, $I_\lambda : X \to \mathbb{R}$ be defined by (6.2.17), (6.3.3), and (6.3.6), respectively, and observe that the regularity conditions required in Theorem 1.1.5 are again satisfied. From the definition of $\mathcal{C}$ in (6.3.8), there exists a sequence $\{x_n\}$ of positive numbers such that $\lim_{n\to\infty} x_n = 0$ and

$$\mathcal{C} = \lim_{n\to\infty} \frac{\sum_{t=1}^{N} \max_{|x|\leq x_n} F(t,x)}{x_n^2}.$$

By the fact that $\inf_X \Phi = 0$ and the definition δ, we have $\delta = \liminf_{r\to 0^+} \varphi(r)$. Then, as in showing (6.3.15) and (6.3.16) in the proof of Theorem 6.3.1, we can prove that $\delta < \infty$ and that if $\lambda \in (\zeta_3, \zeta_4)$ and $\mu \in [0, \overline{\mu}_2)$, then $\lambda \in (0, 1/\delta)$. Let $\lambda \in (\zeta_3, \zeta_4)$ and $\mu \in [0, \overline{\mu}_2)$ be fixed. Then, from Theorem 1.1.5 (c), we see that one of the following alternatives holds

- (c$_1$) either there is a global minimum of Φ which is a local minimum of $I_\lambda = \Phi - \lambda\Psi$, or
- (c$_2$) there exists a sequence $\{u_n\}$ of critical points of I_λ which converges weakly to a global minimum of Φ.

Next, we show that alternative (c$_1$) does not hold. By the definition of $\mathcal{C}$ in (6.3.8), there exists a sequence $\{\eta_n\}$ of positive numbers such that $\lim_{n\to\infty} \eta_n = 0$ and

$$\mathcal{C} = \lim_{n\to\infty} \frac{\sum_{t=1}^{N} F(t, \eta_n)}{\eta_n^2}. \tag{6.3.23}$$

For each $n \in \mathbb{N}$, let $w_n \in X = \mathbb{R}^N$ be defined by (6.3.18) with the above η_n. Then, as in the cases 1 and 2 of the proof of Theorem 6.3.1, we can obtain that, for n large enough, if $\mathcal{C} < \infty$, then

$$I_\lambda(w_n) \leq \left(\frac{1}{2}N\lambda_N - \lambda(\mathcal{C} - \epsilon)\right)\eta_n^2,$$

where

$$\epsilon \in \left(0, \mathcal{D} - \frac{N\lambda_N}{2\lambda}\right);$$

and if $\mathcal{C} = \infty$, then

$$I_\lambda(w_n) \leq \left(\frac{1}{2}N\lambda_N - \lambda M\right)\eta_n^2,$$

where M satisfies (6.3.21). Hence, we always have $I_\lambda(w_n) < 0$ for large n. Thus, since $\lim_{n\to\infty} I_\lambda(w_n) = I_\lambda(0) = 0$, we see that 0 is not a local minimum of I_λ. This, together with the fact that 0 is the only global minimum of Φ, shows that alternative (c$_1$) does not hold. Therefore, there exists a sequence $\{u_n\}$ of critical points of I_λ that converges weakly (and thus also strongly) to 0. Finally, in view of Lemma 6.3.1, the conclusion of the theorem follows. $\qquad\square$

Remark 6.3.1. Notice that in the above theorems, it should be clear that condition (F12) implies $\zeta_1 \geq 0$ and $\zeta_3 \geq 0$. In addition, (6.3.10) and (6.3.22) imply that $\zeta_1 < \zeta_2$ and $\zeta_3 < \zeta_4$, and this ensures that the intervals (ζ_1, ζ_2) and (ζ_3, ζ_4) are nonempty. Note also that the interval $[0, \bar{\mu}_1)$ is well defined since $\bar{\mu}_1 > 0$ under the condition that $\lambda < \zeta_2$ and the interval $[0, \bar{\mu}_2)$ is well defined since $\bar{\mu}_2 > 0$ under the condition that $\lambda < \zeta_4$.

The following corollaries follow easily from Theorems 6.3.1 and 6.3.2.

Corollary 6.3.1. *Assume that (6.3.10) holds. Then, for each $\lambda \in (\zeta_1, \zeta_2)$, the BVP*

$$\begin{cases} \Delta^4 u(t-2) - \Delta\big(p(t-1)\Delta u(t-1)\big) + q(t)u(t) = \lambda f(t, u(t)), & t \in [1, N]_{\mathbb{Z}}, \\ \Delta^i u(-1) = \Delta^i u(N-1), & i = 0, 1, 2, 3, \end{cases}$$

(6.3.24)

has a sequence of solutions that is unbounded in X.

Corollary 6.3.2. *Assume that (6.3.22) holds. Then, for each $\lambda \in (\zeta_3, \zeta_4)$, BVP (6.3.24) has a sequence of solutions converging uniformly to zero in X.*

Corollary 6.3.3. *Assume that $\mathcal{A} = 0$ and $\mathcal{B} = \infty$. Then, for each $\lambda \in (0, \infty)$, each function $g \in C([1, N]_{\mathbb{Z}} \times \mathbb{R}, \mathbb{R})$ with*

$$g(t, x) \leq 0 \text{ on } [1, N]_{\mathbb{Z}} \times \mathbb{R} \quad and \quad G_\infty = \liminf_{\xi \to \infty} \frac{\sum_{t=1}^{N} G(t, \xi)}{\xi^2} = 0, \quad (6.3.25)$$

and each $\mu \in [0, \infty)$, BVP (6.3.1) has a sequence of solutions that is unbounded in X.

Corollary 6.3.4. *Assume that $\mathcal{C} = 0$ and $\mathcal{D} = \infty$. Then, for each $\lambda \in (0, \infty)$, each function $g \in C([1, N]_{\mathbb{Z}} \times \mathbb{R}, \mathbb{R})$ satisfying (6.3.25), and each $\mu \in [0, \infty)$, BVP (6.3.1) has a sequence of solutions converging uniformly to zero in X.*

We conclude this section with an example to illustrate the usefulness of our results. The form of the nonlinear function $f(t, x)$ is partly motivated by [97, Example 3.1].

Example 6.3.1. Let $\{a_n\}$ and $\{b_n\}$ be sequences defined by $b_1 = 2$, $b_{n+1} = b_n^6$, and $a_n = b_n^4$ for $n \in \mathbb{N}$. In BVP (6.3.1), let $N \geq 5$ be a fixed integer,

$$p(t) = -t^2 + Nt + 1, \quad t \in [0, N]_{\mathbb{Z}},$$

$$q(t) = 2 + \sin t, \quad t \in [1, N]_{\mathbb{Z}},$$

$$g(t, x) = -(t^{1/2} + 1)x^{2/5}, \quad (t, x) \in [1, N]_{\mathbb{Z}} \times \mathbb{R},$$

and

$$f(t, x)$$
$$= t^3 \begin{cases} b_1^3 \sqrt{1 - (1 - x)^2} + 1, & \text{for } (t, x) \in [1, N]_{\mathbb{Z}} \times [0, b_1], \\ (a_n - b_n^3)\sqrt{1 - (a_n - 1 - x)^2} + 1, \\ & \text{for } (t, x) \in [1, N]_{\mathbb{Z}} \times \cup_{n=1}^{\infty}[a_n - 2, a_n], \\ (b_{n+1}^3 - a_n)\sqrt{1 - (b_{n+1} - 1 - x)^2} + 1, \\ & \text{for } (t, x) \in [1, N]_{\mathbb{Z}} \times \cup_{n=1}^{\infty}[b_{n+1} - 2, b_{n+1}], \\ 1, & \text{otherwise.} \end{cases}$$

Then, we claim that for each $\lambda \in (0, \infty)$ and $\mu \in [0, \infty)$, BVP (6.3.1) has a sequence of solutions that is unbounded in X.

In fact, with the above f, g, p, and q, it is clear that $p \in C([0, N]_{\mathbb{Z}}, \mathbb{R})$ with $p(0) = p(N)$, $q \in C([1, N]_{\mathbb{Z}}, \mathbb{R})$, $f, g \in C([1, N]_{\mathbb{Z}} \times \mathbb{R}, \mathbb{R})$, and (F11) and (6.3.25) hold. Moreover, for $F(t, x)$ defined by (6.3.4), by simple computations, we see that

$$F(t, a_n) = t^3 \left(\int_0^{a_n} 1 ds + b_1^3 \int_0^2 \sqrt{1 - (1 - s)^2} \, ds \right.$$
$$+ \sum_{i=1}^{n} \int_{a_i - 2}^{a_i} (a_i - b_i^3)\sqrt{1 - (a_i - 1 - s)^2} \, ds$$
$$\left. + \sum_{i=1}^{n-1} \int_{b_{i+1} - 2}^{b_{i+1}} (b_{i+1}^3 - a_i)\sqrt{1 - (b_{i+1} - 1 - s)^2} \, ds \right)$$
$$= t^3 \left(\frac{\pi}{2} a_n + a_n \right)$$

and

$$F(t, b_n) = t^3 \left(\int_0^{b_n} 1 ds + b_1^3 \int_0^2 \sqrt{1 - (1 - s)^2} \, ds \right.$$
$$+ \sum_{i=1}^{n-1} \int_{a_i - 2}^{a_i} (a_i - b_i^3)\sqrt{1 - (a_i - 1 - s)^2} \, ds$$
$$\left. + \sum_{i=1}^{n-1} \int_{b_{i+1} - 2}^{b_{i+1}} (b_{i+1}^3 - a_i)\sqrt{1 - (b_{i+1} - 1 - s)^2} \, ds \right)$$
$$= t^3 \left(\frac{\pi}{2} b_n^3 + b_n \right)$$

for $t \in [1, N]_{\mathbb{Z}}$. Thus,

$$\lim_{n \to \infty} \frac{F(t, a_n)}{a_n^2} = 0 \quad \text{and} \quad \lim_{n \to \infty} \frac{F(t, b_n)}{b_n^2} = \infty \quad \text{for } t \in [1, N]_{\mathbb{Z}}.$$

Then, for $\mathcal{A}$ and $\mathcal{B}$ defined in (6.3.7), it is easy to see that

$$\mathcal{A} = \liminf_{\xi \to \infty} \frac{F(t, \xi) \sum_{t=1}^{N} t^3}{\xi^2} = 0 \quad \text{and} \quad \mathcal{B} = \limsup_{\xi \to \infty} \frac{F(t, \xi) \sum_{t=1}^{N} t^3}{\xi^2} = \infty.$$

Thus, all the conditions of Corollary 6.3.3 are satisfied. The claim then follows directly from Corollary 6.3.3.

6.4   Multi-point Problems with Several Parameters

In this section, we will examine a discrete boundary value problem with four parameters with the intention of proving the existence of three or even infinitely many solutions. The equation to be studied can be viewed as a discrete version of the generalized beam equation.

Consider the four-parameter fourth order BVP consisting of the generalized discrete beam equation

$$\Delta^4 u(t-2) - \alpha \Delta^2 u(t-1) + \beta u(t) = \lambda f(t, u(t)), \quad t \in [1, T]_{\mathbb{Z}}, \quad (6.4.1)$$

and the boundary condition (BC)

$$u(0) = \Delta u(-1) = \Delta^2 u(T) = 0, \quad \Delta^3 u(T-1) - \alpha \Delta u(T) = \mu g(u(T+1)), \quad (6.4.2)$$

where $T \geq 2$ is an integer, $\alpha, \beta \in \mathbb{R}$ and $\lambda, \mu \in \mathbb{R}^+ := [0, \infty)$ are parameters, $f \in C([1, T]_{\mathbb{Z}} \times \mathbb{R}, \mathbb{R})$, and $g \in C(\mathbb{R}, \mathbb{R})$. By a *solution* of (6.4.1), (6.4.2), we mean a function $u \in C([-1, T+2]_{\mathbb{Z}}, \mathbb{R})$ such that u satisfies (6.4.1) and (6.4.2). We assume throughout, and without further mention, that the following condition holds:

(F13) α and β satisfy that

$$1 + \alpha_-(T+1)^2 + \beta_- T^2(T+1)^2 > 0,$$

where $\alpha_- = \min\{\alpha, 0\}$ and $\beta_- = \min\{\beta, 0\}$.

Depending on the values of the parameters α, β, λ, and μ, BVP (6.4.1), (6.4.2) covers many problems as special cases. For instance, if $\alpha = \beta = 0$ and $\mu = 1$, BVP (6.4.1), (6.4.2) becomes

$$\Delta^4 u(t-2) = \lambda f(t, u(t)), \quad t \in [1, T]_{\mathbb{Z}}, \quad (6.4.3)$$

and

$$u(0) = \Delta u(-1) = \Delta^2 u(T) = 0, \ \Delta^3 u(T-1) = g(u(T+1)). \quad (6.4.4)$$

We remark that the continuous version of BVP (6.4.3), (6.4.4), i.e., the BVP consisting of the equation

$$u^{(4)}(t) = \lambda f(t, u), \quad t \in (0, 1),$$

and the BC

$$u(0) = u'(0) = u''(1) = 0, \ u'''(1) = g(u(1)),$$

has been recently investigated in [284], and existence results for multiple solutions are obtained there.

Here, we wish to give criteria for the existence of three and infinitely many solutions for BVP (6.4.1), (6.4.2). We also present a number of consequences of our main results.

We define the real vector space

$$X = \left\{ u : [-1, T+2]_{\mathbb{Z}} \to \mathbb{R} \ : \ u(-1) = u(0) = 0, \ \Delta^2 u(T) = 0 \right\}, \quad (6.4.5)$$

where for any $u \in X$, we let

$$\|u\|_X = \left(\sum_{t=1}^{T+1} \left(|\Delta^2 u(t-2)|^2 + \alpha |\Delta u(t-1)|^2 \right) + \beta \sum_{t=1}^{T} |u(t)|^2 \right)^{1/2}. \quad (6.4.6)$$

We set

$$\rho = (T+1)^{3/2} \left(1 + \alpha_-(T+1)^2 + \beta_- T^2 (T+1)^2 \right)^{-1/2}, \quad (6.4.7)$$

and note that condition (F13) clearly implies $\rho > 0$.

We will need some preliminary lemmas in order to prove our main results in this section.

Lemma 6.4.1. *For any $u \in X$, we have*

$$\sum_{t=1}^{T+1} \left(|\Delta^2 u(t-2)|^2 + \alpha |\Delta u(t-1)|^2 \right) + \beta \sum_{t=1}^{T} |u(t)|^2 \geq 0 \quad (6.4.8)$$

and

$$|u(t)| \leq \rho \|u\|_X \quad \text{for } t \in [1, T+1]_{\mathbb{Z}}. \quad (6.4.9)$$

Hence, $\| \cdot \|_X$ is a norm on X with which X becomes a $T+1$ dimensional separable and reflexive Banach space.

Proof. Let $u \in X$ and $t \in [1, T+1]_\mathbb{Z}$. Note that

$$\Delta u(t-1) = \Delta u(-1) + \sum_{i=1}^{t} \Delta^2 u(i-2) = \sum_{i=1}^{t} \Delta^2 u(i-2).$$

Then, by Hölder's inequality,

$$|\Delta u(t-1)| \leq \sum_{i=1}^{t} |\Delta^2 u(i-2)| \leq \sum_{i=1}^{T+1} |\Delta^2 u(i-2)|$$

$$\leq (T+1)^{1/2} \left(\sum_{i=1}^{T+1} |\Delta^2 u(i-2)|^2 \right)^{1/2}.$$

Consequently,

$$|\Delta u(t-1)|^2 \leq (T+1) \sum_{i=1}^{T+1} |\Delta^2 u(i-2)|^2,$$

which in turn implies that

$$\sum_{t=1}^{T+1} |\Delta u(t-1)|^2 \leq (T+1)^2 \sum_{t=1}^{T+1} |\Delta^2 u(t-2)|^2. \tag{6.4.10}$$

Similarly, note that

$$u(t) = u(0) + \sum_{i=1}^{t} \Delta u(i-1) = \sum_{i=1}^{t} \Delta u(i-1).$$

Then, from Hölder's inequality and (6.4.10), we have that for $t \in [1, T+1]_\mathbb{Z}$,

$$|u(t)| \leq \sum_{i=1}^{T+1} |\Delta u(i-1)| \leq (T+1)^{1/2} \left(\sum_{i=1}^{T+1} |\Delta u(i-1)|^2 \right)^{1/2}$$

$$\leq (T+1)^{3/2} \left(\sum_{i=1}^{T+1} |\Delta^2 u(i-2)|^2 \right)^{1/2}, \tag{6.4.11}$$

and for $t \in [1, T]_\mathbb{Z}$,

$$|u(t)| \leq \sum_{i=1}^{T} |\Delta u(i-1)| \leq T^{1/2} \left(\sum_{i=1}^{T} |\Delta u(i-1)|^2 \right)^{1/2}.$$

Then,

$$|u(t)|^2 \leq T \sum_{i=1}^{T} |\Delta u(i-1)|^2 \quad \text{for } t \in [1, T].$$

Thus,

$$\sum_{t=1}^{T} |u(t)|^2 \leq T^2 \sum_{i=1}^{T} |\Delta u(i-1)|^2 \leq T^2 \sum_{i=1}^{T+1} |\Delta u(i-1)|^2. \qquad (6.4.12)$$

From (6.4.10),

$$\sum_{t=1}^{T} |u(t)|^2 \leq T^2(T+1)^2 \sum_{t=1}^{T+1} |\Delta^2 u(t-2)|^2. \qquad (6.4.13)$$

Using (6.4.10), (6.4.13), and condition (F13), we have

$$\sum_{t=1}^{T+1} \left(|\Delta^2 u(t-2)|^2 + \alpha |\Delta u(t-1)|^2 \right) + \beta \sum_{t=1}^{T} |u(t)|^2$$

$$\geq \left(1 + \alpha_-(T+1)^2 + \beta_- T^2(T+1)^2 \right) \sum_{t=1}^{T+1} \left(|\Delta^2 u(t-2)|^2 \geq 0, \quad (6.4.14) \right.$$

i.e., (6.4.8) holds.

From (6.4.6) and (6.4.14), we have

$$||u||_X \geq \left(1 + \alpha_-(T+1)^2 + \beta_- T^2(T+1)^2 \right)^{1/2} \left(\sum_{t=1}^{T+1} \left(|\Delta^2 u(t-2)|^2 \right) \right)^{1/2}.$$

This, together with (6.4.11), shows that

$$||u||_X \geq \rho^{-1} |u(t)|,$$

for $t \in [1, T+1]_{\mathbb{Z}}$, where ρ is defined by (6.4.7). Therefore, (6.4.9) holds. This competes the proof of the lemma. $\square$

We will need to define some functionals, so for any $u \in X$, let the functionals Φ, J_1, J_2, Ψ_1, Ψ_2, and Ψ be defined by

$$\Phi(u) = \frac{1}{2} ||u||_X^2, \qquad (6.4.15)$$

$$J_1(u) = \Psi_2(u) = -G(u(T+1)), \qquad (6.4.16)$$

$$\Psi_1(u) = J_2(u) = \sum_{t=1}^{T} F(t, u(t)), \qquad (6.4.17)$$

and

$$\Psi(u) = \sum_{t=1}^{T} F(t, u(t)) - \frac{\mu}{\lambda} G(u(T+1)), \qquad (6.4.18)$$

where

$$F(t, u(t)) = \int_0^{u(t)} f(t, s)ds \qquad (6.4.19)$$

and

$$G(x) = \int_0^x g(s)ds, \quad x \in \mathbb{R}. \qquad (6.4.20)$$

Then, Φ, J_i, Ψ_i, $i = 1, 2$, and Ψ are well defined and continuously differentiable and their derivatives at $u \in X$ are the functionals $\Phi'(u)$, $J_i'(u)$, $\Psi_i'(u)$, and $\Psi'(u)$ given by

$$\Phi'(u)(v) = \sum_{t=1}^{T+1} \left(\Delta^2 u(t-2)\Delta^2 v(t-2) + \alpha \Delta u(t-1)\Delta v(t-1) \right)$$

$$+ \beta \sum_{t=1}^{T} u(t)v(t), \qquad (6.4.21)$$

$$J_1'(u)(v) = \Psi_2'(u)(v) = -g(u(T+1))v(T+1), \qquad (6.4.22)$$

$$\Psi_1'(u)(v) = J_2'(u)(v) = \sum_{t=1}^{T} f(t, u(t))v(t), \qquad (6.4.23)$$

and

$$\Psi'(u)(v) = \sum_{t=1}^{T} f(t, u(t))v(t) - \frac{\mu}{\lambda} g(u(T+1))v(T+1)$$

for any $v \in X$.

Our next lemma is arithmetic in nature and will help us in rewriting our functionals.

Lemma 6.4.2. *For any $u, v \in X$, we have*

$$\sum_{t=1}^{T+1} \Delta^2 u(t-2)\Delta^2 v(t-2) = -\Delta^3 u(T-1)v(T+1) + \sum_{t=1}^{T} \Delta^4 u(t-2)v(t)$$

$$(6.4.24)$$

and

$$\sum_{t=1}^{T+1} \Delta u(t-1)\Delta v(t-1) = \Delta u(T)v(T+1) - \sum_{t=1}^{T} \Delta^2 u(t-1)v(t). \quad (6.4.25)$$

Proof. We first prove (6.4.24). For any $u, v \in X$, by the summation by parts formula and the fact that $\Delta^2 u(T) = \Delta v(-1) = v(0) = 0$, it follows that

$$\sum_{t=1}^{T+1} \Delta^2 u(t-2) \Delta^2 v(t-2)$$

$$= \Delta^2 u(T) \Delta v(T) - \Delta^2 u(-1) \Delta v(-1) - \sum_{t=1}^{T+1} \Delta^3 u(t-2) \Delta v(t-1)$$

$$= -\sum_{t=1}^{T+1} \Delta^3 u(t-2) \Delta v(t-1)$$

$$= -\Delta^3 u(T-1) \Delta v(T) - \sum_{t=1}^{T} \Delta^3 u(t-2) \Delta v(t-1)$$

$$= -\Delta^3 u(T-1) \Delta v(T) - \Delta^3 u(T-1) v(T) + \Delta^3 u(-1) v(0)$$

$$+ \sum_{t=1}^{T} \Delta^4 u(t-2) v(t)$$

$$= -\Delta^3 u(T-1) v(T+1) + \sum_{t=1}^{T} \Delta^4 u(t-2) v(t),$$

i.e., (6.4.24) holds.

Now, we show (6.4.25). Again, by the summation by parts formula and the fact that $v(0) = 0$, we have

$$\sum_{t=1}^{T+1} \Delta u(t-1) \Delta v(t-1)$$

$$= \Delta u(T) \Delta v(T) + \sum_{t=1}^{T} \Delta u(t-1) \Delta v(t-1)$$

$$= \Delta u(T) \Delta v(T) + \Delta u(T) v(T) - \Delta u(0) v(0) - \sum_{t=1}^{T} \Delta^2 u(t-1) v(t)$$

$$= \Delta u(T) v(T+1) - \sum_{t=1}^{T} \Delta^2 u(t-1) v(t),$$

i.e., (6.4.25) holds. This completes the proof of the lemma. $\quad\square$

The next three lemmas relate the critical points of our functionals to the solutions of our problem.

Lemma 6.4.3. *If $u \in X$ is a critical point of the functional $\Phi - \lambda\Psi_1 - \mu J_1$, then u is a solution of BVP (6.4.1), (6.4.2).*

Proof. Let $u \in X$ be a critical point of the functional $\Phi - \lambda\Psi_1 - \mu J_1$, i.e., $\Phi'(u)(v) - \lambda\Psi_1'(u)(v) - \mu J_1'(u)(v) = 0$ for any $v \in X$. Then, from (6.4.21)–(6.4.23) and Lemma 6.4.2,

$$\Phi'(u)(v) - \lambda\Psi_1'(u)(v) - \mu J_1'(u)(v)$$
$$= -\big(\Delta^3 u(T-1) - \alpha\Delta u(T) - \mu g(u(T+1))\big)v(T+1)$$
$$+ \sum_{t=1}^{T} \big(\Delta^4 u(t-2) - \alpha\Delta^2 u(t-1) + \beta(t)u(t) - \lambda f(t, u(t))\big)v(t)$$
$$= 0.$$

Thus, by the arbitrariness of $v \in X$, we see that

$$\Delta^3 u(T-1) - \alpha\Delta u(T) = \mu g(u(T+1))$$

and

$$\Delta^4 u(t-2) - \alpha\Delta^2 u(t-1) + \beta(t)u(t) - \lambda f(t, u(t)) = 0 \quad \text{for } t \in [1, T].$$

Note that u also satisfies $u(0) = \Delta u(-1) = \Delta^2 u(T) = 0$. Then, u is a solution of BVP (6.4.1), (6.4.2). This completes the proof of the lemma. $\square$

By (6.4.16), (6.4.17), (6.4.18), and Lemma 6.4.3, the following results are obvious.

Lemma 6.4.4. *If $u \in X$ is a critical point of the functional $\Phi - \lambda J_2 - \mu\Psi_2$, then u is a solution of BVP (6.4.1), (6.4.2).*

Lemma 6.4.5. *The function $u \in X$ is a critical point of the functional $\Phi - \lambda\Psi$ if any only if u is a solution of BVP (6.4.1), (6.4.2).*

We are now ready for our first result on the existence of three solutions.

Theorem 6.4.1. *Assume that the following conditions hold:*

(F14) *there exists $A > 0$ such that $\max\{g^0, g^\infty\} < A$, where*

$$g^0 = \limsup_{y \to 0} \frac{-\int_0^y g(s)ds}{y^2} \quad \text{and} \quad g^\infty = \limsup_{|y| \to \infty} \frac{-\int_0^y g(s)ds}{y^2};$$

(F15) *there exists $\kappa > 0$ such that*

$$2AB\rho^2 < -\int_0^{\kappa(T+1)^2} g(s)ds, \qquad (6.4.26)$$

where ρ is defined by (6.4.7) and

$$B = \frac{\kappa^2}{2} \left(\sum_{t=1}^{T+1} \left(4 + \alpha(2t-1)^2 \right) + \beta \sum_{t=1}^{T} t^4 \right) > 0. \qquad (6.4.27)$$

Then, for each compact interval $[a,b]$ satisfying

$$[a,b] \subset \left(\frac{B}{-\int_0^{\kappa(T+1)^2} g(s)ds}, \frac{1}{2A\rho^2} \right),$$

there exists $K > 0$ such that for every $\mu \in [a,b]$, there exists $\zeta > 0$ such that for each $\lambda \in [0,\zeta]$, BVP (6.4.1), (6.4.2) has at least three solutions in X whose norms are less than K.

Proof. Let the functionals Φ, J_1, $\Psi_1 : X \to \mathbb{R}$ be defined by (6.4.15)–(6.4.17), respectively. Then, Φ, J_1, and Ψ_1 are continuously differentiable and their derivatives $\Phi'(u)$, $J_1'(u)$, and $\Psi_1'(u)$ at $u \in X$ are given by (6.4.21)–(6.4.23), respectively. By Lemma 6.4.3, a critical point of the functional $\Phi - \lambda\Psi_1 - \mu J_1$ is a solution of BVP (6.4.1), (6.4.2). Thus, to prove Theorem 6.4.1, we will apply Theorem 1.1.12 with $E = X$, $J = J_1$, $\Psi = \Psi_1$, $x = \mu$, and $y = \lambda$.

First, we show that some fundamental assumptions of Theorem 1.1.12 are satisfied. Clearly, Φ is a coercive and sequentially weakly lower semicontinuous functional and is bounded on each bounded subset of X, and in view of Remark 1.1.1, $\Phi \in \mathcal{W}_E$. For $u \in X \setminus \{0\}$, from (6.4.6) and (6.4.21), $\Phi'(u)(u) = \|u\|_X^2$, and so $\lim_{\|u\|_X \to \infty} \Phi'(u)(u)/\|u\|_X = \infty$. This shows that Φ' is coercive. Moreover, from (6.4.21),

$$(\Phi'(u) - \Phi'(v))(u-v) = \|u-v\|_X^2 \quad \text{for any } u, v \in X.$$

Then, Φ' is uniformly monotone. Hence, by [289, Theorem 26.A (d)], $(\Phi')^{-1} : X^* \to X$ exists and is continuous. Suppose that $u_n \rightharpoonup u \in X$. Then, by Lemma 6.4.1, $u_n \to u$ on $C([1, T+1]_{\mathbb{Z}})$. Thus, in virtue of (6.4.22) and $g \in C(\mathbb{R}, \mathbb{R})$, we have $J_1'(u_n) \to J_1'(u)$, i.e., J_1' is strongly continuous. Therefore, J_1' is a compact operator by [289, Proposition 26.2]. Similarly, note that since $f \in C([1, T]_{\mathbb{Z}} \times \mathbb{R}, \mathbb{R})$ and (6.4.23) holds, we can show that Ψ_1' is also a compact operator. Let $u_0 \equiv 0$. Then, Φ has a strict local minimum and $\Phi(u_0) = J_1(u_0) = 0$.

Next, we show that

$$\delta \leq 2A\rho^2 \quad \text{and} \quad \frac{-\int_0^{\kappa(T+1)^2} g(s)ds}{B} \leq \eta. \qquad (6.4.28)$$

From (F14), there exist $0 < \gamma_1 < \gamma_2$ such that

$$-\int_0^y g(s)ds \le A|y|^2 \quad \text{for all } |y| \in [0, \gamma_1) \cup (\gamma_2, \infty). \tag{6.4.29}$$

This, together with the continuity of g, implies that there exists $c_1 > 0$ and $\theta > 2$ such that

$$-\int_0^y g(s)ds \le A|y|^2 + c_1|y|^\theta \quad \text{for all } y \in \mathbb{R}.$$

Then, from (6.4.9) and (6.4.16), it follows that

$$J_1(u) \le A|u(T+1)|^2 + c_1|u(T+1)|^\theta$$
$$\le A\rho^2 \|u\|_X^2 + c_1\rho^\theta \|u\|_X^\theta \quad \text{for all } u \in X.$$

Hence,

$$\limsup_{u \to 0} \frac{J_1(u)}{\Phi(u)} \le 2A\rho^2. \tag{6.4.30}$$

For any $u \in X$, note that if $u(T+1) \le \gamma_2$, we have $J_1(u) = -\int_0^{u(T+1)} g(s)ds \le c_2$ for some $c_2 > 0$, and if $u(T+1) > \gamma_2$, (6.4.29) implies that $J_1(u) = -\int_0^{u(T+1)} g(s)ds \le A|u(T+1)|^2$. Thus,

$$J_1(u) \le c_2 + A|u(T+1)|^2 \le c_2 + A\rho^2 \|u\|_X.$$

Thus,

$$\limsup_{u \to \infty} \frac{J_1(u)}{\Phi(u)} \le 2A\rho^2. \tag{6.4.31}$$

By (6.4.30) and (6.4.31), we see that

$$\delta = \max\left\{0, \ \limsup_{\|u\|_X \to \infty} \frac{J_1(u)}{\Phi(u)}, \ \limsup_{\|u\|_X \to u_0} \frac{J_1(u)}{\Phi(u)}\right\} \le 2A\rho^2. \tag{6.4.32}$$

Let κ be given in (F15). Choose $u_1 \in X$ satisfying $u_1(t) = \kappa t^2$ on $[1, T+1]_{\mathbb{Z}}$. Then, by a simple computation, $\Delta u_1(t-1) = \kappa(2t-1)$ and $\Delta^2 u_2(t-2) = 2\kappa$. Combining this with (6.4.6), (6.4.15), and (6.4.27) yields

$$\Phi(u_1) = \frac{\kappa^2}{2}\left(\sum_{t=1}^{T+1}(4 + \alpha(2t-1)^2) + \beta\sum_{t=1}^{T} t^4\right) = B > 0. \tag{6.4.33}$$

Then, $u_1 \in \Phi^{-1}(0, \infty)$. By (6.4.16), we have

$$\eta = \sup_{u \in \Phi^{-1}(0,\infty)} \frac{J_1(u)}{\Phi(u)} \ge \frac{J_1(u_1)}{\Phi(u_1)} = \frac{-\int_0^{\kappa(T+1)^2} g(s)ds}{B}. \tag{6.4.34}$$

Clearly, (6.4.32) and (6.4.34) imply that (6.4.28) holds. Then, from (6.4.26), we have $\delta < \eta$. Hence, all the assumptions of Theorem 1.1.12 are satisfied. The conclusion then follows from Theorem 1.1.12 and Lemma 6.4.3. This completes the proof of the theorem. $\square$

The following corollaries are somewhat easy consequences of Theorem 6.4.1.

Corollary 6.4.1. *Assume that (F14) holds and there exists $\kappa > 0$ such that*

$$\frac{B}{-\int_0^{\kappa(T+1)^2} g(s)ds} < 1 < \frac{1}{2A\rho^2}, \qquad (6.4.35)$$

where A is given in (F14) and ρ and B are defined by (6.4.7) and (6.4.27), respectively. Then, there exist $K > 0$ and $\zeta > 0$ such that for each $\lambda \in [0, \zeta]$, the BVP

$$\Delta^4 u(t-2) - \alpha\Delta^2 u(t-1) + \beta u(t) = \lambda f(t, u(t)), \quad t \in [1, T]_{\mathbb{Z}},$$

$$u(0) = \Delta u(-1) = \Delta^2 u(T) = 0, \quad \Delta^3 u(T-1) - \alpha\Delta u(T) = g(u(T+1)),$$

has at least three solutions in X whose norms are less than K.

Proof. Since (6.4.35) implies (6.4.26), the conclusion follows from Theorem 6.4.1. □

Corollary 6.4.2. *Assume that (F14) holds and there exists $\kappa > 0$ such that*

$$\frac{2(T+1)\kappa^2}{-\int_0^{\kappa(T+1)^2} g(s)ds} < 1 < \frac{1}{2A(T+1)}, \qquad (6.4.36)$$

where A is given in (F14). Then, there exist $K > 0$ and $\zeta > 0$ such that for each $\lambda \in [0, \zeta]$, BVP (6.4.3), (6.4.4) has at least three solutions in X whose norms are less than K.

Proof. When $\alpha = \beta = 0$, from (6.4.7) and (6.4.27), we have $\rho = (T+1)^{1/2}$ and $B = 2(T+1)\kappa^2$. Then, (6.4.36) implies (6.4.35). The conclusion now follows from Corollary 6.4.1. □

Our next main result is contained in the following theorem.

Theorem 6.4.2. *Assume that the following conditions hold:*

(F16) *there exists $C > 0$ such that*

$$\max\left\{\limsup_{y \to 0} \frac{\max\limits_{t \in [1,T]_{\mathbb{Z}}} F(t, y)}{y^2}, \ \limsup_{|y| \to \infty} \frac{\max\limits_{t \in [1,T]_{\mathbb{Z}}} F(t, y)}{y^2}\right\} < C,$$

where F is defined by (6.4.19);

(F17) *there exists $\sigma > 0$ such that*

$$2CDT\rho^2 < \sum_{t=1}^{T} \int_0^{\sigma t^2} f(t,s)\,ds, \qquad (6.4.37)$$

where ρ is defined by (6.4.7) and

$$D = \frac{\sigma^2}{2}\left(\sum_{t=1}^{T+1}\left(4 + \alpha(2t-1)^2\right) + \beta\sum_{t=1}^{T}t^4\right) > 0. \qquad (6.4.38)$$

Then, for each compact interval $[a,b]$ satisfying

$$[a,b] \subset \left(\frac{D}{\sum_{t=1}^{T}\int_0^{\sigma t^2} f(t,s)\,ds}, \frac{1}{2CT\rho^2}\right),$$

there exists $K > 0$ such that for every $\lambda \in [a,b]$, there exists $\zeta > 0$ such that for each $\mu \in [0,\zeta]$, BVP (6.4.1), (6.4.2) has at least three solutions in X whose norms are less than K.

Proof. Let the functionals $\Phi, J_2, \Psi_2 : X \to \mathbb{R}$ be defined by (6.4.15)–(6.4.17), respectively. By Lemma 6.4.4, a critical point of the functional $\Phi - \lambda J_2 - \mu\Psi_2$ is a solution of BVP (6.4.1), (6.4.2). Thus, to prove Theorem 6.4.2, we will apply Theorem 1.1.12 with $E = X$, $J = J_2$, $\Psi = \Psi_2$, $x = \lambda$, and $y = \mu$.

As in the proof of Theorem 6.4.1, we see that Φ, J_2, and Ψ_2 satisfy the regularity assumptions of Theorem 1.1.12. Moreover, for $u_0 \equiv 0$, Φ has a strict local minimum and $\Phi(u_0) = J_2(u_0) = 0$.

In the following, we show that

$$\delta \leq 2CT\rho^2 \quad \text{and} \quad \frac{\sum_{t=1}^{T}\int_0^{\sigma t^2} f(t,s)\,ds}{D} \leq \eta. \qquad (6.4.39)$$

From (F16), there exist $0 < \tau_1 < \tau_2$ such that

$$F(t,y) \leq C|y|^2 \quad \text{for all } t \in [1,T]_{\mathbb{Z}} \text{ and } |y| \in [0,\tau_1) \cup (\tau_2,\infty). \qquad (6.4.40)$$

This, together with the continuity of f, implies that there exists $c_3 > 0$ and $\vartheta > 2$ such that

$$F(t,y) \leq C|y|^2 + c_3|y|^\vartheta \quad \text{for all } t \in [1,T]_{\mathbb{Z}} \text{ and } |y| \in [0,\tau_1) \cup (\tau_2,\infty).$$

Then, from (6.4.9) and (6.4.17), it follows that

$$J_2(u) \leq \sum_{t=1}^{T}\left(C|u(t)|^2 + c_3|u(t)|^\vartheta\right)$$

$$\leq CT\rho^2\|u\|_X^2 + c_1 T\rho^\vartheta\|u\|_X^\vartheta \quad \text{for all } u \in X.$$

Hence,

$$\limsup_{u \to 0} \frac{J_2(u)}{\Phi(u)} \le 2CT\rho^2. \tag{6.4.41}$$

For any $u \in X$ and $t \in [1, T]_{\mathbb{Z}}$, note that if $u(t) \le \tau_2$, we have $F(t, u(t)) \le c_4$ for some $c_4 > 0$, and if $u(t) > \gamma_2$, (6.4.40) implies that $F(t, u(t)) \le C|u(t)|^2$. Thus, in view of (6.4.9) and (6.4.17),

$$J_2(u) \le c_4 T + CT|u(t)|^2 \le c_4 T + CT\rho^2 \|u\|_X.$$

Thus,

$$\limsup_{u \to \infty} \frac{J_2(u)}{\Phi(u)} \le 2CT\rho^2. \tag{6.4.42}$$

By (6.4.41) and (6.4.42), we see that

$$\delta = \max \left\{ 0, \ \limsup_{\|u\|_X \to \infty} \frac{J_1(u)}{\Phi(u)}, \ \limsup_{\|u\|_X \to u_0} \frac{J_1(u)}{\Phi(u)} \right\} \le 2CT\rho^2. \tag{6.4.43}$$

Let σ be given in (F17). Choose $u_2 \in X$ satisfying $u_2(t) = \sigma t^2$ on $[1, T+1]_{\mathbb{Z}}$. Then, as obtaining (6.4.33), we have

$$\Phi(u_2) = \frac{\sigma^2}{2} \left(\sum_{t=1}^{T+1} \left(4 + \alpha(2t-1)^2 \right) + \beta \sum_{t=1}^{T} t^4 \right) = D > 0. \tag{6.4.44}$$

Then, $u_2 \in \Phi^{-1}(0, \infty)$. By (6.4.17) and (6.4.19), we have

$$\eta = \sup_{u \in \Phi^{-1}(0, \infty)} \frac{J_2(u)}{\Phi(u)} \ge \frac{J_2(u_2)}{\Phi(u_2)} = \frac{\sum_{t=1}^{T} \int_0^{\sigma t^2} f(t, s) ds}{D}. \tag{6.4.45}$$

Clearly, (6.4.43) and (6.4.45) imply that (6.4.39) holds. Then, from (6.4.37), we have $\delta < \eta$. Hence, all the assumptions of Theorem 1.1.12 are satisfied. The conclusion then follows from Theorem 1.1.12 and Lemma 6.4.4. This completes the proof of the theorem. $\square$

The next two corollaries follow from Theorem 6.4.2.

Corollary 6.4.3. *Assume that (F16) holds and there exists $\sigma > 0$ such that*

$$\frac{D}{\sum_{t=1}^{T} \int_0^{\sigma t^2} f(t, s) ds} < 1 < \frac{1}{2CT\rho^2} \tag{6.4.46}$$

where C is given in (F16) and ρ and D are defined by (6.4.7) and (6.4.38), respectively. Then, there exist $K > 0$ and $\zeta > 0$ such that for each $\mu \in [0, \zeta]$, the BVP

$$\Delta^4 u(t-2) - \alpha \Delta^2 u(t-1) + \beta u(t) = f(t, u(t)), \quad t \in [1, T]_{\mathbb{Z}},$$

$$u(0) = \Delta u(-1) = \Delta^2 u(T) = 0, \quad \Delta^3 u(T-1) - \alpha \Delta u(T) = \mu g(u(T+1)),$$

has at least three solutions in X whose norms are less than K.

Proof. Since (6.4.46) implies (6.4.37), the conclusion follows from Theorem 6.4.2. $\qquad\square$

Corollary 6.4.4. *Assume that (F16) holds and there exists $\sigma > 0$ such that*

$$\frac{2(T+1)\sigma^2}{\sum_{t=1}^{T} \int_0^{\sigma t^2} f(t,s)ds} < 1 < \frac{1}{2CT(T+1)} \tag{6.4.47}$$

where C is given in (F16). Then, there exist $K > 0$ and $\zeta > 0$ such that for each $\mu \in [0, \zeta]$, the BVP consisting of the equation

$$\Delta^4 u(t-2) = f(t, u(t)), \quad t \in [1, T]_{\mathbb{Z}}, \tag{6.4.48}$$

and the BC

$$u(0) = \Delta u(-1) = \Delta^2 u(T) = 0, \ \Delta^3 u(T-1) = \mu g(u(T+1)), \tag{6.4.49}$$

has at least three solutions in X whose norms are less than K.

Proof. If $\alpha = \beta = 0$, from (6.4.7) and (6.4.38), we have $\rho = (T+1)^{1/2}$ and $D = 2(T+1)\sigma^2$. Then, (6.4.47) implies (6.4.46) and so the conclusion follows from Corollary 6.4.3. $\qquad\square$

We now turn our attention to results on the existence of infinitely many solutions. let X and ρ be defined by (6.4.5) and (6.4.7), and let $\mathcal{A}$, $\mathcal{B}$, $\mathcal{C}$, and $\mathcal{D}$ be defined in (6.3.7) and (6.3.8), i.e.,

$$\mathcal{A} = \liminf_{\xi \to \infty} \frac{\sum_{t=1}^{T} \max_{|x| \leq \xi} F(t, x)}{\xi^2}, \quad \mathcal{B} = \limsup_{\xi \to \infty} \frac{\sum_{t=1}^{T} F(t, \xi)}{\xi^2}, \tag{6.4.50}$$

$$\mathcal{C} = \liminf_{\xi \to 0^+} \frac{\sum_{t=1}^{T} \max_{|x| \leq \xi} F(t, x)}{\xi^2}, \quad \mathcal{D} = \limsup_{\xi \to 0^+} \frac{\sum_{t=1}^{T} F(t, \xi)}{\xi^2} \tag{6.4.51}$$

and in what follows, we assume that

(F18) $\mathcal{A}$, $\mathcal{B}$, $\mathcal{C}$, $\mathcal{D} \geq 0$.

We define the constants

$$\theta_1 = \frac{2 + \alpha + \beta T}{2\mathcal{B}}, \quad \theta_2 = \frac{1}{2\rho^2 \mathcal{A}}, \tag{6.4.52}$$

$$\theta_3 = \frac{2 + \alpha + \beta T}{2\mathcal{D}}, \quad \theta_4 = \frac{1}{2\rho^2 \mathcal{C}},$$

and adopt the convention that $1/a = \infty$ when $a = 0$.

Theorem 6.4.3. *Assume that*

$$\mathcal{A} < \frac{\mathcal{B}}{\rho^2(2 + \alpha + \beta T)}. \tag{6.4.53}$$

Then, for each $\lambda \in (\theta_1, \theta_2)$ *and each function* $g \in C(\mathbb{R}, \mathbb{R})$ *with*

$$g(x) \leq 0 \text{ on } \mathbb{R} \quad \text{and} \quad G_\infty = \liminf_{\xi \to \infty} \frac{G(\xi)}{\xi^2} > -\infty, \tag{6.4.54}$$

and for each $\mu \in [0, \overline{\mu}_1)$ *with*

$$\overline{\mu}_1 = \frac{1 - 2\rho^2 \lambda \mathcal{A}}{-2\rho^2 G_\infty}, \tag{6.4.55}$$

BVP (6.4.1), (6.4.2) *has a sequence of solutions which is unbounded in* X.

The proof of Theorem 6.4.3 relies on Theorem 1.1.5 (b).

Proof. We are going to apply Theorem 1.1.5 (b), so let the functionals Φ, $\Psi : X \to \mathbb{R}$ be defined as in (6.4.15) and (6.4.18). As before, Φ and Ψ satisfy all the regularity conditions in Theorem 1.1.5. From the definition of $\mathcal{A}$ in (6.4.50), there exists a sequence $\{x_n\}$ of positive numbers such that $\lim_{n \to \infty} x_n = \infty$ and

$$\mathcal{A} = \lim_{n \to \infty} \frac{\sum_{t=1}^T \max_{|x| \leq x_n} F(t, x)}{x_n^2}. \tag{6.4.56}$$

Let $r_n = \frac{x_n^2}{2\rho^2}$. Then, for any $u \in X$ with $\Phi(u) < r_n$, from (6.4.9), we have

$$\max_{t \in [1, T+1]_\mathbb{Z}} |u(t)| \leq \rho ||u||_X < \rho(2r_n)^{1/2} = x_n. \tag{6.4.57}$$

Note that $0 \in \Phi^{-1}(-\infty, r_n)$ and $\Psi(0) = 0$. Then, by (1.1.4) and (6.4.54),

$$\begin{aligned}
\varphi(r_n) &= \inf_{u \in \Phi^{-1}(-\infty, r_n)} \frac{\left(\sup_{v \in \Phi^{-1}(-\infty, r_n)} \Psi(v) \right) - \Psi(u)}{r_n - \Phi(u)} \\
&\leq \inf_{u \in \Phi^{-1}(-\infty, r_n)} \frac{\sup_{v \in \Phi^{-1}(-\infty, r_n)} \Psi(v)}{r_n} \\
&\leq \frac{\sum_{t=1}^T \max_{|x| \leq x_n} F(t, x) - \frac{\mu}{\lambda} \min_{|s| \leq x_n} G(s)}{r_n} \\
&= 2\rho^2 \frac{\sum_{t=1}^T \max_{|x| \leq x_n} F(t, x) - \frac{\mu}{\lambda} G(x_n)}{x_n^2}.
\end{aligned}$$

Thus, from (6.4.54) and (6.4.56), we see that, for γ defined in Theorem 1.1.5,

$$\gamma \leq \liminf_{n \to \infty} \varphi(r_n) \leq 2\rho^2 \left(A - \frac{\mu}{\lambda} G_\infty \right) < \infty. \tag{6.4.58}$$

We claim that

$$\text{if } \lambda \in (\theta_1, \theta_2) \text{ and } \mu \in [0, \bar{\mu}_1), \text{ then } \lambda \in (0, 1/\gamma). \tag{6.4.59}$$

In fact, it is clear that $\lambda > 0$. Now, when $\lambda \in (\theta_1, \theta_2)$ and $\mu \in [0, \bar{\mu}_1)$, from (6.4.55) and (6.4.58), we have

$$\gamma \leq 2\rho^2 \left(\mathcal{A} - \frac{\bar{\mu}_1}{\lambda} G_\infty \right) = 2\rho^2 \left(\mathcal{A} + \frac{1 - 2\rho^2 \lambda \mathcal{A}}{2\rho^2 \lambda} \right) = \frac{1}{\lambda},$$

and so, $\lambda < 1/\gamma$. Thus, (6.4.59) holds.

Let $\lambda \in (\theta_1, \theta_2)$ and $\mu \in [0, \bar{\mu}_1)$ be fixed. Then, in view of (6.4.58) and (6.4.59), by Theorem 1.1.5 (b), it follows that one of the following alternatives holds

(b$_1$) either $I_\lambda := \Phi - \lambda \Psi$ has a global minimum, or
(b$_2$) there exists a sequence $\{u_n\}$ of critical points of I_λ such that $\lim_{n \to \infty} \|u_n\|_X = \infty$.

In what follows, we show that alternative (b$_1$) does not hold. By the definition of $\mathcal{B}$ in (6.4.50), there exists a sequence $\{\eta_n\}$ of positive numbers such that $\lim_{n \to \infty} \eta_n = \infty$ and

$$\mathcal{B} = \lim_{n \to \infty} \frac{\sum_{t=1}^T F(t, \eta_n)}{\eta_n^2}. \tag{6.4.60}$$

For each $n \in \mathbb{N}$, define a function $w_n : [-1, T+2]_{\mathbb{Z}} \to \mathbb{R}$ by

$$w_n(t) = \begin{cases} 0, & t = -1, 0, \\ \eta_n, & t \in [1, T+2]_{\mathbb{Z}}. \end{cases} \tag{6.4.61}$$

Then, $w_n \subseteq X$. Moreover, from (6.4.15) and (6.4.18), it is easy to see that

$$\Phi(w_n) = \frac{1}{2}(2 + \alpha + \beta T)\eta_n^2$$

and

$$\Psi(w_n) = \sum_{t=1}^T F(t, \eta_n) - \frac{\mu}{\lambda} G(\eta_n).$$

Note that $G(\eta_n) \leq 0$ by (6.4.54). Then, we have

$$I_\lambda(w_n) = \Phi(w_n) - \lambda \Psi(w_n)$$

$$= \frac{1}{2}(2 + \alpha + \beta T)\eta_n^2 - \lambda \sum_{t=1}^T F(t, \eta_n) + \mu G(\eta_n)$$

$$\leq \frac{1}{2}(2 + \alpha + \beta T)\eta_n^2 - \lambda \sum_{t=1}^T F(t, \eta_n). \tag{6.4.62}$$

Now, we consider two cases.

Case 1: $\mathcal{B} < \infty$. By $\lambda > \theta_1$ and the definition of θ_1 in (6.4.52), we have $\mathcal{B} - \frac{2+\alpha+\beta T}{2\lambda} > 0$. Let

$$\epsilon \in \left(0, \mathcal{B} - \frac{2+\alpha+\beta T}{2\lambda}\right). \tag{6.4.63}$$

From (6.4.60), there exists $N_1 \in \mathbb{N}$ such that

$$\sum_{t=1}^{T} F(t, \eta_n) > (\mathcal{B} - \epsilon)\eta_n^2 \quad \text{for } n \geq N_1.$$

This, together with (6.4.34), implies that

$$I_\lambda(w_n) \leq \left(\frac{1}{2}(2 + \alpha + \beta T)\eta_n^2 - \lambda(\mathcal{B} - \epsilon)\right)\eta_n^2.$$

Thus, from (6.4.63) and $\lim_{n\to\infty} \eta_n = \infty$, it follows that $\lim_{n\to\infty} I_\lambda(w_n) = -\infty$.

Case 2: $\mathcal{B} = \infty$. Choose

$$M > \frac{2+\alpha+\beta T}{2\lambda}. \tag{6.4.64}$$

Then, (6.4.60) implies that there exists $N_2 \in \mathbb{N}$ such that

$$\sum_{t=1}^{T} F(t, \eta_n) > M\eta_n^2 \quad \text{for } n \geq N_2.$$

Thus, from (6.4.62),

$$I_\lambda(w_n) \leq \left(\frac{1}{2}(2 + \alpha + \beta T)\eta_n^2 - \lambda M\right)\eta_n^2.$$

Then, from (6.4.64) and $\lim_{n\to\infty} \eta_n = \infty$, we have $\lim_{n\to\infty} I_\lambda(w_n) = -\infty$.

Combining the above two cases, we see that the functional I_λ is always unbounded from below. Hence, the alternative (b_1) does not hold. Therefore, there exists a sequence $\{u_n\}$ of critical points of I_λ such that $\lim_{n\to\infty} \|u_n\|_X = \infty$. Finally, taking into account Lemma 6.4.5, we finish the proof of the theorem. $\qquad\square$

In our next theorem we prove the existence of a sequence of solutions that converges to zero.

Theorem 6.4.4. *Assume that*

$$\mathcal{C} < \frac{\mathcal{D}}{\rho^2(2 + \alpha + \beta T)}. \tag{6.4.65}$$

Then, for each $\lambda \in (\theta_3, \theta_4)$ and each function $g \in C(\mathbb{R}, \mathbb{R})$ satisfying (6.4.53), and for each $\mu \in [0, \bar{\mu}_2)$ with

$$\bar{\mu}_2 = \frac{1 - 2\lambda\rho^2 \mathcal{C}}{-2\rho^2 G_\infty},$$

BVP (6.4.1), (6.4.2) has a sequence of solutions converging uniformly to zero in X.

Proof. The proof of this result is similar to the proof of the above theorem except that now we will apply Theorem 1.1.5 (c). Again take the functionals Φ, $\Psi : X \to \mathbb{R}$ as in the proof of Theorem 6.4.3 and note that the regularity conditions in Theorem 1.1.5 are satisfied. From the definition of $\mathcal{C}$ in (6.4.51), there exists a sequence $\{x_n\}$ of positive numbers such that $\lim_{n\to\infty} x_n = 0$ and

$$\mathcal{C} = \lim_{n \to \infty} \frac{\sum_{t=1}^{T} \max_{|x| \leq x_n} F(t, x)}{x_n^2}.$$

By the fact that $\inf_X \Phi = 0$ and the definition δ, we have $\delta = \liminf_{r \to 0^+} \varphi(r)$. Then, as showing (6.4.58) and (6.4.59) in the proof of Theorem 6.4.3, we can prove that $\delta < \infty$ and that if $\lambda \in (\theta_3, \theta_4)$ and $\mu \in [0, \bar{\mu}_2)$, then $\lambda \in (0, 1/\delta)$. Let $\lambda \in (\theta_3, \theta_4)$ and $\mu \in [0, \bar{\mu}_2)$ be fixed. Then, by Theorem 1.1.5 (c), we see that one of the following alternatives holds

(c₁) either there is a global minimum of Φ which is a local minimum of $I_\lambda = \Phi - \lambda\Psi$, or

(c₂) there exists a sequence $\{u_n\}$ of critical points of I_λ which converges weakly to a global minimum of Φ.

In the following, we show that alternative (c₁) does not hold. By the definition of $\mathcal{C}$ in (6.4.51), there exists a sequence $\{\eta_n\}$ of positive numbers such that $\lim_{n\to\infty} \eta_n = 0$ and

$$\mathcal{C} = \lim_{n \to \infty} \frac{\sum_{t=1}^{T} F(t, \eta_n)}{\eta_n^2}. \tag{6.4.66}$$

For each $n \in \mathbb{N}$, let $w_n : [-1, T+2]_{\mathbb{Z}} \to \mathbb{R}$ be defined by (6.4.61) with the above η_n. Then, as in the cases 1 and 2 of the proof of Theorem 6.4.3, we can get that, when n is large enough, if $\mathcal{C} < \infty$, then

$$I_\lambda(w_n) \leq \left(\frac{1}{2}(2 + \alpha + \beta T)\eta_n^2 - \lambda(\mathcal{C} - \epsilon) \right)\eta_n^2,$$

where
$$\epsilon \in \left(0, \mathcal{C} - \frac{2 + \alpha + \beta T}{2\lambda}\right),$$
and if $\mathcal{C} = \infty$, then
$$I_\lambda(w_n) \leq \left(\frac{1}{2}(2 + \alpha + \beta T)\eta_n^2 - \lambda M\right)\eta_n^2,$$
where M satisfies (6.4.64). Therefore, we always have $I_\lambda(w_n) < 0$ for large n. Then, since $\lim_{n\to\infty} I_\lambda(w_n) = I_\lambda(0) = 0$, we see that 0 is not a local minimum of I_λ. This, together with the fact that 0 is the only global minimum of Φ, shows that alternative (c_1) does not hold. Therefore, there exists a sequence $\{u_n\}$ of critical points of I_λ which converges weakly (and thus also strongly) to 0. Finally, taking into account Lemma 6.4.5, we finish the proof of the theorem. $\qquad\square$

Remark 6.4.1. It is easy to see that condition (F13) implies $2 + \alpha + \beta T > 0$, so $\theta_1 \geq 0$ and $\theta_3 \geq 0$. Also, conditions (6.4.53) and (6.4.65) imply that $\theta_1 < \theta_2$ and $\theta_3 < \theta_4$ thus ensuring that the intervals (θ_1, θ_2) and (θ_3, θ_4) are nonempty. The interval $[0, \overline{\mu}_1)$ is well defined since $\overline{\mu}_1 > 0$ under the condition that $\lambda < \theta_2$, and the interval $[0, \overline{\mu}_2)$ is well defined since $\overline{\mu}_2 > 0$ under the condition that $\lambda < \theta_4$.

The following results are direct consequences of Theorems 6.4.3 and 6.4.4. In particular, Corollaries 6.4.5, 6.4.7, and 6.4.9 follow from Theorem 6.4.3, and Corollaries 6.4.6, 6.4.8, and 6.4.10 follow from Theorem 6.4.4.

Corollary 6.4.5. *Assume that* (6.4.53) *hold. Then, for each* $\lambda \in (\theta_1, \theta_2)$, *the BVP*
$$\begin{cases} \Delta^4 u(t-2) - \alpha\Delta^2 u(t-1) + \beta u(t) = \lambda f(t, u(t)), & t \in [1, T]_{\mathbb{Z}}, \\ u(0) = \Delta u(-1) = \Delta^2 u(T) = 0, \ \Delta^3 u(T-1) - \alpha\Delta u(T) = 0, \end{cases} \tag{6.4.67}$$
has a sequence of solutions which is unbounded in X.

Corollary 6.4.6. *Assume that* (6.4.65) *hold. Then, for each* $\lambda \in (\theta_3, \theta_4)$, *BVP* (6.4.67) *has a sequence of solutions converging uniformly to zero in* X.

Corollary 6.4.7. *Assume that* $\mathcal{A} = 0$ *and* $\mathcal{B} = \infty$. *Then, for each* $\lambda \in (0, \infty)$ *and each function* $g \in C(\mathbb{R}, \mathbb{R})$ *with*
$$g(x) \leq 0 \quad on \ \mathbb{R} \quad and \quad G_\infty = \liminf_{\xi\to\infty} \frac{G(\xi)}{\xi^2} = 0, \tag{6.4.68}$$
and for each $\mu \in [0, \infty)$, *BVP* (6.4.1), (6.4.2) *has a sequence of solutions which is unbounded in* X.

Corollary 6.4.8. *Assume that* $C = 0$ *and* $\mathcal{D} = \infty$. *Then, for each* $\lambda \in (0, \infty)$ *and each function* $g \in C(\mathbb{R}, \mathbb{R})$ *satisfying* (6.4.68), *and for each* $\mu \in [0, \infty)$, *BVP* (6.4.1), (6.4.2) *has a sequence of solutions converging uniformly to zero in* X.

Corollary 6.4.9. *Assume that* $\mathcal{A} < \frac{\mathcal{B}}{2(T+1)^3}$. *Then, for each* $\lambda \in \left(\frac{1}{\mathcal{B}}, \frac{1}{2\mathcal{A}(T+1)^3} \right)$ *and each function* $g \in C(\mathbb{R}, \mathbb{R})$ *satisfying* (6.4.68), *BVP* (6.4.3), (6.4.4) *has a sequence of solutions which is unbounded in* X.

Corollary 6.4.10. *Assume that* $\mathcal{C} < \frac{\mathcal{D}}{2(T+1)^3}$. *Then, for each* $\lambda \in \left(\frac{1}{\mathcal{D}}, \frac{1}{2\mathcal{C}(T+1)^3} \right)$ *and each function* $g \in C(\mathbb{R}, \mathbb{R})$ *satisfying* (6.4.68), *BVP* (6.4.3), (6.4.4) *has a sequence of solutions converging uniformly to zero in* X.

We conclude this section with some examples to illustrate our results.

Example 6.4.1. Let $g \in C(\mathbb{R}, \mathbb{R})$ be defined by

$$
g(y) = \begin{cases} 3, & y < 1, \\ 3y^2, & -1 \leq y < 0, \\ -3y^2, & 0 \leq y \leq 1, \\ -3, & y > 1. \end{cases} \tag{6.4.69}
$$

We claim that for each compact interval $[a, b]$ satisfying

$$
[a, b] \subset \left(\left(\frac{16}{27} \right)^2, \infty \right),
$$

there exists $K > 0$ such that for every $\mu \in [a, b]$, there exists $\zeta > 0$ such that for each $\lambda \in [0, \zeta]$, the BVP consisting of the equation

$$
\Delta^4 u(t - 2) - \Delta^2 u(t - 1) + u(t) = \lambda f(t, u(t)), \quad t \in [1, 2]_{\mathbb{Z}}, \tag{6.4.70}
$$

and the BC

$$
u(0) = \Delta u(-1) = \Delta^2 u(2) = 0, \quad \Delta^3 u(1) - \Delta u(2) = \mu g(u(3)), \tag{6.4.71}
$$

where $f \in C([1, 2]_{\mathbb{Z}} \times \mathbb{R}, \mathbb{R})$, has at least three solutions in X whose norms are less than K.

In fact, with $\alpha = \beta = 1$ and $T = 2$, we see that BVP (6.4.70), (6.4.71) is of the form of BVP (6.4.1), (6.4.2). From (6.4.69), it is easy to see that

$$
\int_0^y g(s)ds = \begin{cases} -|y|^3, & |y| \leq 1, \\ -3|y| + 2, & |y| > 1. \end{cases}
$$

Then,

$$g^0 = \limsup_{y \to 0} \frac{-\int_0^y g(s)ds}{y^2} = 0 \quad \text{and} \quad g^\infty = \limsup_{|y| \to \infty} \frac{-\int_0^y g(s)ds}{y^2} = 0.$$

Thus, (F14) holds for any $A > 0$ and (F15) is satisfied for any $\kappa > 0$.

In what follows, we choose a $\kappa > 0$ such that

$$h(\kappa) := \frac{B}{-\int_0^{(T+1)^2\kappa} g(s)ds} = \frac{B}{-\int_0^{9\kappa} g(s)ds}$$

has the smallest value, where B is defined by (6.4.27). For any $\kappa > 0$, from (6.4.27),

$$B = \frac{\kappa^2}{2}\left(\sum_{t=1}^{3}\left(4 + (2t-1)^2\right) + \sum_{t=1}^{2} t^4\right) = 32\kappa^2.$$

By (6.4.69), it follows that if $0 < 9\kappa \leq 1$,

$$-\int_0^{9\kappa} g(s)ds = \int_0^{9\kappa} 3s^2 ds = 9^3\kappa^3;$$

and if $9\kappa > 1$,

$$-\int_0^{9\kappa} g(s)ds = \int_0^1 3s^2 ds + \int_1^{9\kappa} 3ds = 27\kappa - 2.$$

Thus, we have

$$h(\kappa) = \frac{B}{-\int_0^{9\kappa} g(s)ds} = \begin{cases} \frac{32}{9^3\kappa} & \text{if } 0 < 9\kappa \leq 1, \\ \frac{32\kappa^2}{27\kappa - 2} & \text{if } 9\kappa > 1. \end{cases} \tag{6.4.72}$$

From (6.4.72) and some basic calculus, it is easy to see that $h(\kappa)$ has the smallest value $(16/27)^2$ at $\kappa = 4/27$.

Finally, by the arbitrariness of A in (F14) and choosing $\kappa = 4/27$ in (F15), we see that the claim readily follows from Theorem 6.4.1.

Example 6.4.2. We claim that there exist $K > 0$ and $\zeta > 0$ such that for each $\mu \in [0, \zeta]$, the BVP consisting of the equation

$$\Delta^4 u(t-2) = \frac{1}{32}y, \quad t \in [1,5]_\mathbb{Z}, \tag{6.4.73}$$

$$u(0) = \Delta u(-1) = \Delta^2 u(5) = 0, \quad \Delta^3 u(4) = \mu g(u(6)), \tag{6.4.74}$$

where $g \in C(\mathbb{R}, \mathbb{R})$, has at least three solutions in X whose norms are less than K.

In fact, with $f(t, y) = y/32$ and $T = 5$, we see that BVP (6.4.73), (6.4.74) is of the form of (6.4.48), (6.4.49). Let $C = 1/62$. Since

$$F(t, y) = \int_0^y f(t, s)ds = \frac{1}{64}y^2 \quad \text{for } (t, y) \in [1, 5]_\mathbb{Z} \times \mathbb{R},$$

we have

$$\max \left\{ \lim_{y \to 0} \frac{\max\limits_{t \in [1, T]_\mathbb{Z}} F(t, y)}{y^2}, \ \lim_{|y| \to \infty} \frac{\max\limits_{t \in [1, T]_\mathbb{Z}} F(t, y)}{y^2} \right\} = \frac{1}{64} < C,$$

i.e., (F16) holds.

For any $\sigma > 0$, by a simple calculation, we see that

$$\frac{2(T + 1)\sigma^2}{\sum_{t=1}^T \int_0^{\sigma t^2} f(t, s)ds} = \frac{768}{979} < 1$$

and

$$\frac{1}{2CT(T + 1)} = \frac{31}{30} > 1,$$

i.e., (6.4.47) holds. Then, the claim readily follows from Corollary 6.4.4.

The form of the nonlinear function $f(t, x)$ in our next example is motivated by [97, Example 3.1] as well as by Example 6.3.1 above.

Example 6.4.3. Let $T \geq 2$ be an integer, $\{a_n\}$ and $\{b_n\}$ be sequences defined by $b_1 = 2$, $b_{n+1} = b_n^6$, and $a_n = b_n^4$ for $n \in \mathbb{N}$. Let $f : [0, T]_\mathbb{Z} \times \mathbb{R} \to \mathbb{R}$ be a positive continuous function defined by

$$f(t, x) = t^2 \begin{cases} b_1^3 \sqrt{1 - (1 - x)^2} + 1, & x \in [0, b_1], \\ (a_n - b_n^3)\sqrt{1 - (a_n - 1 - x)^2} + 1, & x \in \cup_{n=1}^\infty [a_n - 2, a_n], \\ (b_{n+1}^3 - a_n)\sqrt{1 - (b_{n+1} - 1 - x)^2} + 1, & x \in \cup_{n=1}^\infty [b_{n+1} - 2, b_{n+1}], \\ 1, & \text{otherwise.} \end{cases}$$

Let $\alpha, \beta \in \mathbb{R}$ satisfy (F13). We claim that for each $\lambda \in (0, \infty)$ and $\mu \in [0, \infty)$, the BVP

$$\begin{cases} \Delta^4 u(t - 2) - \alpha \Delta^2 u(t - 1) + \beta u(t) = \lambda f(t, u(t)), & t \in [1, T]_\mathbb{Z}, \\ u(0) = \Delta u(-1) = \Delta^2 u(T) = 0, & (6.4.75) \\ \Delta^3 u(T - 1) - \alpha \Delta u(T) = -\mu(u(T + 1))^{2/3}, \end{cases}$$

has a sequence of solutions which is unbounded in X.

In fact, with $g(x) = -x^{2/3}$, it is clear that BVP (6.4.75) is a special case of BVP (6.4.1), (6.4.2) and that (6.4.68) holds. Let $F(t,x)$ be defined by (6.4.19). Then, for $t \in [1,T]_{\mathbb{Z}}$, by simple computations,

$$F(t, a_n) = t^2 \left(\int_0^{a_n} 1 ds + b_1^3 \int_0^2 \sqrt{1 - (1-s)^2} \, ds \right.$$
$$+ \sum_{i=1}^n \int_{a_i-2}^{a_i} (a_i - b_i^3) \sqrt{1 - (a_i - 1 - s)^2} \, ds$$
$$\left. + \sum_{i=1}^{n-1} \int_{b_{i+1}-2}^{b_{i+1}} (b_i^3 - a_i) \sqrt{1 - (b_{i+1} - 1 - s)^2} \, ds \right)$$
$$= t^2 \left(\frac{\pi}{2} a_n + a_n \right)$$

and

$$F(t, b_n) = t^2 \left(\int_0^{b_n} 1 ds + b_1^3 \int_0^2 \sqrt{1 - (1-s)^2} \, ds \right.$$
$$+ \sum_{i=1}^{n-1} \int_{a_i-2}^{a_i} (a_i - b_i^3) \sqrt{1 - (a_i - 1 - s)^2} \, ds$$
$$\left. + \sum_{i=1}^{n-1} \int_{b_{i+1}-2}^{b_{i+1}} (b_i^3 - a_i) \sqrt{1 - (b_{i+1} - 1 - s)^2} \, ds \right)$$
$$= t^2 \left(\frac{\pi}{2} b_n^3 + b_n \right).$$

Thus,

$$\lim_{n \to \infty} \frac{F(t, a_n)}{a_n^2} = 0 \quad \text{and} \quad \lim_{n \to \infty} \frac{F(t, b_n)}{b_n^2} = \infty \quad \text{for } t \in [1,T]_{\mathbb{Z}}.$$

Then, for $\mathcal{A}$ and $\mathcal{B}$ defined in (6.4.50), it is easy to see that

$$\mathcal{A} = \liminf_{\xi \to \infty} \frac{F(t, \xi) \sum_{t=1}^T t^2}{\xi^2} = 0 \quad \text{and} \quad \mathcal{B} = \limsup_{\xi \to \infty} \frac{F(t, \xi) \sum_{t=1}^T t^2}{\xi^2} = \infty.$$

Thus, all the conditions of Corollary 6.4.7 are satisfied. The claim then follows directly from Corollary 6.4.7.

6.5 Homoclinic Solutions for Difference Equations

In this section, we consider the existence of solutions of the second order difference equation with a p-Laplacian

$$\begin{cases} -\Delta\big(a(k)\phi_p(\Delta u(k-1))\big) + b(k)\phi_p(u(k)) = \lambda f(k, u(k)), & k \in \mathbb{Z}, \\ u(k) \to 0 \quad \text{as } |k| \to \infty, \end{cases}$$

$$(6.5.1)$$

where $p > 1$ is a real number, $\phi_p(t) = |t|^{p-2}t$ for $t \in \mathbb{R}$, $\lambda > 0$ is a parameter, $a, b : \mathbb{Z} \to (0, \infty)$, and $f : \mathbb{Z} \times \mathbb{R} \to \mathbb{R}$ is a continuous function in the second variable. A solution of problem (6.5.1) is referred to as a *homoclinic solution* of the equation

$$-\Delta\big(a(k)\phi_p(\Delta u(k-1))\big) + b(k)\phi_p(u(k)) = \lambda f(k, u(k))), \quad k \in \mathbb{Z}.$$

Homoclinic solutions for discrete systems have been studied for example in $[83, 152, 153, 183, 193, 220, 231, 261, 298]$.

Let

$$F(k, t) = \int_0^t f(k, s)ds \quad \text{for all } (k, t) \in \mathbb{Z} \times \mathbb{R}. \tag{6.5.2}$$

We will make use of the following conditions in this section:

(F19) $b(k) \geq b_0 > 0$ for all $k \in \mathbb{Z}$, $b(k) \to \infty$ as $|k| \to \infty$;

(F20) $\sup_{|t| \leq T} |F(\cdot, t)| \in \ell^1$ for all $T > 0$;

(F21) $\limsup_{|t| \to \infty} \frac{F(k,t)}{|t|^p} \leq 0$ uniformly for all $k \in \mathbb{Z}$;

(F22) there exists a constant $M > 0$ such that

$$a(k) \leq Mb(k) \quad \text{for all } k \in \mathbb{Z};$$

(F23) there exist a constant $\rho_1 > 0$ and two positive functions $w_1, w_2 \in \ell^1$ such that

$$w_1(k)|t|^p \leq F(k, t) \leq w_2(k)|t|^p \tag{6.5.3}$$

for all $k \in \mathbb{Z}$ and $|t| \leq \rho_1$;

(F24) $f(k, -t) = -f(k, t)$ for all $k \in \mathbb{Z}$ and $t \in \mathbb{R}$;

(F25) there exist $\alpha \geq p$, $C > D > 0$, and $0 < \delta < 1$ such that

$$D|t|^{\alpha-1} < |f(k, t)| < C|t|^{\alpha-1} \quad \text{for } k \in \mathbb{Z} \text{ and } 0 < |t| \leq \delta;$$

(F26) $tf(k, t) \geq 0$ for $k \in \mathbb{Z}$ and $t \in [-\delta, \delta]$, where δ is as given in (F25).

(F27) there exist $d > 0$ and $q > p$ such that

$$|F(k, t)| \leq d|t|^q \quad \text{for all } k \in \mathbb{Z} \text{ and } t \in \mathbb{R};$$

(F28) $\lim_{|t| \to \infty} \frac{f(k,t)t}{|t|^p} = \infty$ uniformly for all $k \in \mathbb{Z}$;

(F29) there exists $\sigma \geq 1$ such that

$$\sigma\mathcal{F}(k, t) \geq \mathcal{F}(k, st) \quad \text{for } k \in \mathbb{Z}, \ t \in \mathbb{R}, \text{ and } s \in [0, 1],$$

where $\mathcal{F}(k, t) = f(k, t)t - pF(k, t)$.

Remark 6.5.1. We wish to point out that in condition (F23) we only need for (6.5.3) to hold for small $|t| \in \mathbb{R}$; there is no restriction on the behavior of $F(k, t)$ when $|t|$ is large.

Remark 6.5.2. If we assume that $f(k,t)/\phi_p(t)$ increases as $|t|$ increases, then, it can be shown that $\mathcal{F}(k,t) \geq \mathcal{F}(k,s)$ if $|t| \geq |s|$ and $ts \geq 0$. Hence, (F29) holds with $\sigma = 1$. To see this, note that for any $t \geq s \geq 0$, we have

$$
\begin{aligned}
\mathcal{F}(k,t) - \mathcal{F}(k,s) &= p\left[\frac{1}{p}f(k,t)t - \frac{1}{p}f(k,s)s - \big(F(k,t) - F(k,s)\big)\right] \\
&= p\left[\int_0^t \frac{f(k,t)}{t^{p-1}}\tau^{p-1}d\tau - \int_0^s \frac{f(k,s)}{s^{p-1}}\tau^{p-1}d\tau \right.\\
&\qquad \left. - \int_s^t \frac{f(k,\tau)}{\tau^{p-1}}\tau^{p-1}d\tau\right] \\
&= p\left[\int_s^t \left(\frac{f(k,t)}{t^{p-1}} - \frac{f(k,\tau)}{\tau^{p-1}}\right)\tau^{p-1}d\tau\right.\\
&\qquad \left. + \int_0^s \left(\frac{f(k,t)}{t^{p-1}} - \frac{f(k,s)}{s^{p-1}}\right)\tau^{p-1}d\tau\right] \\
&\geq 0.
\end{aligned}
$$

A similar argument holds if $t \leq s \leq 0$.

Remark 6.5.3. It is not hard to see that the Ambrosetti–Rabinowitz condition (see 1.1.2) implies (F28). In fact, if condition (AR) holds, then, we can show there exist $r_1, r_2 > 0$ such that

$$
F(k,t) \geq r_1|t|^\mu - r_2 \quad \text{for } (k,t) \in \mathbb{Z} \times \mathbb{R}^N,
$$

so (F28) holds. However, there are many functions that do not satisfy condition (AR). For example, it is easy to see if $\mu > 1$ and $\nu \geq 1$, then the function

$$
f(k,t) = \frac{1}{k^\mu}|t|^{p-2}t\ln\left(1 + |t|^\nu\right), \quad (k,t) \in \mathbb{Z} \times \mathbb{R} \tag{6.5.4}
$$

does not satisfy condition (AR). It does satisfy conditions (F20), (F24), and (F27)–(F29) however.

We begin by setting up the variational framework for problem (6.5.1). For all $1 \leq p < \infty$, let ℓ^p be the set of all functions $u : \mathbb{Z} \to \mathbb{R}$ such that

$$
\|u\|_p = \left(\sum_{k \in \mathbb{Z}} |u(k)|^p\right)^{1/p} < \infty,
$$

and let ℓ^∞ be the set of all functions $u : \mathbb{Z} \to \mathbb{R}$ such that

$$
\|u\|_\infty = \sup_{k \in \mathbb{Z}} |u(k)| < \infty.
$$

Then, for any $1 \le p \le q < \infty$, we have

$$\ell^p \subseteq \ell^q \quad \text{and} \quad \|u\|_q \le \|u\|_p. \tag{6.5.5}$$

In fact, for $u \in \ell^p$, by normalizing, we may assume $\|u\|_p = 1$. Then, $|u(k)| \le 1$ for any $k \in \mathbb{Z}$. Hence, $|u(k)|^q \le |u(k)|^p$. This shows that $\|u\|_q \le \|u\|_p$, and so $\ell^p \subseteq \ell^q$.

The following lemma can be found in [114, pp. 3 and 429] and [183, Proposition 2].

Lemma 6.5.1. *For all $1 \le p < \infty$, $(\ell^p, \|\cdot\|_p)$ is a reflexive and separable Banach space whose dual is $(\ell^q, \|\cdot\|_q)$, where $1/p + 1/q = 1$. Moreover, $(\ell^\infty, \|\cdot\|_\infty)$ is a Banach space, and for all $1 \le p < \infty$, the embedding $\ell^p \hookrightarrow \ell^\infty$ is continuous as*

$$\|u\|_\infty \le \|u\|_p \quad \text{for all } u \in \ell^p. \tag{6.5.6}$$

Let

$$X = \left\{ u : \mathbb{Z} \to \mathbb{R} \ : \ \sum_{k \in \mathbb{Z}} \left[a(k)|\Delta u(k-1)|^p + b(k)|u(k)|^p \right] < \infty \right\} \tag{6.5.7}$$

and

$$\|u\| = \left(\sum_{k \in \mathbb{Z}} \left[a(k)|\Delta u(k-1)|^p + b(k)|u(k)|^p \right] \right)^{1/p}. \tag{6.5.8}$$

From condition (F19), we have

$$\|u\|_\infty \le \|u\|_p \le b_0^{-1/p}\|u\|. \tag{6.5.9}$$

Lemma 6.5.2. *For all $1 < p < \infty$, $(X, \|\cdot\|)$ is a reflexive and separable Banach space, and the embedding $X \hookrightarrow \ell^p$ is compact.*

Proof. By Lemma 6.5.1, $(\ell^p, \|\cdot\|_p)$ is a reflexive and separable Banach space. Then, the Cartesian product space $\ell_2^p = \ell^p \times \ell^p$ is also a reflexive and separable Banach space with respect to the norm

$$\|v\|_{\ell_2^p} = \left(\sum_{i=1}^{2} \|v_i\|_p^p \right)^{1/p}, \quad v = (v_1, v_2) \in \ell_2^p. \tag{6.5.10}$$

Consider the space

$$Y = \left\{ \left((b(k))^{1/p} u(k), (a(k))^{1/p} \Delta u(k-1) \right) \ : \ u \in X, \ k \in \mathbb{Z} \right\}.$$

Then, Y is a closed subset of ℓ_2^p. Hence, Y is also a reflexive and separable Banach space with respect to the norm (6.5.10).

Let the operator $T : X \to Y$ be defined by

$$Tu = \big((b(k))^{1/p} u(k), (a(k))^{1/p} \Delta u(k-1)\big) \quad \text{for any } u \in X.$$

Clearly,

$$\|u\| = \|Tu\|_{\ell_2^p}.$$

Thus, $T : X \to Y$ is an isometric isomorphic mapping and X is isometric isomorphic to Y. Therefore, X is a reflexive and separable Banach space.

Finally, by using a proof similar to the ones used to prove [183, Proposition 3] or [220, Lemma 2.1], it can be shown that the embedding $X \hookrightarrow \ell^p$ is compact. We omit the details. This completes the proof of the lemma. $\square$

Next, we define the functionals needed for our problem. Then we will prove a number of lemmas describing properties of these functionals.

For any $u \in X$ and $\lambda > 0$, let

$$\Phi(u) = \frac{1}{p} \sum_{k \in \mathbb{Z}} \big[a(k)|\Delta u(k-1)|^p + b(k)|u(k)|^p \big],$$

$$\Psi(u) = \sum_{k \in \mathbb{Z}} F(k, u(k)),$$

and

$$I_\lambda(u) = \Phi(u) - \lambda \Psi(u). \tag{6.5.11}$$

Lemma 6.5.3. *For functionals Φ, Ψ, and I_λ, we have the following:*

(a) *Assume that (F19) holds. Then $\Phi \in C^1(X, \mathbb{R})$ with*

$$\langle \Phi'(u), v \rangle = \sum_{k \in \mathbb{Z}} \big[a(k)\phi_p(\Delta u(k-1))\Delta v(k-1) + b(k)\phi_p(u(k))v(k) \big]$$

for all $u, v \in X$.

(b) *Assume that one of (F23), (F25), and (F27) holds. Then $\Psi \in C^1(\ell^p, \mathbb{R})$ with*

$$\langle \Psi'(u), v \rangle = \sum_{k \in \mathbb{Z}} f(k, u(k))v(k) \quad \text{for all } u, v \in \ell^p.$$

(c) *Assume that (F19) and one of (F23), (F25), and (F27) hold. Then, for all $\lambda > 0$, every critical point $u \in X$ of I_λ is a solution of problem (6.5.1).*

Proof. Part (a) with $a(k) \equiv 1$ on $\mathbb{Z}$ was proved in [183, Proposition 5] and part (b) was shown in [183, Proposition 6] under the assumption

$$\lim_{t \to 0} \frac{|f(k,t)|}{|t|^{p-1}} = 0 \quad \text{uniformly for all} \quad k \in \mathbb{Z}. \tag{6.5.12}$$

Part (c) with $a(k) \equiv 1$ on $\mathbb{Z}$ was proved in [183, Proposition 7] under conditions (F19) and (6.5.12). Our form of the lemma can be proved in essentially the same way as in [183], so we only give a proof of part (b) for (F25) holding.

By (F25), we have

$$|f(k,t)| \le C|t|^{\alpha-1} \le C|t|^{p-1} \quad \text{for } k \in \mathbb{Z} \text{ and } |t| \le \delta. \tag{6.5.13}$$

Thus,

$$|F(k,t)| \le \frac{C}{p}|t|^p \quad \text{for } k \in \mathbb{Z} \text{ and } |t| \le \delta.$$

For any $u \in \ell^p$, there exists $h \in \mathbb{N}$ such that $|u(k)| \le \delta$ for all $k \in \mathbb{Z}$ with $|k| > h$. Hence,

$$\left| \sum_{k \in \mathbb{Z}} F(k, u(k)) \right| \le \sum_{|k| \le h} |F(k, u(k))| + \frac{C}{p} \sum_{|k| > h} |u(k)|^p,$$

and so Ψ is well defined.

Noting the similarity between (6.5.13) and inequality (6) in [183], the remainder of the proof is almost the same as that of the proof of [183, Proposition 6]. The details are omitted. □

Lemma 6.5.4. *Assume that (F19), (F20), (F21), (F23), and (F24) hold. Then, the functional I_λ defined by (6.5.11) satisfies condition (A3) of Theorem 1.1.13 with $I = I_\lambda$, i.e., $I_\lambda(u)$ is even, bounded from below, $I_\lambda(0) = 0$, and $I_\lambda(u)$ satisfies the (PS) condition.*

Proof. We see that $I_\lambda(0) = 0$ and $I_\lambda(u)$ is even by (F23). Under (F19)–(F21) and (I1), the rest part of the lemma has been proved in [183, Proposition 9] when $a(k) \equiv 1$ on $\mathbb{Z}$. The proof there can be slightly modified to prove the present version of the lemma. We omit the details. □

Lemma 6.5.5. *Assume that (F19), (F22), and (F23) hold. Then, for each $n \in \mathbb{N}$, there exist $H_n \in \Gamma_n$ and $\underline{\lambda} > 0$ satisfying*

$$\sup_{u \in H_n} I_\lambda(u) < 0 \quad \text{for all } \lambda > \underline{\lambda}.$$

Proof. For any fixed $n \in \mathbb{N}$, define
$$v_i(k) = \delta_{ik} = \begin{cases} 1 & \text{if } i = k, \\ 0 & \text{if } i \neq k, \end{cases} \quad \text{for } i = 1, \ldots, n \text{ and } k \in \mathbb{Z},$$
and
$$F_n = \text{span}\{v_i(k) \; : \; i = 1, \ldots, n\}.$$
Then, $\dim F_n = n$. Let
$$S = \left\{ u \in X \; : \; \|u\| = \rho_1 b_0^{1/p} \right\}, \quad H_n = S \cap F_n, \quad \text{and} \quad H_\infty = \lim_{n \to \infty} H_n$$
where ρ_1 is given in (F23). Then, by the property of genus (see, for example, [187, Lemma 2.6]), $\gamma(H_n) = n$. Let
$$\kappa = \inf_{n \in \mathbb{N}} \inf_{u \in H_n} \sum_{k=1}^{n} w_1(k)|u(k)|^p. \tag{6.5.14}$$
Since $H_n \subset H_{n+1}$ for any $n \in \mathbb{N}$, we have
$$\kappa = \inf_{u \in H_\infty} \sum_{k=1}^{\infty} w_1(k)|u(k)|^p.$$
We claim that $\kappa > 0$. In fact, assume, to the contrary, that $\kappa = 0$, then for any $l \in \mathbb{N}$, there exists and $u_l \in H_\infty$ such that
$$\sum_{k=1}^{\infty} w_1(k)|u_l(k)|^p < \frac{1}{l}. \tag{6.5.15}$$
Thus,
$$\lim_{l \to \infty} u_l(k) = 0 \quad \text{for } k \in \mathbb{Z}.$$
Then, from the fact that $\sum_{k=1}^{\infty} b(k)|u_l(k)|^p \leq \|u_l\|^p = \rho_1^p b_0 < \infty$, we see that there exists $L \in \mathbb{N}$ such that
$$\sum_{k=1}^{\infty} b(k)|u_l(k)|^p < \frac{\rho_1^p b_0}{(2^p M + 1)} \quad \text{for } l \geq L. \tag{6.5.16}$$
By Minkowski's inequality and (F22), we have
$$\sum_{k \in \mathbb{Z}} a(k)|\Delta u_l(k-1)|^p$$
$$= \sum_{k \in \mathbb{Z}} a(k)|u_l(k) - u_l(k-1)|^p$$
$$\leq \left[\left(\sum_{k \in \mathbb{Z}} a(k)|u_l(k)|^p \right)^{1/p} + \left(\sum_{k \in \mathbb{Z}} a(k)|u_l(k-1)|^p \right)^{1/p} \right]^p$$
$$= 2^p \sum_{k=1}^{\infty} a(k)|u_l(k)|^p$$
$$\leq 2^p M \sum_{k=1}^{\infty} a(k)|u_l(k)|^p.$$

Thus, from (6.5.8), we see that

$$\rho_1^p b_0 = \|u_l\|^p = \sum_{k \in \mathbb{Z}} \left[a(k)|\Delta u_l(k-1)|^p + b(k)|u_l(k)|^p \right]$$

$$\leq (2^p M + 1) \sum_{k=1}^{\infty} b(k)|u_l(k)|^p \quad \text{for } l \in \mathbb{N}.$$

This contradicts (6.5.16). Hence, the claim is true.

Now, for any $u \in H_n$, from (6.5.9), we have $\|u\|_\infty \leq b_0^{-1/p}\|u\| = \rho_1$. This, together with (F23) and (6.5.14), implies that

$$\sup_{u \in H_n} I_\lambda(u) = \sup_{u \in H_n} \left\{ \frac{1}{p} \sum_{k \in \mathbb{Z}} \left[a(k)|\Delta u(k-1)|^p + b(k)|u(k)|^p \right] \right. \tag{6.5.17}$$

$$\left. -\lambda \sum_{k \in \mathbb{Z}} F(k, u(k)) \right\}$$

$$\leq \sup_{u \in H_n} \left\{ \frac{1}{p}\|u\|^p - \lambda \sum_{k \in \mathbb{Z}} w_1(k)|u(k)|^p \right\}$$

$$= \sup_{u \in H_n} \left\{ \frac{1}{p}\rho_1^p b_0 - \lambda \sum_{k=1}^{n} w_1(k)|u(k)|^p \right\}$$

$$\leq \sup_{u \in H_n} \left\{ \frac{1}{p}\rho_1^p b_0 - \lambda \kappa \right\} \tag{6.5.18}$$

Let $\underline{\lambda} = \rho_1^p b_0/(\kappa p)$. Then, (6.5.17) implies that

$$\sup_{u \in H_n} I_\lambda(u) < 0 \quad \text{for all } \lambda > \underline{\lambda}.$$

This completes the proof of the lemma. $\qquad \square$

We are now ready to give our first main result in this section.

Theorem 6.5.1. *Assume that (F19)–(F24) hold. Then, there exists a constant $\underline{\lambda} > 0$ such that for all $\lambda > \underline{\lambda}$, problem (6.5.1) has a sequence $\{u_n(k)\}$ of nontrivial solutions satisfying*

$$u_n \to 0 \text{ in } X \quad \text{and} \quad I_\lambda(u_n) \leq 0, \tag{6.5.19}$$

where X and I_λ are defined by (6.5.7) and (6.5.11), respectively.

Proof. For $\underline{\lambda}$ given in Lemma 6.5.5, by Lemmas 6.5.4 and 6.5.5, conditions (A3) and (A4) of Theorem 1.1.13 with $I = I_\lambda$ are satisfied if $\lambda > \underline{\lambda}$. Hence, Theorem 1.1.13 and Lemma 6.5.3 (c) imply that for all $\lambda > \underline{\lambda}$, problem (6.5.1) has a sequence $\{u_n(k)\}$ of nontrivial solutions satisfying the required properties. This completes the proof the theorem. $\qquad \square$

The following corollary is an easy consequence of Theorem 6.5.1.

Corollary 6.5.1. *Assume that (F19) and (F22) hold, $w \in \ell^1$, and $1 < q < p$. Let*

$$g(t) = \begin{cases} \phi_p(t) & \text{if } |t| \leq 1, \\ \phi_q(t) & \text{if } |t| > 1. \end{cases}$$

Then, there exists a constant $\underline{\lambda} > 0$ such that for all $\lambda > \underline{\lambda}$, the problem

$$\begin{cases} -\Delta\big(a(k)\phi_p(\Delta u(k-1))\big) + b(k)\phi_p(u(k)) = \lambda w(k)g(u(k)), & k \in \mathbb{Z}, \\ u(k) \to 0 & \text{as } |k| \to \infty, \end{cases}$$

(6.5.20)

has a sequence $\{u_n(k)\}$ of nontrivial solutions satisfying (6.5.19).

Proof. With $f(k,t) = w(k)g(t)$ and $F(k,t)$ defined by (6.5.2), it is easy to see that (F20), (F21), (F23), and (F24) hold. Moreover, in (F23), we can let $w_1 = w_2 = w$ and $\rho_1 = 1$, The conclusion then follows from Theorem 6.5.1. This completes the proof of the corollary. $\qquad\square$

In order to prove our next main result we need the following lemma.

Lemma 6.5.6. *Assume that (F19)–(F21) and (F25) hold. Then, for any $\lambda > 0$, I_λ is coercive and satisfies the (PS) condition.*

Proof. The conclusion of the lemma with $a(t) \equiv 1$ was proved in [183, Proposition 9] under the assumptions (F19)–(F21) and (6.5.12). The role (6.5.12) is to guarantee that $\Psi \in C^1(X, \mathbb{R})$. In view of Lemma 6.5.3 (b), the in our case is essentially the same and so we omit the details. $\qquad\square$

Theorem 6.5.2. *Assume that (F19)–(F21), (F25), and (F26) hold. Then, there exists $\lambda^* > 0$ such that for any $\lambda > \lambda^*$, problem (6.5.1) has at least one nontrivial solution $\{u_\lambda(k)\}$ which is a global minimizer of the functional I_λ defined by (6.5.11).*

Proof. Fix $\lambda > 0$. By Lemma 6.5.6, I_λ is coercive, i.e., $\lim_{\|u\| \to \infty} I_\lambda(u) = \infty$. Note that I_λ is differentiable and sequentially weakly lower semicontinuous and X is reflexive (by Lemma 6.5.2). Then, from Theorem 1.1.9 with $I = I_\lambda$, I_λ attains its infimum in X at some $u_\lambda \in X$ and $I'_\lambda(u_\lambda) = 0$. In view of Lemma 6.5.3 (c), $u_\lambda(k)$ is a solution of problem (6.5.1).

We now show that $u_\lambda(k) \not\equiv 0$ on $\mathbb{Z}$. Let δ be given as in (F25). Choose a function $w \in X$ satisfying $w(k) \not\equiv 0$ and $|w(k)| \leq \delta$ on $\mathbb{Z}$. Then, by Lemma

6.5.2 and the fact that $1 < p \leq \alpha$, we have $w \in \ell^\alpha$. From (F25) and (F26), we see that

$$F(k,t) \geq \frac{1}{\alpha} D |t|^\alpha \quad \text{for } k \in \mathbb{Z} \text{ and } |t| \leq \delta. \tag{6.5.21}$$

Then, from (6.5.8), (6.5.11), and (6.5.21), we have

$$\begin{aligned}
I_\lambda(w) &= \Phi(w) - \lambda\Psi(w) \\
&< \frac{1}{p}\|w\|^p - \frac{1}{\alpha}\lambda D \sum_{k \in \mathbb{Z}} |w(k)|^\alpha \\
&= \frac{1}{p}\|w\|^p - \frac{1}{\alpha}\lambda \|w\|_\alpha^\alpha D < 0 \quad \text{if } \lambda > \lambda^*,
\end{aligned} \tag{6.5.22}$$

where

$$\lambda^* = \frac{\alpha\|w\|^p}{pD\|w\|_\alpha^\alpha} > 0.$$

Then, $I_\lambda(u_\lambda) < 0$ if $\lambda > \lambda^*$. Hence, $u_\lambda(k) \not\equiv 0$ on $\mathbb{Z}$ for $\lambda > \lambda^*$, i.e, problem (6.5.1) has a nontrivial solution for $\lambda > \lambda^*$. This completes the proof of the theorem. $\qquad\square$

To prove our next main result we need the following lemma.

Lemma 6.5.7. *Assume that (F19), (F25), and (F26) hold. Then, there exists $\overline{\lambda} > 0$ such that for any $\lambda > \overline{\lambda}$ and $n \in \mathbb{N}$, there exists $A_n \in \Gamma_n$ such that $\sup_{u \in A_n} I_\lambda(u) < 0$.*

Proof. Let $\lambda > 0$ be fixed. For any $n \in \mathbb{N}$, we can choose an n-dimensional subspace $Y_n \subset X$. Let

$$S_{n-1} = \{u \in Y_n \; : \; \|u\| = 1\}.$$

Since S_{n-1} is finite dimensional and compact, we have

$$\kappa_{n-1} := \inf_{w \in S_{n-1}} \|w\|_\alpha^\alpha > 0.$$

Define

$$\kappa = \inf_{n \in \mathbb{N}} \kappa_{n-1}.$$

To show that $\kappa > 0$, assume to the contrary that $\kappa = 0$. Then, for any $l \in \mathbb{N}$, there exists $w_l \in S_{l-1}$ such that

$$\sum_{k \in \mathbb{Z}} |w_l(k)|^\alpha < \frac{1}{l}.$$

Thus,

$$\lim_{l \to \infty} w_l(k) \to 0 \quad \text{for } k \in \mathbb{Z}.$$

Then, from the facts that

$$\sum_{k \in \mathbb{Z}} b(k)|w_l(k)|^p \le \|w_l\|^p = 1$$

and

$$\sum_{k \in \mathbb{Z}} a(k)|\Delta w_l(k-1)|^p \le \|w_l\|^p = 1,$$

we see that there exists $L \in \mathbb{N}$ such that

$$\sum_{k \in \mathbb{Z}} b(k)|w_l(k)|^p \le \frac{1}{3} \quad \text{and} \quad \sum_{k \in \mathbb{Z}} a(k)|\Delta w_l(k-1)|^p \le \frac{1}{3} \quad \text{for } l \ge L.$$

Hence,

$$\|w_l\|^p = \sum_{k \in \mathbb{Z}} \left[a(k)|\Delta w_l(k-1)|^p + b(k)|w_l(k)|^p \right] \le \frac{2}{3} \quad \text{for } l \ge L.$$

On the other hand, since $w_l \in S_{l-1}$, we have $\|w_l\| = 1$ for any $l \in \mathbb{N}$, which is a contradiction. Therefore, $\kappa > 0$.

Now, choose $\mu > 0$ small enough so that $\mu b_0^{-1/p} \le \delta$, where δ is given in (F25). Then, for any $w \in S_{n-1}$, in view of (6.5.9), we see that

$$\|\mu w\|_\infty \le \mu b_0^{-1/p} \|w\| \le \delta.$$

Then, as in (6.5.22), from (6.5.8), (6.5.11), and (6.5.6), we have that, for any $w \in S_{n-1}$,

$$
\begin{aligned}
I_\lambda(\mu w) &= \Phi(\mu w) - \lambda \Psi(\mu w) \\
&< \frac{1}{p}\|w\|^p \mu^p - \frac{1}{\alpha} \lambda \mu^\alpha D \sum_{k \in \mathbb{Z}} |w(k)|^\alpha \\
&\le \frac{1}{p} \mu^p - \frac{1}{\alpha} \lambda \kappa_{n-1} \mu^\alpha D \\
&\le \frac{1}{p} \mu^p - \frac{1}{\alpha} \lambda \kappa \mu^\alpha D < 0 \quad \text{if } \lambda > \overline{\lambda},
\end{aligned}
$$

where

$$\overline{\lambda} = \frac{\alpha \mu^{p-\alpha}}{p \kappa D}.$$

Let $A_n = \mu S_{n-1}$. Then, $\gamma(A_n) = n$ and $\sup_{u \in A_n} I_\lambda(u) < 0$ for $\lambda > \overline{\lambda}$. This completes the proof of the lemma. $\qquad \square$

Theorem 6.5.3. *Assume that (F19)–(F21) and (F24)–(F26) hold. Then, for any $\lambda > 0$, problem (6.5.1) has a sequence $\{u_{\lambda,n}(k)\}$ of nontrivial solutions satisfying*

$$u_{\lambda,n} \to 0 \text{ in } X \text{ as } n \to \infty \quad \text{and} \quad I_\lambda(u_{\lambda,n}) \le 0,$$

where X and I_λ are defined by (6.5.7) and (6.5.11), respectively.

Proof. Let $\lambda > 0$ be fixed. Then, by (F24) and Lemmas 6.5.6 and 6.5.7, conditions (A3) and (A4) of Theorem 1.1.13 with $I = I_\lambda$ are satisfied. Hence, Theorem 1.1.13 and Lemma 6.5.9 (c) imply that, for all $\lambda > 0$, problem (6.5.1) has a sequence $\{u_n(k)\}$ of nontrivial solutions satisfying the required properties. This completes the proof of the theorem.   □

Remark 6.5.4. It is not hard to see that if $\alpha = p$, then the results here are independent from those of Iannizzotto and Tersian [183]. That is, our condition (F25) does not imply their condition (F_1) and vice-versa. If $\alpha > p$, then our condition (F25) does imply their (F_1). On the other hand, our Theorem 6.5.3 guarantees the existence of an infinite number of solutions, whereas in [183] the authors obtain the existence of only two solutions.

In the remainder of this section, let the Banach space X be defined by (6.5.7), and Y_n and Z_n be given in (1.1.8). By Lemma 6.5.2, X is reflexive and separable.

Lemma 6.5.8. *Let $q > p$. For $n \in \mathbb{N}$, define*

$$\beta_n = \sup_{u \in Z_n, \|u\|=1} \|u\|_q.$$

Then, $\lim_{n \to \infty} \beta_n = 0$.

Proof. Obviously, $0 < \beta_{n+1} < \beta_n$. Then, there exists $\beta \ge 0$ such that $\lim_{n \to \infty} \beta_n = \beta$. For any $n \in \mathbb{N}$, let $u_n \in \mathbb{Z}_n$ be such that

$$\|u_n\| = 1 \quad \text{and} \quad 0 \le \beta_n - \|u_n\|_q < \frac{1}{n}. \tag{6.5.23}$$

Since X is reflexive, there exists a subsequence of $\{u_n\}$ (which is still denoted by $\{u_n\}$) such that $u_n \rightharpoonup u$. We claim $u = 0$. If fact, for any $e_m^* \in \{e_n^* : n \in \mathbb{N}\}$, we have $\langle e_m^*, u_n \rangle = 0$ when $n > m$. Hence,

$$\langle e_m^*, u \rangle = \lim_{n \to \infty} \langle e_m^*, u_n \rangle = 0 \quad \text{for } m \in \mathbb{N}.$$

Therefore, $u = 0$, i.e., $u_n \rightharpoonup 0$. By (6.5.10) and Lemma 6.5.2, $\ell^p \hookrightarrow \ell^q$ is continuous and $X \hookrightarrow \ell^p$ is compact. Thus, $X \hookrightarrow \ell^q$ is compact. Hence, $u_n \to 0$ in ℓ^p. Then, from (6.5.23), $\lim_{n \to \infty} \beta_n = 0$. This completes the proof of the lemma.   □

Using the same approach as in the proof of [183, Lemma 4], we can prove the following result.

Lemma 6.5.9. *If S is a compact subset of ℓ^p, then, for all $\delta > 0$, there exists $h > 0$ such that*

$$|u(k)| \leq \left(\sum_{|k|>h} |u(k)|^p \right)^{1/p} < \delta \quad \text{for all } u \in S \text{ and } |k| > h.$$

Lemma 6.5.10. *Assume that (F19) and (F27)–(F29) hold. Then, for any $\lambda > 0$, I_λ satisfies Cerami's condition.*

In the sequel, we let C_i denote a generic positive constant.

Proof. Let $\lambda > 0$ be fixed. We first show that I_λ satisfies part (i) of Cerami's condition. Let $\{u_n\} \subset X$ be a bounded sequence such that $I_\lambda(u_n) \to c$ and $I_\lambda'(u_n) \to 0$. Then, passing to a subsequence if necessary, it can be assumed that $u_n \rightharpoonup u$ weakly in X. Hence, by Lemma 6.5.2, $u_n \to u$ in ℓ^p. Note that (F27) implies (I1). Then, for any $\epsilon > 0$, by Lemma 6.5.9, we see that there exist $K \in \mathbb{N}$, independent of n, and $N \in \mathbb{N}$ such that

$$|f(k, u_n(k))| < \epsilon |u_n(k)|^{p-1} \quad \text{and} \quad |f(k, u(k))| < \epsilon |u(k)|^{p-1} \quad \text{for all } k > K,$$
$$(6.5.24)$$

and

$$\|u_n - u\|_p < \epsilon \quad \text{for all } n > N. \tag{6.5.25}$$

It is obvious that

$$\sum_{k \in \mathbb{Z}} (f(k, u_n(k)) - f(k, u(k)))(u_n - u)$$
$$= \sum_{|k| \leq K} (f(k, u_n(k)) - f(k, u(k)))(u_n - u)$$
$$+ \sum_{|k| > K} (f(k, u_n(k)) - f(k, u(k)))(u_n - u)$$
$$= J_1(u_n) + J_2(u_n),$$

where

$$J_1(u_n) = \sum_{|k| \leq K} (f(k, u_n(k)) - f(k, u(k)))(u_n - u)$$

and

$$J_2(u_n) = \sum_{|k| > K} (f(k, u_n(k)) - f(k, u(k)))(u_n - u).$$

By the continuity of $f(k,t)$ in t and $u_n \to u$ in ℓ^p, we can see that $J_1(u_n) \to 0$ as $n \to \infty$. Moreover, from (6.5.24), (6.5.25), and Hölder's inequality, it follows that

$$
\begin{aligned}
|J_2(u_n)| &\leq \sum_{|k|>K} |f(k, u_n(k))|\, |u_n - u| + \sum_{|k|>K} |f(k, u(k))|\, |u_n - u| \\
&\leq \epsilon \sum_{|k|>K} |u_n(k)|^{p-1} |u_n - u| + \epsilon \sum_{|k|>K} |u(k)|^{p-1} |u_n - u| \\
&\leq \epsilon \left(\sum_{|k|>K} |u_n(k)|^p \right)^{(p-1)/p} \left(\sum_{|k|>K} |u_n - u|^p \right)^{1/p} \\
&\quad + \epsilon \left(\sum_{|k|>K} |u(k)|^p \right)^{(p-1)/p} \left(\sum_{|k|>K} |u_n - u|^p \right)^{1/p} \\
&\leq \epsilon(\|u_n\|^{p-1} + \|u\|^{p-1}) \|u_n - u\|_p \\
&\leq C_1 \epsilon \|u_n - u\|_p < C_1 \epsilon^2 \quad \text{if } n > N.
\end{aligned}
$$

Hence, $J_2(u_n) \to 0$ as $n \to \infty$. Then, we have

$$
\sum_{k \in \mathbb{Z}} (f(k, u_n(k)) - f(k, u(k)))(u_n - u) \to 0 \quad \text{as } n \to \infty. \tag{6.5.26}
$$

Moreover, it is easy to see that

$$
\langle I'_\lambda(u_n) - I'_\lambda(u), u_n - u \rangle \to 0 \quad \text{as } n \to \infty. \tag{6.5.27}
$$

Now, by (6.5.11) and parts (a) and (b) of Lemma 6.5.3,

$$
\begin{aligned}
&\langle I'_\lambda(u_n) - I'_\lambda(u), u_n - u \rangle \\
&= \sum_{k \in \mathbb{Z}} \big[a(k)\big(\phi_p(\Delta u_n(k-1)) - \phi_p(\Delta u(k-1))\big)\big(\Delta u_n(k-1) - \Delta u(k-1)\big) \\
&\quad + b(k)\big(\phi_p(u_n(k)) - \phi_p(u(k))\big)\big(u_n(k) - u(k)\big) \big] \\
&\quad + \lambda \sum_{k \in \mathbb{Z}} (f(k, u_n(k)) - f(k, u(k)))(u_n - u). \tag{6.5.28}
\end{aligned}
$$

Recall the following well known inequalities (see, for example (2.2) in [253])

$$
(\phi(x) - \phi(y))(x - y) \geq
\begin{cases}
\dfrac{1}{2^p} |x - y|^p & \text{if } p \geq 2, \\[2mm]
\dfrac{(p-1)|x-y|^2}{(|x|+|y|)^{2-p}} & \text{if } 1 < p < 2,
\end{cases}
\quad \text{for } x, y \in \mathbb{R}. \tag{6.5.29}
$$

If $p \geq 2$, then in view of (6.5.28) and (6.5.29), we have

$$\langle I'_\lambda(u_n) - I'_\lambda(u), u_n - u \rangle$$
$$\geq \frac{1}{2^p} \sum_{k \in \mathbb{Z}} \left[a(k)|\Delta u_n(k-1) - \Delta u(k-1)|^p + b(k)|u_n(k) - u(k)|^p \right]$$
$$+ \lambda \sum_{k \in \mathbb{Z}} (f(k, u_n(k)) - f(k, u(k)))(u_n - u)$$
$$= \frac{1}{2^p} \|u_n - u\|^p + \lambda \sum_{k \in \mathbb{Z}} (f(k, u_n(k)) - f(k, u(k)))(u_n - u).$$

Hence, from (6.5.26) and (6.5.27), we see that $\|u_n - u\|^p \to 0$ as $n \to \infty$, i.e., $u_n \to u$ in X.

If $1 < p < 2$, then, by Hölder's inequality, we obtain

$$\sum_{k \in \mathbb{Z}} b(k)|u_n(k) - u(k)|^p$$
$$\leq \left(\sum_{k \in \mathbb{Z}} \frac{b(k)|u_n(k) - u(k)|^2}{(|u_n(k)| + |u(k)|)^{2-p}} \right)^{p/2} \left(\sum_{k \in \mathbb{Z}} b(k)(|u_n(k)| + |u(k)|)^p \right)^{(2-p)/2}$$
$$\leq \left(\sum_{k \in \mathbb{Z}} \frac{b(k)|u_n(k) - u(k)|^2}{(|u_n(k)| + |u(k)|)^{2-p}} \right)^{p/2}$$
$$\times 2^{p(2-p)/2} \left(\sum_{k \in \mathbb{Z}} b(k)(|u_n(k)|^p + |u(k)|^p) \right)^{(2-p)/2}$$
$$\leq C_2 \left(\sum_{k \in \mathbb{Z}} \frac{b(k)|u_n(k) - u(k)|^2}{(|u_n(k)| + |u(k)|)^{2-p}} \right)^{p/2} (\|u_n\| + \|u\|)^{(2-p)p/2}. \qquad (6.5.30)$$

Similarly, we have

$$\sum_{k \in \mathbb{Z}} a(k)|\Delta u_n(k-1) - \Delta u(k-1)|^p$$
$$\leq C_3 \left(\sum_{k \in \mathbb{Z}} \frac{a(k)|\Delta u_n(k-1) - \Delta u(k-1)|^2}{(|\Delta u_n(k-1)| + |\Delta u(k-1)|)^{2-p}} \right)^{p/2}$$
$$\times (\|u_n\| + \|u\|)^{(2-p)p/2}. \qquad (6.5.31)$$

Then, from (6.5.28)–(6.5.31), we have

$$\langle I_\lambda'(u_n) - I_\lambda'(u), u_n - u \rangle$$

$$\geq (p-1) \sum_{k\in\mathbb{Z}} \left[\frac{a(k)|\Delta u_n(k-1) - \Delta u(k-1)|^2}{(|\Delta u_n(k-1)| + |\Delta u(k-1)|)^{2-p}} + \frac{b(k)|u_n(k) - u(k)|^2}{(|u_n(k)| + |u(k)|)^{2-p}} \right]$$

$$+ \lambda \sum_{k\in\mathbb{Z}} (f(k,u_n(k)) - f(k,u(k)))(u_n - u)$$

$$\geq \left[C_4 \left(\left(\sum_{k\in\mathbb{Z}} a(k)|\Delta u_n(k-1) - \Delta u(k-1)|^p \right)^{2/p} \right. \right.$$

$$\left. \left. + \left(\sum_{k\in\mathbb{Z}} b(k)|u_n(k) - u(k)|^p \right)^{2/p} \right) \right] \times \left[(\|u_n\| + \|u\|)^{(2-p)} \right]^{-1}$$

$$+ \lambda \sum_{k\in\mathbb{Z}} (f(k,u_n(k)) - f(k,u(k)))(u_n - u)$$

$$\geq \frac{C_5 \left(\sum_{k\in\mathbb{Z}} [a(k)|\Delta u_n(k-1) - \Delta u(k-1)|^p + b(k)|u_n(k) - u(k)|^p] \right)^{2/p}}{(\|u_n\| + \|u\|)^{(2-p)}}$$

$$+ \lambda \sum_{k\in\mathbb{Z}} (f(k,u_n(k)) - f(k,u(k)))(u_n - u)$$

$$= \frac{C_6 \|u_n - u\|^2}{(\|u_n\| + \|u\|)^{(2-p)}} + \lambda \sum_{k\in\mathbb{Z}} (f(k,u_n(k)) - f(k,u(k)))(u_n - u). \quad (6.5.32)$$

Now, from (6.5.26), (6.5.27), and (6.5.32), it follows that $\|u_n - u\|^2 \to 0$ as $n \to \infty$, i.e., $u_n \to u$ in X. This shows that I_λ satisfies part (i) of Cerami's condition.

In the following, we show that, for any fixed $\lambda > 0$, I_λ satisfies part (ii) of Cerami's condition. Assume, to the contrary, that there exist $c \in \mathbb{R}$ and $\{u_n\} \subset X$ such that

$$I_\lambda(u_n) \to c, \quad \|u_n\| \to \infty, \quad \text{and} \quad \|I_\lambda'(u_n)\|\|u_n\| \to 0 \quad \text{as } n \to \infty. \quad (6.5.33)$$

Then, we have

$$\lim_{n\to\infty} \left(\frac{\lambda}{p} \sum_{k\in\mathbb{Z}} f(k,u_n(k))u_n(k) - \lambda \sum_{k\in\mathbb{Z}} F(k,u_n(k)) \right)$$

$$= \lim_{n\to\infty} \left(I_\lambda(u_n) - \frac{1}{p} \langle I_\lambda'(u_n), u_n \rangle \right) = c. \quad (6.5.34)$$

Let $w_n(k) = u_n(k)/\|u_n\|$. Then, $\|w_n\| = 1$. In view of Lemmas 6.5.1 and 6.5.2, up to subsequences, we can assume that, for some $w \in X$,

$$\begin{cases} w_n \rightharpoonup w & \text{in } X, \\ w_n \to w & \text{in } \ell^p, \\ w_n(k) \to w(k) & \text{in } \mathbb{Z}. \end{cases} \tag{6.5.35}$$

We now consider two cases:

Case 1: $w(k) \equiv 0$ on $\mathbb{Z}$. For this case, we define a sequence $\{t_n\}$ of real numbers such that

$$I_\lambda(t_n u_n) = \max_{t \in [0,1]} I_\lambda(t u_n). \tag{6.5.36}$$

If, for some $n \in \mathbb{N}$, there exist more than one t_n satisfying (6.5.36), we can arbitrarily choose one of them. For any $m > 0$, let $\overline{w}_n = \sqrt[p]{2pm}\, w_n$. By (6.5.35), $w_n(k) \to 0$ on $\mathbb{Z}$. Then,

$$\lim_{n \to \infty} \sum_{k \in \mathbb{Z}} F(k, \overline{w}_n(k)) = \lim_{n \to \infty} \sum_{k \in \mathbb{Z}} F\left(k, \sqrt[p]{2pm}\, w_n(k)\right) = 0.$$

Thus, for n large enough,

$$I_\lambda(t_n u_n) \geq I_\lambda(\overline{w}_n) = 2m - \lambda \lim_{n \to \infty} \sum_{k \in \mathbb{Z}} F(k, \overline{w}_n(k)) \geq m.$$

Hence, $\lim_{n \to \infty} I_\lambda(t_n u_n) = \infty$. Since $I_\lambda(u_n) \to c$ (see (6.5.33)) and $I_\lambda(0) = 0$, we have $t_n \in (0,1)$. Then, for n large enough, we have

$$\sum_{k \in \mathbb{Z}} [a(k)|\Delta(t_n u_n(k-1))|^p + b(k)|t_n u_n(k)|^p]$$

$$-\lambda \sum_{k \in \mathbb{Z}} f(k, t_n u_n(k)) t_n u_n(k) = \langle I_\lambda'(t_n u_n), t_n u_n \rangle = t_n \frac{d}{dt}\bigg|_{t=t_n} I_\lambda(t u_n) = 0.$$

Thus,

$$\lim_{n \to \infty} \left(\frac{\lambda}{p} \sum_{k \in \mathbb{Z}} f(k, t_n u_n(k)) t_n u_n(k) - \lambda \sum_{k \in \mathbb{Z}} F(k, t_n u_n(k)) \right)$$

$$= \lim_{n \to \infty} \left(\frac{1}{p} \sum_{k \in \mathbb{Z}} [a(k)|\Delta(t_n u_n(k-1))|^p + b(k)|t_n u_n(k)|^p] \right.$$

$$\left. -\lambda \sum_{k \in \mathbb{Z}} F(k, t_n u_n(k)) \right)$$

$$= \lim_{n \to \infty} I_\lambda(t_n u_n) = \infty. \tag{6.5.37}$$

On the other hand, since $t_n \in (0,1)$, by (F29), it follows that

$$\sigma \mathcal{F}(k, u_n(k)) \geq \mathcal{F}(k, t_n u_n(k)).$$

This, together with (6.5.37), implies that

$$\frac{\lambda}{p} \sum_{k \in \mathbb{Z}} f(k, u_n(k)) u_n(k) - \lambda \sum_{k \in \mathbb{Z}} F(k, u_n(k))$$

$$= \frac{\lambda}{p} \sum_{k \in \mathbb{Z}} \mathcal{F}(k, u_n(k))$$

$$\geq \frac{\lambda}{p\sigma} \sum_{k \in \mathbb{Z}} \mathcal{F}(k, t_n u_n(k))$$

$$= \frac{1}{\sigma} \left(\frac{\lambda}{p} \sum_{k \in \mathbb{Z}} f(k, t_n u_n(k)) t_n u_n(k) - \lambda \sum_{k \in \mathbb{Z}} F(k, t_n u_n(k)) \right) \to \infty$$

as $n \to \infty$, which contradicts (6.5.34).

Case 2: $w(k) \not\equiv 0$ on $\mathbb{Z}$. From (6.5.33), we see that

$$\sum_{k \in \mathbb{Z}} [a(k)|\Delta(u_n(k-1))|^p + b(k)|u_n(k)|^p] - \lambda \sum_{k \in \mathbb{Z}} f(k, u_n(k)) u_n(k)$$

$$= \langle I'_\lambda(u_n), u_n \rangle = o(1) \quad \text{as } n \to \infty. \tag{6.5.38}$$

Hence,

$$1 - o(1) = \lambda \sum_{k \in \mathbb{Z}} \frac{f(k, u_n(k)) u_n(k)}{\|u_n\|^p}$$

$$= \lambda \left(\sum_{k \in \mathbb{Z}_1} + \sum_{k \in \mathbb{Z}_2} \right) \frac{f(k, u_n(k)) u_n(k)}{|u_n(k)|^p} |w_n(k)|^p, \quad (6.5.39)$$

where $Z_1 = \{k \in \mathbb{Z} : w(k) \neq 0\}$ and $Z_2 = \mathbb{Z} \setminus Z_1$. For $k \in Z_1$, we have $u_n(k) \to \infty$. Then, by (F28),

$$\lim_{n \to \infty} \frac{f(k, u_n(k)) u_n(k)}{|u_n(k)|^p} |w_n(k)|^p = \infty.$$

Consequently,

$$\lim_{n \to \infty} \sum_{k \in Z_1} \frac{f(k, u_n(k)) u_n(k)}{|u_n(k)|^p} |w_n(k)|^p = \infty. \tag{6.5.40}$$

Again, by (F28), there exists $A_1 > -\infty$ such that

$$\frac{f(k,t)t}{|t|^p} > A_1 \quad \text{for } (k,t) \in \mathbb{Z} \times \mathbb{R}.$$

Now, it is easy to see that $\lim_{n\to\infty} \sum_{k\in\mathbb{Z}_2} |w_n(k)|^p = 0$. Thus, we have

$$\liminf_{n\to\infty} \sum_{k\in\mathbb{Z}_2} \frac{f(k,u_n(k))u_n(k)}{|u_n(k)|^p} |w_n(k)|^p \geq A \lim_{n\to\infty} \sum_{k\in\mathbb{Z}_2} |w_n(k)|^p = 0.$$

(6.5.41)

From (6.5.39), (6.5.40), and (6.5.41), we see that there is a contradiction. Thus, we have proved that I_λ satisfies part (ii) of Cerami's condition. This completes the proof of the lemma. □

We are now ready to give the last main result in this section.

Theorem 6.5.4. *Assume that (F19), (F20), (F24), and (F27)–(F29) hold. Then, for any $\lambda > 0$, problem (6.5.1) has a sequence $\{u_n(k)\}$ of nontrivial solutions such that $I_\lambda(u_n) \to \infty$ as $n \to \infty$, where I_λ is defined by (6.5.11).*

Moreover, whenever $u : \mathbb{Z} \to \mathbb{R}$ is a nontrivial solution of problem (6.5.1), there exist integers k_- and k_+ such that both sequences $\{u(k)\}_{k\leq k_-}$ and $\{u(k)\}_{k\geq k_+}$ are strictly monotone.

Proof. Let $\lambda > 0$ be fixed. By (F24) and Lemma 6.5.10, I_λ is even and satisfies Cerami's condition. In the following, we show that, for any $n \in \mathbb{N}$, there exist $\rho_n > r_n > 0$ such that conditions (a) and (b) of Theorem 1.1.14 with $I = I_\lambda$ are satisfied.

For β_n defined in Lemma 6.5.8, we have $\lim_{n\to\infty} \beta_n = 0$ by Lemma 6.5.8. Let $r_n = (\lambda dq\beta_n^q)^{1/(p-q)}$, where $d > 0$ and $q > p$ are given in (F27). Then, $r_n > 0$ and $\lim_{n\to\infty} r_n = \infty$. For $u \in Z_n$ with $\|u\| = r_n$, by the definition of β_n, $\|u\|_q \leq \beta_n \|u\|$. Combining this with (F27), we obtain that

$$\begin{aligned} I_\lambda(u) &= \frac{1}{p} \sum_{k\in\mathbb{Z}} \left[a(k)|\Delta u(k-1)|^p + b(k)|u(k)|^p \right] - \lambda \sum_{k\in\mathbb{Z}} F(k,u(k)) \\ &\geq \frac{1}{p}\|u\|^p - \lambda d\|u\|_q^q \\ &\geq \frac{1}{p}\|u\|^p - \lambda d\beta_n^q \|u\|^q \\ &= \left(\frac{1}{p} - \frac{1}{q}\right)\|u\|^p + \frac{1}{q}\|u\|^p - \lambda d\beta_n^q \|u\|^q \\ &= \left(\frac{1}{p} - \frac{1}{q}\right)(\lambda dq\beta_n^q)^{p/(p-q)}. \end{aligned}$$

(6.5.42)

Since $q > p$, from (6.5.42), it follows that

$$a_n = \inf_{u\in Z_n, \|u\|=r_n} I_\lambda(u) \to \infty \quad \text{as } n \to \infty,$$

i.e., condition (a) of Theorem 1.1.14 with $I = I_\lambda$ holds.

Since the dimension of Y_n is finite, all norms of Y_n are equivalent. So there exists $C_n > 0$ such that

$$\frac{1}{p}\|u\|^p \leq \lambda C_n \|u\|_\infty^p \quad \text{for all } u \in Y_n. \tag{6.5.43}$$

By (F20) and (F28), we see that there exists $T \in \mathbb{N}$ and a nonnegative $w \in \ell^1$ such that

$$F(k, t) \geq 2C_n |t|^p \quad \text{for } (k, t) \in \mathbb{Z} \times \mathbb{R} \text{ with } |t| > T \tag{6.5.44}$$

and

$$F(k, t) \geq -w(k) \quad \text{for } (k, t) \in \mathbb{Z} \times \mathbb{R} \text{ with } |t| \leq T. \tag{6.5.45}$$

For any $u \in Y_n$ with $\|u\|_\infty > T$, let

$$K_{u,1} = \{k \in \mathbb{Z} : |u(k)| > T\} \quad \text{and} \quad K_{u,2} = \mathbb{Z} \setminus K_{u,1}.$$

Then, as the inequality (6.5.6) in Lemma 6.5.1, one can see that

$$\sup_{k \in K_{u,1}} |u(k)| \leq \left(\sum_{k \in K_{u,1}} |u(k)|^p\right)^{1/p}. \tag{6.5.46}$$

Note that $\sup_{k \in K_{u,1}} |u(k)| = \sup_{k \in \mathbb{Z}} |u(k)| = \|u\|_\infty$. Then, from (6.5.46), we have

$$\|u\|_\infty^p \leq \sum_{k \in K_{u,1}} |u(k)|^p. \tag{6.5.47}$$

Thus, from (6.5.43)–(6.5.45) and (6.5.47), it follows that

$$\begin{aligned}
I_\lambda(u) &= \frac{1}{p}\|u\|^p - \lambda \sum_{k \in K_{u,1}} F(k, u(k)) - \lambda \sum_{k \in K_{u,2}} F(k, u(k)) \\
&\leq \frac{1}{p}\|u\|^p - 2\lambda C_n \sum_{k \in K_{u,1}} |u(k)|^p + \sum_{k \in K_{u,2}} |w(k)| \\
&\leq \lambda C_n \|u\|_\infty^p - 2\lambda C_n \|u\|_\infty^p + \|w\|_1 \\
&= -\lambda C_n \|u\|_\infty^p + \|w\|_1 \\
&\leq -\frac{1}{p}\|u\|^p + \|w\|_1.
\end{aligned}$$

Thus, we always can choose ρ_n large enough so that $\rho_n > r_n > 0$ and

$$b_n = \max_{u \in Y_n, \|u\| = \rho_n} I(u) \leq 0,$$

i.e., condition (b) of Theorem 1.1.14 with $I = I_\lambda$ holds.

We have verified that all the conditions of Theorem 1.1.14 with $I = I_\lambda$ are satisfied. Therefore, the first part of the theorem follows from Theorem 1.1.14. Note that (F27) implies (I1). Then, the "moreover" part of the theorem can be proved essentially by the same way as in the proof of the "moreover" part of [183, Theorem 1]. The details are omitted here. This completes the proof of the theorem. $\qquad \square$

The following result is an immediate consequence of Theorem 6.5.4.

Corollary 6.5.2. *Assume that (F19) holds, $\mu > 1$, and $\nu \geq 1$. Then, for all $\lambda > 0$, the problem*

$$
\begin{cases}
-\Delta\big(a(k)\phi_p(\Delta u(k-1))\big) + b(k)\phi_p(u(k)) \\
\qquad\qquad = \lambda k^{-\mu}|u(k)|^{p-2}u(k)\ln(1+|u(k)|^\nu), \ k \in \mathbb{Z}, \quad (6.5.48) \\
u(k) \to 0 \quad as \ |k| \to \infty,
\end{cases}
$$

has a sequence $\{u_n(k)\}$ of nontrivial solutions such that $I_\lambda(u_n) \to \infty$ as $n \to \infty$, where I_λ is defined by (6.5.11) with $f(k,t)$ given by (6.5.4). The "moreover" part of Theorem 6.5.4 also holds for problem (6.5.48).

Proof of Corollary 6.5.2. In view of the comments in Remark 6.5.3 on the function $f(k,t)$ defined by (6.5.4), the conclusion then follows directly from Theorem 6.5.4. □

6.6 Anti-periodic Solutions of Higher Order Difference Equations

As was pointed out in Section 4.4, anti-periodic problems have attracted attention because of their many application especially in the area of interpolation problems. Various functional-analytic techniques such as topological degree, lower and upper solutions, and the maximal monotone method have been used to study such problems.

In this section, for $n \geq 3$, we will study a $2n$th order nonlinear difference equation

$$\Delta^n(r(t-n)\Delta^n x(t-n)) + f(t,x(t)) = 0, \quad t \in \mathbb{Z}, \quad (6.6.1)$$

where $\mathbb{Z}$ is the integers. We will study the existence of an anti-periodic solution, i.e. $x(t+T) = -x(t)$ for $t \in \mathbb{Z}$. Throughout this section and without further mention, $r(t)$ and f are assumed to satisfy the following conditions:

(F30) $r(t+T) = r(t) > 0$, for a given positive integer T and for all $t \in \mathbb{Z}$.
(F31) $f : \mathbb{Z} \times \mathbb{R} \to \mathbb{R}$ is a continuous function in the second variable, $f(t+T,z) = f(t,z)$ and $f(t,-z) = -f(t,z)$, for all $(t,z) \in \mathbb{Z} \times \mathbb{R}$.

First we will construct a suitable functional J, and then we show that critical points of J are just the anti-periodic solutions of the difference equation (6.6.1). By using various variational approaches, we establish existence results for anti-periodic solutions when the nonlinearity satisfies different assumptions.

We begin with the following lemma.

Lemma 6.6.1. ([227, Theorem 1.1]) *If the functional J is weakly lower semicontinuous on a reflexive Banach space X and has a bounded minimizing sequence, then J has a minimum on X.*

In Definition 1.1.1 we defined the concept of a Palais-Smale condition. Below we have a variation of that notion that will be used in what is called the Linking Theorem.

Definition 6.6.1. *We say that J satisfies $(PS)_c$ condition if the existence of a sequence $\{u_n\} \subset E$ such that $J(u_n) \to c$ and $J'(u_n) \to 0$ (strongly in E^*) as $n \to \infty$, implies that $\{u_n\}$ has a convergent subsequence.*

Lemma 6.6.2. (Linking Theorem [283]) *Let $E = V \oplus X$ be a real Banach space with $\dim V < \infty$. Let $\rho > r > 0$ and let $z \in X$ be such that $\|z\| = r$. Define*

$$M = \{u = y + \lambda z : \|u\| \leq \rho, \lambda \geq 0, y \in V\},$$

$$M_0 = \{u = y + \lambda z : y \in V, \|u\| = \rho, \lambda \geq 0, \ or \ \|u\| \leq \rho, \lambda = 0\},$$

$$N = \{u \in X : \|u\| = r\}.$$

Let $J \in C^1(J, \mathbb{R})$ be such that

$$b = \inf_{u \in N} J(u) > a = \max_{u \in M_0} J(u).$$

If J satisfies the $(PS)_c$ condition with

$$c = \inf_{\gamma \in \Gamma} \max_{u \in M} J(\gamma(u)), \quad \Gamma = \{\gamma \in C(M, E) : \gamma|_{M_0} = id\},$$

then c is a critical value of J.

Now we recall some concepts and results from Morse theory.

Definition 6.6.2. ([214]) *Let $J^c = \{u \in E : J(u) \leq c\}$, let u be an isolated critical point of J with $J(u) = c$ and let U be a neighborhood of u, containing the unique critical point. We call*

$$C_q(J, u) = H_q(J^c \cap U, J^c \cap U\setminus\{u\})$$

the q-th critical group of J at u, $q = 0, 1, 2, \ldots$, where $H_q(\cdot, \cdot)$ stands for the qth singular relative homology group with integer coefficients. We say that u is nontrivial critical point of J, if at least one of its critical groups is nontrivial.

Lemma 6.6.3. [214]) *Let 0 be a critical point of J with $J(0) = 0$. Assume that J has a local linking at 0 with respect to $E = V_1 \oplus V_2, k = \dim V_1 < \infty$, that is, there exists $\rho > 0$ sufficiently small, such that*

$$J(u) \leq 0, \ u \in V_1, \|u\| \leq \rho;$$

$$J(u) > 0, \ u \in V_2, 0 < \|u\| \leq \rho.$$

Then $C_k(J,0) \ncong 0$. That is, 0 is a homological nontrivial critical point of J.

Lemma 6.6.4. ([226]) *Assume that J satisfies (PS) condition and is bounded from below. If J has a critical point that is homological nontrivial and does not minimize J, then J has at least three critical points.*

Next, we present some notations and lemmas that will be used in the proofs of our main results. For a positive integer T, let $Z(1,T) = \{1, 2, \ldots, T\}$. Let S be the set of sequences

$$x = (\ldots, x(-t), \ldots, x(-1), x(0), x(1), \ldots, x(t), \ldots) = \{x(t)\}_{t=-\infty}^{\infty},$$

i.e., $S = \{x = \{x(t) : x(t) \in \mathbb{R}, t \in \mathbb{Z}\}$. For a given positive integer T, E_{-T} is defined as the subsequence of S given by

$$E_{-T} = \{x = \{x(t)\} \in S : x(t+T) = -x(t), \ t \in \mathbb{Z}\}.$$

For any $x, y \in S$ and $a, b \in \mathbb{R}$, set $ax + by = \{ax(t) + by(t)\}_{t=-\infty}^{\infty}$; then S becomes a vector space. Clearly, S is isomorphic to $\mathbb{R}^T$. In addition, if E_{-T} is equipped with the inner product

$$\langle x, y \rangle = \sum_{t=1}^{T} x(t)y(t), \quad \forall x, y \in E_{-T},$$

then it is a finite dimensional Hilbert space and is homeomorphic to $\mathbb{R}^T$. Define the functional J on E_{-T} by

$$J(x) = \frac{1}{2} \sum_{t=1}^{T} r(t)(\Delta^n x(t))^2 + (-1)^n \sum_{t=1}^{T} F(t, x(t)),$$

where $F(t,z) = \int_0^z f(t,s)ds$. Clearly $J \in C^1(E_{-T}, \mathbb{R})$. For any $v \in E_{-T}$,

$$\langle J'(x), v \rangle = \sum_{t=1}^{T} r(t)\Delta^n x(t)\Delta^n v(t) + (-1)^n \sum_{t=1}^{T} f(t, x(t))v(t).$$

The next lemma relates critical points of J to solutions of our problem.

Lemma 6.6.5. *If $x \in E_{-T}$ is a critical point of J, then $x \in E_{-T}$ is a solution of* (6.6.1).

Proof. If $x \in E_{-T}$ is a critical point of J, then for any $v \in E_{-T}$, $\langle J'(x), v \rangle = 0$, i.e.,

$$\sum_{t=1}^{T} r(t) \Delta^n x(t) \Delta^n v(t) + (-1)^n \sum_{t=1}^{T} f(t, x(t)) v(t) = 0. \qquad (6.6.2)$$

Since

$$\sum_{t=1}^{T} r(t) \Delta y(t) \Delta z(t) = r(t-1) \Delta y(t-1) z(t)|_{t=1}^{T+1} - \sum_{t=1}^{T} \Delta(r(t-1) \Delta y(t-1)) z(t)$$

for any $y, v \in C(\mathbb{Z}, \mathbb{R})$, we have

$$\sum_{t=1}^{T} r(t) \Delta^n x(t) \Delta^n v(t) = r(t-1) \Delta^n x(t-1) \Delta^{n-1} v(t)|_{t=1}^{T+1}$$

$$- \sum_{t=1}^{T} \Delta(r(t-1) \Delta^n x(t-1)) \Delta^{n-1} v(t).$$

Since $x(t+T) = -x(t)$, $v(t+T) = -v(t)$, $r(t+T) = r(t)$, and $\Delta^n x(t-1) = \sum_{k=0}^{n} (-1)^k \binom{n}{k} x(t+n-k-1)$ for $t \in \mathbb{Z}$, we have $r(t-1) \Delta^n x(t-1) \Delta^{n-1} v(t)|_{t=1}^{T+1} = 0$. Hence,

$$\sum_{t=1}^{T} r(t) \Delta^n x(t) \Delta^n v(t) = - \sum_{t=1}^{T} \Delta(r(t-1) \Delta^n x(t-1)) \Delta^{n-1} v(t)$$

$$= \cdots = (-1)^n \sum_{t=1}^{T} \Delta^n (r(t-n) \Delta^n x(t-n)) v(t).$$

Thus, (6.6.2) becomes

$$(-1)^n \sum_{t=1}^{T} \Delta^n (r(t-n) \Delta^n x(t-n)) v(t) + (-1)^n \sum_{t=1}^{T} f(t, x(t)) v(t) = 0,$$

i.e., $\sum_{t=1}^{T} \left[\Delta^n (r(t-n) \Delta^n x(t-n)) + f(t, x(t)) \right] v(t) = 0$ for all $v \in E_{-T}$. So

$$\Delta^n (r(t-n) \Delta^n x(t-n)) + f(t, x(t)) = 0, \ t \in Z(1, T).$$

Since $r(t+T) = r(t)$, $f(t+T, z) = f(t, z)$, and $f(t, -z) = -f(t, z)$, $z \in \mathbb{R}$, we have

$$\Delta^n (r(t+T-n) \Delta^n x(t+T-n)) + f(t+T, x(t+T))$$
$$= -\Delta^n (r(t-n) \Delta^n x(t-n)) - f(t, x(t)) = 0.$$

Therefore x satisfies the difference equation (6.6.1) for $t \in \mathbb{Z}$. $\qquad \square$

Lemma 6.6.6. *For $x \in E_{-T}$, we have $\sum\limits_{t=1}^{T}(\Delta^n x(t))^2 = y^T A y$, where $y = (\Delta^{n-1}x(1), \ldots, \Delta^{n-1}x(T))$,*

$$A = \begin{pmatrix} 2 & -1 & 0 & \cdots & 0 & 1 \\ -1 & 2 & -1 & \cdots & 0 & 0 \\ 0 & -1 & 2 & \cdots & 0 & 0 \\ \cdots & \cdots & \cdots & \cdots & \cdots & \cdots \\ 0 & 0 & 0 & \cdots & 2 & -1 \\ 1 & 0 & 0 & \cdots & -1 & 2 \end{pmatrix}.$$

Proof. Clearly,

$$\sum_{t=1}^{T}(\Delta^n x(t))^2 = \sum_{t=1}^{T}(\Delta^{n-1}x(t+1) - \Delta^{n-1}x(t))^2$$

$$= \sum_{t=1}^{T}(\Delta^{n-1}x(t+1))^2 + (\Delta^{n-1}x(t))^2 - 2\Delta^{n-1}x(t+1)\Delta^{n-1}x(t)$$

$$= 2\sum_{t=2}^{T}(\Delta^{n-1}x(t))^2 + (\Delta^{n-1}x(T+1))^2 + (\Delta^{n-1}x(1))^2$$

$$- 2\sum_{t=1}^{T}\Delta^{n-1}x(t+1)\Delta^{n-1}x(t).$$

Since $x(t+T) = -x(t)$, we have

$$\sum_{t=1}^{T}(\Delta^n x(t))^2 = 2\sum_{t=1}^{T}\left[(\Delta^{n-1}x(t))^2 - \Delta^{n-1}x(t+1)\Delta^{n-1}x(t)\right]. \quad (6.6.3)$$

Note that

$$y^T A y = 2\sum_{t=1}^{T}(\Delta^{n-1}x(t))^2 - 2\sum_{t=1}^{T-1}\Delta^{n-1}x(t+1)\Delta^{n-1}x(t)$$

$$+ 2\Delta^{n-1}x(T)\Delta^{n-1}x(1)$$

and $x(t+T) = -x(t)$, we have

$$y^T A y = 2\sum_{t=1}^{T}(\Delta^{n-1}x(t))^2 - 2\sum_{t=1}^{T-1}\Delta^{n-1}x(t+1)\Delta^{n-1}x(t) \quad (6.6.4)$$

$$- 2\Delta^{n-1}x(T)\Delta^{n-1}x(T+1)$$

$$= 2\sum_{t=1}^{T}\left[(\Delta^{n-1}x(t))^2 - \Delta^{n-1}x(t+1)\Delta^{n-1}x(t)\right]. \quad (6.6.5)$$

By (6.6.3) and (6.6.4), we have the desired result. $\qquad\square$

Remark 6.6.1. It is easy to verify that A is a positive definite matrix. We assume that the eigenvalues $\lambda_1, \lambda_2, \ldots, \lambda_T$ are ordered as $0 < \lambda_1 \leq \lambda_2 \leq \cdots \leq \lambda_T$, and the corresponding orthonormal eigenvectors are $\xi_1, \xi_2, \ldots, \xi_T$.

Lemma 6.6.7. For $x \in E_{-T}$, we have $\lambda_1^n \|x\|^2 \leq \sum_{t=1}^{T}(\Delta^n x(t))^2 \leq \lambda_T^n \|x\|^2$.

Proof. By Lemma 6.6.6 and Remark 6.6.1,

$$\lambda_1 \|y\|^2 \leq \sum_{t=1}^{T}(\Delta^n x(t))^2 \leq \lambda_T \|y\|^2. \tag{6.6.6}$$

$$\|y\|^2 = \sum_{t=1}^{T}(\Delta^{n-1} x(t))^2 = \sum_{t=1}^{T}(\Delta^{n-2} x(t+1) - \Delta^{n-2} x(t))^2$$

$$= (\Delta^{n-2} x(1), \ldots, \Delta^{n-2} x(T))A \begin{pmatrix} \Delta^{n-2} x(1) \\ \vdots \\ \Delta^{n-2} x(T) \end{pmatrix} \tag{6.6.7}$$

$$\geq \lambda_1 \sum_{t=1}^{T}(\Delta^{n-2} x(t))^2 \geq \cdots \geq \lambda_1^{n-1} \sum_{t=1}^{T} |x(t)|^2 = \lambda_1^{n-1} \|x\|^2.$$

Similarly

$$\|y\|^2 \leq \lambda_T^{n-1} \|x\|^2. \tag{6.6.8}$$

By (6.6.6), (6.6.7) and (6.6.8), we obtain the stated result. $\qquad\square$

Remark 6.6.2. If x is an anti-periodic solution of (6.6.1), then $-x$ is also an anti-periodic solution of (6.6.1) by condition (F31). That is, the solution set of (6.6.1) is symmetric with respect to the origin in the space E_{-T}.

For any continuous function $h \in C(Z(1,T))$, we set $\underline{h} = \min_{t \in Z(1,T)}\{h(t)\}$ and $\overline{h} = \max_{t \in Z(1,T)}\{h(t)\}$. We will need the following conditions.

(F32) there exists $a(t) > 0$ such that $\limsup_{|z| \to \infty} \frac{F(t,z)}{|z|^2} < a(t)$, where $\overline{a} < \frac{1}{2} r \lambda_1^n$;

(F33) there exist $\mu > 2$ and $M > 0$ such that $0 < \mu F(t,z) \leq z f(t,z)$, for $|z| \geq M$;

(F34) there exists $b(t) > 0$ such that $\limsup_{|z| \to 0} \frac{F(t,z)}{|z|^2} < b(t)$, where $\overline{b} < \frac{1}{2} r \lambda_1^n$;

(F35) there exists $c(t) > 0$, $t \in Z(1,T)$, such that $\limsup_{|z| \to 0} \frac{F(t,z)}{|z|^2} \leq c(t)$, where $\bar{c} < \frac{1}{2}\underline{r}\lambda_{i+1}^n$, $i \in Z(1,T)$;

(F36) there exists $d(t) > 0$, $t \in Z(1,T)$, such that $F(t,z) \geq d(t)|z|^2$, where $\underline{d} > \frac{1}{2}\bar{r}\lambda_i^n$, $i \in Z(1,T)$;

(F37) there exist $k_1 > 0$ and $k_2 > 0$ such that $k_2|z|^2 \leq F(t,z) \leq k_1|z|^2$, where $k_2 > \frac{1}{2}\bar{r}\lambda_i^n$, $k_1 < \frac{1}{2}\underline{r}\lambda_{i+1}^n$, $i \in Z(1,T)$;

(F38) there exists $e(t) > 0, t \in Z(1,T)$, such that $\liminf_{|z| \to 0} \frac{F(t,z)}{|z|^2} \geq e(t)$, where $\underline{e} > \frac{1}{2}\bar{r}\lambda_i^n$, $i \in Z(1,T)$.

Next, we have some lemmas about the functional J.

Lemma 6.6.8. *Let f satisfy (F32). Then for n odd:*

   (i) *J is coercive on $\mathbb{R}^T$, i.e. $J(x) \to \infty$ as $|x| \to \infty$;*

   (ii) *J satisfies the (PS) condition.*

Proof. (i) Since $\limsup_{|x| \to \infty} \frac{F(t,z)}{|z|^2} < a(t)$, there exists $M_1 > 0$ such that $F(t,z) < a(t)|z|^2$ for $|z| > M_1, t \in Z(1,T)$. Since $F(t,z) - a(t)|z|^2$ is continuous on $[-M_1, M_1]$, $F(t,z) - a(t)|z|^2 < M_2$ for some constant $M_2 > 0$ and all $x \in [-M_2, M_2]$. So $F(t,z) < a(t)|z|^2 + M_2$ for all $z \in \mathbb{R}, t \in Z(1,T)$. So by Lemma 6.6.6 and Lemma 6.6.7,

$$
\begin{aligned}
J(x) &= \frac{1}{2}\sum_{t=1}^{T} r(t)(\Delta^n x(t))^2 - \sum_{t=1}^{T} F(t, x(t)) \\
&\geq \frac{1}{2}\underline{r}\sum_{t=1}^{T}(\Delta^n x(t))^2 - \sum_{t=1}^{T}[a(t)|x(t)|^2 + M_2] \\
&\geq \frac{1}{2}\underline{r}\lambda_1^n\|x\|^2 - \bar{a}\|x\|^2 - M_2 T = (\frac{1}{2}\underline{r}\lambda_1^n - \bar{a})\|x\|^2 - M_2 T \to \infty
\end{aligned}
$$

as $\|x\| \to \infty$.

(ii) From (i), J is bounded from below, which means $\{x_m(t)\}$ is bounded when $J(x_m)$ is bounded. Because E_{-T} is homeomorphic to $\mathbb{R}^T$, $\{x_m(t)\}$ has a convergent subsequence. This completes the proof of the lemma. $\square$

Lemma 6.6.9. *If f satisfies (F32), then J satisfies the (PS) condition.*

Proof. Since $\{x_m(t)\}$ is such that $J(x_m)$ is bounded and $J'(x_m) \to 0$ as $m \to \infty$, there exists $M_3 > 0$ such that

$$
M_3 \geq \mu J(x_m) - \langle J'(x_m), x_m \rangle. \tag{6.6.9}
$$

By Lemma 6.6.7 and (F32), we have

$$\mu J(x_m) - \langle J'(x_m), x_m \rangle = \left(\frac{\mu}{2} - 1\right) \sum_{t=1}^{T} r(t) \left(\Delta^n x_m(t)\right)^2$$
$$- \sum_{t=1}^{T} \mu F(t, x_m(t)) + \sum_{t=1}^{T} f(t, x_m(t)) x_m(t)$$
$$\geq \left(\frac{\mu}{2} - 1\right) \underline{r} \lambda_1^n \|x_m\|^2.$$

(6.6.10)

Now, (6.6.9) and (6.6.10) imply that $\{x_m(t)\}$ is bounded. Because E_{-T} is homeomorphic to $\mathbb{R}^T$, $\{x_m(t)\}$ has a convergent subsequence. □

Our first result on the existence of anti-periodic solutions is the following.

Theorem 6.6.1. *If f satisfies (F32), then the difference equation (6.6.1) has at least two anti-periodic solutions for n odd.*

Proof. We shall verify that the functional J satisfies the conditions of Lemma 6.6.1. E_{-T} is a reflexive Banach space. J is weakly lower semicontinuous. By Lemma 6.6.8 J is coercive on $\mathbb{R}^T$. So J has a bounded minimizing sequence. By Lemma 6.6.1, J has a minimum on E_{-T}. By Lemma 6.6.5, the difference equation (6.6.1) has at least one anti-periodic solution for n odd. By Remark 6.6.2, the difference equation (6.6.1) has at least two anti-periodic solutions for n odd. □

Another result on the existence of at least two anti-periodic solutions is contained in the following theorem.

Theorem 6.6.2. *If f satisfies (F33) and (F34), then the difference equation (6.6.1) has at least two nontrivial anti-periodic solutions for n odd.*

Proof. We shall verify that functional J satisfies the conditions of Theorem 1.1.8. By Lemma 6.6.9, J satisfies the (PS) condition. By (F34), for $\varepsilon \in (0, \frac{1}{2}\underline{r}\lambda_1^n - \overline{b})$, there exists $\delta > 0$ satisfying $F(t, z) \leq (b(t) + \varepsilon)|z^2|$ for $|z| < \delta$.

So by Lemma 6.6.7,

$$J(x) = \frac{1}{2} \sum_{t=1}^{T} r(t)(\Delta^n x(t))^2 - \sum_{t=1}^{T} F(t, x(t))$$

$$\geq \frac{1}{2} \underline{r} \lambda_1^n \|x\|^2 - \sum_{t=1}^{T} (b(t) + \varepsilon)|x(t)|^2$$

$$\geq (\frac{1}{2} \underline{r} \lambda_1^n - \overline{b} - \varepsilon)\|x\|^2,$$

for $x \in \partial B_\delta$. This implies $\inf_{x \in \partial B_\delta} J(x) \geq (\frac{1}{2} \underline{r} \lambda_1^n - \overline{b} - \varepsilon)\delta^2 > 0$. It is obvious that $J(0) = 0$. By (F33), we have

$$F(t, z) \geq M_4 |z|^\mu - M_5 \quad \text{for some } M_4, M_5 > 0, z \in \mathbb{R}. \tag{6.6.11}$$

Let $s \in \mathbb{R}$, we have by (6.6.11) and the proof of Lemma 6.6.7,

$$J(s\xi_i) = \frac{1}{2} \sum_{t=1}^{T} r(t)(\Delta^n s\xi_i(t))^2 - \sum_{t=1}^{T} F(t, s\xi_i(t))$$

$$\leq \frac{1}{2} \overline{r} s^2 \sum_{t=1}^{T} (\Delta^n \xi_i(t))^2 - \sum_{t=1}^{T} [M_4 |s\xi_i(t)|^\mu - M_5]$$

$$= \frac{1}{2} \overline{r} s^2 \lambda_i^n \sum_{t=1}^{T} (\xi_i(t))^2 - M_4 |s|^\mu \sum_{t=1}^{T} \xi_i(t)|^\mu + M_5 T \to -\infty,$$

as $|s| \to \infty$. Then there exists a sufficiently large $s_0 > \delta$ such that $x = s_0\xi_i \notin \overline{B}_\delta$ and $J(x) < 0$.

By Theorem 1.1.8, J has a critical point on E_{-T}. By Lemma 6.6.5, the difference equation (6.6.1) has at least one anti-periodic solution for n odd. By Remark 6.6.2, the difference equation (6.6.1) has at least two anti-periodic solutions for n odd. $\qquad\square$

Theorem 6.6.3. *If f satisfies (F33), (F35), and (F36), then the difference equation (6.6.1) has at least two nontrivial anti-periodic solutions for n odd.*

Proof. We shall verify that the functional J satisfies the conditions of Lemma 6.6.2. By Lemma 6.6.9, J satisfies the (PS) condition. So J satisfies the (PS)$_c$ condition. Let $V_1 = \text{span}\{\xi_1, \xi_2, \ldots, \xi_i\}, (i \in Z(1, T-1)), V_2 = V_1^T$. For $\sigma > 0$, now we shall show $\inf_{x \in V_2 \cap \partial B_\sigma} J(x) > 0$. The condition (F35) means that for $\varepsilon \in (0, \frac{1}{2} \underline{r} \lambda_1^n - \overline{b})$, there exists $\delta_1 > 0$ satisfying

$F(t, z) \leq (c(t) + \varepsilon)|z|^2$ for $|z| < \sigma$. For $x \in V_2$ and $x \in \partial B_\sigma$, we have $|x(t)| \leq \sigma$. By the proof of Lemma 6.6.7, one has

$$J(x) = \frac{1}{2} \sum_{t=1}^{T} r(t)(\Delta^n x(t))^2 - \sum_{t=1}^{T} F(t, x(t))$$

$$\geq \frac{1}{2} \underline{r} \lambda_{i+1}^n \|x\|^2 - \sum_{t=1}^{T} (c(t) + \varepsilon)|x(t)|^2$$

$$\geq (\frac{1}{2} \underline{r} \lambda_{i+1}^n - \overline{c} - \varepsilon)\|x\|^2 = (\frac{1}{2} \underline{r} \lambda_{i+1}^n - \overline{c} - \varepsilon)\sigma^2,$$

for $x \in \partial B_\sigma$. So $\inf_{x \in V_2 \in \partial B_\sigma} J(x) > 0$.

Next, we show that $\inf_{x \in M_0} J(x) < 0$, where

$$M_0 = \{u = y + \lambda z : y \in V_1, \|u\| = \rho, \lambda \geq 0, \text{ or } \|u\| \leq \rho, \lambda = 0\},$$

and where $\rho > \sigma, z = \xi_{i+1}, \|z\| = \sigma$. For $u \in M_0$ and $\|u\| \leq \rho$, and $\lambda = 0$, we have $u = y \in V_1$. By (F36) and the proof of Lemma 6.6.7,

$$J(u) = J(y) = \frac{1}{2} \sum_{t=1}^{T} r(t)(\Delta^n y(t))^2 - \sum_{t=1}^{T} F(t, y(t))$$

$$\leq \frac{1}{2} \overline{r} \lambda_i^n \|y\|^2 - \sum_{t=1}^{T} d(t)|y(t)|^2$$

$$\leq (\frac{1}{2} \overline{r} \lambda_i^n - \underline{d})\|y\|^2 < 0.$$

For $u \in M_0$ and $\|u\| = \rho$, and $\lambda \geq 0$, we have $u = y + \lambda z = y + \lambda \xi_i$. By (6.6.11),

$$J(u) = \frac{1}{2} \sum_{t=1}^{T} r(t)(\Delta^n u(t))^2 - \sum_{t=1}^{T} F(t, u(t))$$

$$\leq \frac{1}{2} \overline{r} \lambda_T^n \|u\|^2 - \sum_{t=1}^{T} (M_4|u(t)|^\mu - M_5)$$

$$\leq \frac{1}{2} \overline{r} \lambda_T^n \|u\|^2 - M_6\|u\|^\mu + M_5 T \to -\infty$$

as $\|u\| = \rho \to \infty$, $M_4, M_5, M_6 > 0$.

So there exists $\rho > \sigma$ such that $\max_{u \in M_0} J(u) = 0 < \inf_{u \in V_2 \cap \partial B_\sigma} J(u)$.

By Lemma 6.6.2, J has a critical value $c^* > 0$, i.e. there exists $x^* \in R^T$ such that $J(x^*) = c^*$ and $J'(x^*) = 0$. It is obvious $x^* \neq 0$ since $J(0) = 0$. By Lemma 6.6.5, x^* is a nontrivial anti-periodic solution of the difference equation (6.6.1) when n is odd. By Remark 6.6.2, the difference equation (6.6.1) has at least two anti-periodic solutions for n odd.   □

Theorem 6.6.4. *If f satisfies (F32) and (F37), then the difference equation (6.6.1) has at least four nontrivial anti-periodic solutions for n odd.*

Proof. We shall first verify that the functional J satisfies the conditions of Lemma 6.6.3. By Lemma 6.6.8, J satisfies the (PS) condition and is bounded from below. It is clear that $J(0) = 0$. Let $V_1 = span\{\xi_1, \xi_2, \ldots, \xi_i\}, V_2 = span\{\xi_{i+1}, \ldots, \xi_T\}$, $i \in Z(1, T-1)$. We shall show that 0 is a homological nontrivial critical point of J. For $x \in V_2, \|x\| \leq \rho$, we have $|x(t)| \leq \rho$. So by the proof of Lemma 6.6.7 and (F37) we have

$$J(x) = \frac{1}{2}\sum_{t=1}^{T} r(t)(\Delta^n x(t))^2 - \sum_{t=1}^{T} F(t, x(t))$$

$$\geq \frac{1}{2}\underline{r}\lambda_{i+1}^n \|x\|^2 - k_1 \sum_{t=1}^{T} |x(t)|^2$$

$$= \left(\frac{1}{2}\underline{r}\lambda_{i+1}^n - k_1\right)\|x\|^2 > 0.$$

For $x \in V_1, \|x\| \leq \rho$, we have $|x(t)| \leq \rho$. So by the proof of Lemma 6.6.7 and (F37)

$$J(x) \leq \frac{1}{2}\overline{r}\lambda_i^n \|x\|^2 - \sum_{t=1}^{T} B|x(t)|^2 = \left(\frac{1}{2}\overline{r}\lambda_i^n - k_2\right)\|x\|^2 \leq 0.$$

It follows from Lemma 6.6.3 that 0 is a homological nontrivial critical point of J.

If $\inf_{x \in R^T} J(x) \geq 0$, then $J(x) = \inf_{x \in V_1} J(x) = 0$ for $x \in V_1, \|x\| < \rho$. So $x \in V_1$ with $\|x\| \leq \rho$ are solutions of (6.6.1).

If $\inf_{x \in R^T} J(x) < 0$, then 0 is not a homological nontrivial critical point of J. By Lemma 6.6.4, J has at least three critical points. So the difference equation (6.6.1) has at least two nontrivial anti-periodic solutions.

Therefore, difference equation (6.6.1) has at least two nontrivial anti-periodic solutions. By Remark 6.6.2, the difference equation (6.6.1) has at least four anti-periodic solutions for n odd. $\square$

Theorem 6.6.5. *If f satisfies (F32) and (F38), then the difference equation (6.6.1) has at least $2i$ ($i \in Z(1, T-1)$) nontrivial anti-periodic solutions for n odd.*

Proof. We shall verify that the functional J satisfies the conditions of Theorem 1.1.11. By Lemma 6.6.8, J satisfies the (PS) condition and is bounded from below. By condition (F31), J is even. Let

$$K = \left\{ \sum_{j=1}^{i} a_j \xi_j : \sum_{j=1}^{i} a_j^2 = a^2 \right\}, \quad i \in Z(1, T-1).$$

It is clear that K is homeomorphic to S^{i-1} by an odd map. We shall show $J|_K < 0$ if a is sufficiently small. For $x \in K$, we have by the proof of Lemma 6.6.7 and (F38),

$$J(x) \leq \frac{1}{2} \overline{r} \lambda_i^n \|x\|^2 - \sum_{t=1}^{T} e(t) |x(t)|^2 \leq (\frac{1}{2} \overline{r} \lambda_i^n - \underline{e}) \|x\|^2 < 0$$

as a is sufficiently small. By Theorem 1.1.11, J has at least i critical points. By Lemma 6.6.5 and Remark 6.6.2, the difference equation (6.6.1) has at least $2i$ anti-periodic solutions for n odd. $\qquad\square$

Similar to the above theorems, we can establish the existence of anti-periodic solutions for (6.6.1) when n is even. To this end, we assume the following conditions hold.

(F39) there exists $a(t) > 0$ such that $\liminf_{|z| \to \infty} \frac{F(t,z)}{|z|^2} > a(t)$, where $\underline{a} > -\frac{1}{2} \overline{r} \lambda_1^n$;

(F40) there exist $\mu > 2$ and $M > 0$ such that $0 > \mu F(t,z) \geq z f(t,z)$, for $|z| \geq M$;

(F41) there exists $b(t) > 0$ such that $\liminf_{|z| \to 0} \frac{F(t,z)}{|z|^2} \geq b(t)$, where $\underline{b} > -\frac{1}{2} \underline{r} \lambda_1^n$;

(F42) there exists $c(t) > 0$, $t \in Z(1, T)$, such that $\liminf_{|z| \to 0} \frac{F(t,z)}{|z|^2} \geq c(t)$, where $\underline{c} > -\frac{1}{2} \underline{r} \lambda_{i+1}^n$, $i \in Z(1, T)$;

(F43) there exists $d(t) > 0$, $t \in Z(1, T)$, such that $F(t,z) \leq d(t)|z|^2$, where $\overline{d} < -\frac{1}{2} \overline{r} \lambda_i^n$, $i \in Z(1, T)$;

(F44) there exist $k_1 > 0$ and $k_2 > 0$ such that $k_2 |z|^2 \leq F(t,z) \leq k_1 |z|^2$, where $k_2 > -\frac{1}{2} \underline{r} \lambda_{i+1}^n$, $k_1 < -\frac{1}{2} \overline{r} \lambda_i^n$, $i \in Z(1, T)$;

(F45) there exists $e(t)$ satisfying $\limsup_{|x| \to 0} \frac{F(t,z)}{|z|^2} \leq e(t)$, where $\overline{e} > -\frac{1}{2} \overline{r} \lambda_i^n$, $i \in Z(1, T)$.

Theorem 6.6.6. *If f satisfies (F39), then the difference equation (6.6.1) has at least two anti-periodic solutions for n even.*

Theorem 6.6.7. *If f satisfies (F40) and (F41), then the difference equation (6.6.1) has at least two nontrivial anti-periodic solutions for n even.*

Theorem 6.6.8. *If f satisfies (F40), (F42), and (F43), then the difference equation (6.6.1) has at least two nontrivial anti-periodic solutions for n even.*

Theorem 6.6.9. *If f satisfies (F39) and (F44), then the difference equation (6.6.1) has at least four nontrivial anti-periodic solutions for n even.*

Theorem 6.6.10. *If f satisfies (F39) and (F45), then the difference equation (6.6.1) has at least 2i nontrivial anti-periodic solutions for n even.*

Remark 6.6.3. If $x \in E_{-T}$ is an anti-periodic solution of (6.6.1), then x is a $2T$−periodic solution of (6.6.1). So under the assumptions of Theorems 6.6.1–6.6.5 and Theorems 6.6.6–6.6.10, we obtain corresponding results for each of those theorems for the existence of $2T$-periodic solutions.

Chapter 7

Notes

Chapter 1: While the forms of the variational principles presented in this chapter are primarily taken from the works of [22], [39], [40], [42], [54], [66], [75], [77], [89], [185], [187], [213], [227], [241], [244], [248], [250], [251], [287], [289], [299], various forms of these statements can be found throughout the papers listed in the bibliography.

Chapter 2: The results in this chapter are based primarily on the papers of J. R. Graef, S. Heidarkhani, and L. Kong [137, 138, 140]

Chapter 3: The source of the results in this chapter are mainly the papers of J. R. Graef, S. Heidarkhani, and L. Kong [130, 132, 134, 135], and Y. Tian, J. R. Graef, L. Kong, and M. Wang [270]

Chapter 4: Results in this chapter are based on the papers of J. R. Graef, S. Heidarkhani, and L. Kong [141, 142] and the paper of Heidarkhani, G. A. Afrouzi, A. Hadjian, and J. Henderson [166]

Chapter 5: The papers of J. R. Graef, S. Heidarkhani, L. Kong [133, 136], L. Kong [195], S. Heidarkhani and J. Henderson [168] formed the basis for much of the content in this chapter.

Chapter 6: The results in this chapter are taken from the papers of J. R. Graef, L. Kong, and Q. Kong [146, 147], J. R. Graef, L. Kong, Y. Tian, and M. Wang [148], J. R. Graef, L. Kong, and M. Wang [149–153], L. Kong [193], and Y. Tian and J. Henderson [274].

Bibliography

[1] A. Abdurahman, F. Anton, and J. Bordes, Half-string oscillator approach to string field theory (Ghost sector: I), *Nuclear Phys. B* **397** (1993), 260–282.

[2] G. A. Afrouzi and S. Heidarkhani, Three solutions for a Dirichlet boundary value problem involving the p-Laplacian, *Nonlinear Anal.* **66** (2007), 2281–2288.

[3] G. A. Afrouzi and S. Heidarkhani, Existence of three solutions for a class of Dirichlet quasilinear elliptic systems involving the $(p_1, \ldots, p_n)$-Laplacian, *Nonlinear Anal.* **70** (2009), 135–143.

[4] G. A. Afrouzi and S. Heidarkhani, Multiplicity theorems for a class of Dirichlet quasilinear elliptic systems involving the $(p_1, \ldots, p_n)$-Laplacian, *Nonlinear Anal.* **73** (2010), 2594–2602.

[5] G. A. Afrouzi, S. Heidarkhani, and D. O'Regan, Three solutions to a class of Neumann doubly eigenvalue elliptic systems driven by a $(p_1, \ldots, p_n)$-Laplacian, *Bull. Korean Math. Soc.* **47** (2010), 1235–1250.

[6] G. A. Afrouzi, S. Heidarkhani, and D. O'Regan, Existence of three solutions for a doubly eigenvalue fourth-order boundary value problem, *Taiwanese J. Math.* **15** (2011), 201–210.

[7] A. R. Aftabizadeh, S. Aizicovici, and N. H. Pavel, On a class of second-order anti-periodic boundary value problems, *J. Math. Anal. Appl.* **171** (1992), 301–320.

[8] A. R. Aftabizadeh, S. Aizicovici, and N. H. Pavel, Anti-periodic boundary value problems for higher order differential equations in Hilbert spaces, *Nonlinear Anal.* **18** (1992), 253–267.

[9] R. P. Agarwal, *Difference Equations and Inequalities, Theory, Methods, and Applications*, Second Edition, Marcel Dekker, New York, 2000.

[10] R. P. Agarwal and D. O'Regan, Multiple nonnegative solutions for second order impulsive differential equations, *Appl. Math. Comput.* **114** (2000), 51–59.

[11] B. Ahmad and J. J. Nieto, Existence of solutions for anti-periodic boundary value problems involving fractional differential equations via Leray-Schauder degree, *Topol. Methods Nonlinear Anal.* **35** (2010), 295–304.

[12] C. Ahn and C. Rim, Boundary flows in general coset theories, *J. Phys. A.* **32** (1999), 2509–2525.

[13] S. Aizicovici, M. McKibben, and S. Reich, Anti-periodic solutions to non-monotone evolution equations with discontinuous nonlinearities, *Nonlinear Anal.* **43** (2001), 233–251.

[14] S. Aizicovici and N. H. Pavel, Anti-periodic solutions to a class of nonlinear differential equations in Hilbert space, *J. Funct. Anal.* **99** (1991), 387–408.

[15] C. O. Alves, F. S. J. A. Correa, and T. F. Ma, Positive solutions for a quasilinear elliptic equation of Kirchhoff type, *Comput. Math. Appl.* **49** (2005), 85–93.

[16] C. Amrouche, Singular boundary conditions and regularity for the biharmonic problem in the half-space, *Commun. Pure Appl. Anal.* **6** (2007), 957–982.

[17] D. R. Anderson and R. I. Avery, Existence of a periodic solution for continuous and discrete periodic second-order equations with variable potentials, *J. Appl. Math. Comput.* **37** (2011), 297–312.

[18] D. R. Anderson and F. Minhós, A discrete fourth-order Lidstone problem with parameters, *Appl. Math. Comput.* **214** (2009), 523–533.

[19] S. Aouaoui, Existence of three solutions for some equation of Kirchhoff type involving variable exponents, *Appl. Math. Comput.* **218** (2012), 7184–7192.

[20] A. Arosio and S. Panizzi, On the well-posedness of the Kirchhoff string, *Trans. Amer. Math. Soc.* **348** (1996), 305–330.

[21] F. M. Atici and G. Sh. Guseinov, Positive periodic solutions for nonlinear difference equations with periodic coefficients, *J. Math. Anal. Appl.* **232** (1999), 166–182.

[22] D. Averna and G. Bonanno, A three critical points theorem and its applications to the ordinary Dirichlet problem, *Topol. Methods Nonlinear Anal.* **22** (2003), 93–103.

[23] D. Averna and G. Bonanno, Three solutions for a quasilinear two-point boundary-value problem involving the one-dimensional p-Laplacian, *Proc. Edinburgh Math. Soc.* **47** (2004), 257–270.

[24] D. Averna and G. Bonanno, A mountain pass theorem for a suitable class of functions, *Rocky Mountain J. Math.* **39** (2009), 707–727.

[25] M. B. Ayed and M. Hammami, On a fourth order elliptic equation with critical nonlinearity in dimension six, *Nonlinear Anal.* **64** (2006), 924–957.

[26] M. Ayed and A. Selmi, Asymptotic behavior and existence results for a biharmonic equation involving the critical Sobolev exponent in a five-dimensional domain, *Commun. Pure Appl. Anal.* **9** (2012), 1705–1722.

[27] A. Ayoujil and A. R. El Amrouss, On the spectrum of a fourth order elliptic equation with variable exponent, *Nonlinear Anal.* **71** (2009), 4916–4926.

[28] A. Ayoujil and A. R. El Amrouss, Continuous spectrum of a fourth order nonhomogenous elliptic equation with variable exponent, *Electron. J. Differential Equations* **2011** (2011), No. 24, 12pp.

[29] Z. Bai, Iterative solutions for some fourth-order periodic boundary value problems, *Taiwanese J. Math.* **12** (2008), 1681–1690.

[30] L. Bai and B. Dai, Application of variational method to a class of Dirichlet

boundary value problems with impulsive effects, *J. Franklin Inst.* **348** (2011), 2607–2624.

[31] Z. Bai and H. Wang, On positive solutions of some nonlinear fourth-order beam equations, *J. Math. Anal. Appl.* **270** (2002), 357–368.

[32] D. Bainov and P. Simeonov, *Systems with Impulse Effect*, Ellis Horwood Series: Mathematics and Its Applications, Ellis Horwood, Chichester, 1989.

[33] M. Benchohra, J. Henderson, and S. Ntouyas, *Theory of Impulsive Differential Equations*, Contemporary Mathematics and Its Applications, Vol. 2, Hindawi, New York, 2006.

[34] J. Benedikt and P. Drábek, Estimates of the principal eigenvalue of the *p*-biharmonic operator, *Nonlinear Anal.* **75** (2012), 5374–5379.

[35] A. Benmezai, J. R. Graef, and L. Kong, Positive solutions to a two point singular boundary value problem, *Differ. Equ. Appl.* **3** (2011), 347–373.

[36] C. Bereanu, Periodic solutions of some fourth-order nonlinear differential equations, *Nonlinear Anal.* **71** (2009), 53–57.

[37] L. Boccardo and D. Figueiredo, Some remarks on a system of quasilinear elliptic equations, *NoDEA Nonlinear Differential Equations Appl.* **9** (2002), 309–323.

[38] G. Bonanno, Some remarks on a three critical points theorem, *Nonlinear Anal.* **54** (2003), 651–665.

[39] G. Bonanno, A critical point theorem via the Ekeland variational principle, *Nonlinear Anal.* **75** (2012), 2992–3007.

[40] G. Bonanno and P. Candito, Non-differentiable functionals and applications to elliptic problems with discontinuous nonlinearities, *J. Differential Equations* **244** (2008), 3031–3059.

[41] G. Bonanno and A. Chinnì, Existence of three solutions for a perturbed two-point boundary value problem, *Appl. Math. Lett.* **23** (2010), 807–811.

[42] G. Bonanno and G. D'Aguì, A Neumann boundary value problem for the Sturm-Liouville equation, *Appl. Math. Comput.* **208** (2009), 318–327.

[43] G. Bonanno and G. D'Aguì, Multiplicity results for a perturbed elliptic Neumann problem, *Abstr. Appl. Anal.* **2010** (2010), doi:10.1155/2010/564363, 10 pp.

[44] G. Bonanno and B. Di Bella, A boundary value problem for fourth-order elastic beam equations, *J. Math. Anal. Appl.* **343** (2008), 1166–1176.

[45] G. Bonanno and B. Di Bella, Infinitely many solutions for a fourth-order elastic beam equation, *NoDEA Nonlinear Differential Equations Appl.* **18** (2011), 357–368.

[46] G. Bonanno, B. Di Bella, and J. Henderson, Existence of solutions to second-order boundary-value problems with small perturbations of impulses, *Electron. J. Differential Equations* **2013** (2013), No. 126, 14pp.

[47] G. Bonanno, S. Heidarkhani, and D. O'Regan, Multiple solutions for a class of Dirichlet quasilinear elliptic systems driven by a (p, q)-Laplacian operator, *Dynam. Systems Appl.* **20** (2011), 89–99.

[48] G. Bonanno, S. Heidarkhani, and D. O'Regan, Nontrivial solutions for Sturm-Liouville systems via a local minimum theorem for functionals, *Bull. Austral. Math. Soc.*, to appear.

[49] G. Bonanno and R. Livrea, Multiplicity theorems for the Dirichlet problem involving the p-Laplacian, *Nonlinear Anal.* **54** (2003), 1–7.

[50] G. Bonanno and R. Livrea, Periodic solutions for a class of second-order Hamiltonian systems, *Electron. J. Differential Equations* **2005** (2005), No. 115, 13pp.

[51] G. Bonanno and R. Livrea, Multiple periodic solutions for Hamiltonian systems with not coercive potential, *J. Math. Anal. Appl.* **363** (2010), 627–638.

[52] G. Bonanno and R. Livrea, Existence and multiplicity of periodic solutions of second order Hamiltonian systems depending on a parameter, *J. Convex Anal.* **20** (2013), 1075–1094.

[53] G. Bonanno and S. A. Marano, On the structure of the critical set of non-differentiable functions with a weak compactness condition, *Appl. Anal.* **89** (2010), 1–10.

[54] G. Bonanno and G. Molica Bisci, Infinitely many solutions for a boundary value problem with discontinuous nonlinearities, *Bound. Value Probl.* **2009** (2009), 1–20.

[55] G. Bonanno and G. Molica Bisci, Infinitely many solutions for a Dirichlet problem involving the p-Laplacian, *Proc. Roy. Soc. Edinburgh Sect. A* **140** (2010), 737–742.

[56] G. Bonanno and G. Molica Bisci, A remark on perturbed elliptic Neumann problems, *Stud. Univ. Babeş-Bolyai Math.* **55** (2010), 17–25.

[57] G. Bonanno and G. Molica Bisci, Three weak solutions for elliptic Dirichlet problems, *J. Math. Anal. Appl.* **382** (2011), 1–8.

[58] G. Bonanno, G. Molica Bisci, and D. O'Regan, Infinitely many weak solutions for a class of quasilinear elliptic systems, *Math. Comput. Modelling* **52** (2010), 152–160.

[59] G. Bonanno, G. Molica Bisci, and V. Rădulescu, Existence of three solutions for a non-homogeneous Neumann problem through Orlicz-Sobolev spaces, *Nonlinear Anal.* **74** (2011), 4785–4795.

[60] G. Bonanno, G. Molica Bisci, and V. Rădulescu, Multiple solutions of generalized Yamabe equations on Riemannian manifolds and applications to Emden-Fowler problems, *Nonlinear Anal. Real World Appl.* **12** (2011), 2656–2665.

[61] G. Bonanno, G. Molica Bisci, and V. Rădulescu, Arbitrarily small weak solutions for a nonlinear eigenvalue problem in Orlicz-Sobolev spaces, *Monatsh. Math.* **165** (2012), 305–318.

[62] G. Bonanno and P. F. Pizzimenti, Existence results for nonlinear elliptic problems, *Appl. Anal.* **92** (2013), 411–423.

[63] G. Bonanno and G. Riccobono, Multiplicity results for Sturm-Liouville boundary value problems, *Appl. Math. Comput.* **210** (2009), 294–297.

[64] Y. Bozhkov and E. Mitidieri, Existence of multiple solutions for quasilinear systems via fibering method, *J. Differential Equations* **190** (2003), 239–267.

[65] H. Brézis, *Analyse Functionelle — Théorie et Applications*, Masson, Paris, 1983.

[66] H. Brézis and L. Nirenberg, Remarks on finding critical points, *Comm.*

Pure Appl. Math. **44** (1991), 939–963.

[67] A. Cabada, J. A. Cid and L. Sanchez, Positivity and lower and upper solutions for fourth-order boundary value problems, *Nonlinear Anal.* **67** (2007), 1599–1612.

[68] A. Cabada and N. Dimitrov, Multiplicity results for nonlinear periodic fourth order difference equations with parameter dependence and singularities, *J. Math. Anal. Appl.* **371** (2010), 518–533.

[69] A. Cabada and J. B. Ferreiro, Existence of positive solutions for nth-order periodic difference equations, *J. Difference Equ. Appl.* **17** (2011), 935–954.

[70] A. Cabada and S. Lois, Maximum principles for fourth and sixth order periodic boundary value problems. *Nonlinear Anal.* **29** (1997), 1161–1171.

[71] X. Cai and Z. Guo, Existence of solutions of nonlinear fourth order discrete boundary value problem, *J. Difference Equ. Appl.* **12** (2006), 459–466.

[72] X. Cai, J. Yu, Existence of periodic solutions for a 2nth-order nonlinear difference equation, *J. Math. Anal. Appl.* **329** (2007), 870–878.

[73] P. Candito and G. D'Agui, Three solutions to a perturbed nonlinear Dirichlet problem, *J. Math. Anal. Appl.* **375** (2011), 594–601.

[74] P. Candito and R. Livrea, Infinitely many solutions for a nonlinear Navier boundary value problem involving the *p*-biharmonic, *Stud. Univ. Babeş-Bolyai Math.* **55** (2010), 41–51.

[75] G. Cerami, An existence criterion for the critical points on unbounded manifolds, *Istit. Lombardo Accad. Sci. Lett. Rend. A* **112** (1978), 332–336.

[76] J. Chabrowski and J. Marcos do Ó, On some fourth-order semilinear elliptic problems in R^N, *Nonlinear Anal.* **49** (2002), 861–884.

[77] K. C. Chang, *Critical Point Theory and Applications*, Shanghai Scientific and Technology Press, Shanghai, 1986.

[78] G. Chen and S. Ma, Periodic solutions for Hamiltonian systems without Ambrosetti-Rabinowitz condition and spectrum 0, *J. Math. Anal. Appl.* **379** (2011), 842–851.

[79] H. Chen, Antiperiodic wavelets, *J. Comput. Math.* **14** (1996), 32–39.

[80] H. Chen and Z. He, New results for perturbed Hamiltonian systems with impulses, *Appl. Math. Comput.* **218** (2012), 9489–9497.

[81] P. Chen and X. Tang, Existence of homoclinic orbits for 2nth-order nonlinear difference equations containing both many advances and retardations, *J. Math. Anal. Appl.* **381** (2011), 485–505.

[82] P. Chen and X. Tang, Existence of solutions for a class of *p*-Laplacian systems with impulsive effects, *Taiwanese J. Math.* **16** (2012), 803–828.

[83] P. Chen, X. Tang, and R. P. Agarwal, Existence of homoclinic solutions for *p*(*n*)-Laplacian Hamiltonian systems on Orlicz sequence spaces, *Math. Comput. Modelling* **55** (2012), 989–1002.

[84] Y. Q. Chen, Note on Masseras theorem on anti-periodic solution, *Adv. Math. Sci. Appl.* **9** (1999), 125–128.

[85] Y. Q. Chen, J. J. Nieto, and D. O'Regan, Anti-periodic solutions for fully nonlinear first-order differential equations, *Math. Comput. Model.* **46** (2007), 1183–1190.

[86] Y. Q. Chen, X. D. Wang, and H. X. Xu, Anti-periodic solutions for

semilinear evolution equations, *J. Math. Anal. Appl.* **273** (2002), 627–636.

[87] B. Cheng and X. Xu, Existence results of positive solutions of Kirchhoff type problems, *Nonlinear Anal.* **71** (2009), 4883–4892.

[88] M. Chipot and B. Lovat, Some remarks on nonlocal nonlinear elliptic and parabolic problems, *Nonlinear Anal.* **30** (1997), 4619–4627.

[89] D. C. Clark, A variant of the Lusternik–Schnirelmann theory, *Indiana Univ. Math. J.* **22** (1972), 65–74.

[90] M. Conti, S. Terracini, and G. Verzini, Infinitely many solutions to fourth order superlinear periodic problems, *Trans. Amer. Math. Soc.* **356** (2004), 3283–3300.

[91] G. Cordaro, Three periodic solutions to an eigenvalue problem for a class of second order Hamiltonian systems, *Abstr. Appl. Anal.* **18** (2003), 1037–1045.

[92] G. Cordaro and G. Rao, Three periodic solutions for perturbed second order Hamiltonian systems, *J. Math. Anal. Appl.* **359** (2009), 780–785.

[93] V. Coti-Zelati, I. Ekeland, and E. Sere, A variational approach to homoclinic orbits in Hamiltonian systems, *Math. Ann.* **288** (1990), 133–160.

[94] C. Cowan, P. Esposito, and N. Ghoussoub, Regularity of extremal solutions in fourth order nonlinear eigenvalue problems on general domains, *Discrete Contin. Dyn. Syst.* **28** (2010), 1033–1050.

[95] G. D'Aguì, Multiplicity results for nonlinear mixed boundary value problem, *Bound. Value Prob.* **2012** (2012), No. 134.

[96] G. D'Aguì and G. Molica Bisci, Three non-zero solutions for elliptic Neumann problems, *Anal. Appl.* **9** (2011), 383–394.

[97] G. D'Aguì and A. Sciammetta, Infinitely many solutions to elliptic problems with variable exponent and nonhomogeneous Neumann conditions, *Nonlinear Anal.* **75** (2012), 5612–5619.

[98] G. Dai and R. Hao, Existence of solutions for a $p(x)$-Kirchhoff-type equation, *J. Math. Anal. Appl.* **359** (2009), 275–284.

[99] J. M. Davis, L. H. Erbe, and J. Henderson, Multiplicity of positive solutions for higher order Sturm-Liouville problems, *Rocky Mountain J. Math.* **31** (2001), 169–184.

[100] F. J. Delvos and L. Knoche, Lacunary interpolation by antiperiodic trigonometric polynomials, *BIT* **39** (1999), 439–450.

[101] X. Deng, X. Liu, Y. Zhang, and H. Shi, Periodic and subharmonic solutions for a 2nth-order difference equation involving p-Laplacian, *Indag. Math.* **24** (2013), 613–625.

[102] Y. Ding and C. Lee, Periodic solutions for Hamiltonian systems, *SIAM J. Math. Anal.* **32** (2000), 555–571.

[103] A. Djellit and S. Tas, On some nonlinear elliptic systems, *Nonlinear Anal.* **59** (2004), 695–706.

[104] A. Djellit and S. Tas, Quasilinear elliptic systems with critical Sobolev exponents in R^N, *Nonlinear Anal.* **66** (2007), 1485–1497.

[105] P. Drábek, N. M. Stavrakakis, and N. B. Zographopoulos, Multiple non-semitrivial solutions for quasilinear elliptic systems, *Differential Integral Equations* **16** (2003), 1519–1531.

[106] J. Y. Du, H. L. Han, and G. X. Jin, On trigonometric and paratrigonometric Hermite interpolation, *J. Approx. Theory* **131** (2004), 74–99.

[107] Z. Du, W. Ge, and M. Zhou, Singular perturbations for third-order nonlinear multi-point boundary value problems, *J. Differential Equations* **218** (2005), 69–90.

[108] Z. Du and L. Kong, Existence of three solutions for systems of multi-point boundary value problems, *Electron. J. Qual. Theory Diff. Equ.* **Spec. Ed. I**, (2009), No. 10, 17 pp. (electronic).

[109] Z. Du, W. Liu, and X. Lin, Multiple solutions to a three-point boundary value problem for higher-order ordinary differential equations, *J. Math. Anal. Appl.* **335** (2007), 1207–1218.

[110] D. Edmunds and J. Rákosník, Sobolev embeddings with variable exponent, *Studia Math.* **143** (2000), 267–293.

[111] A. R. El Amrouss and A. Ourraoui, Existence of solutions for a boundary value problem involving $p(x)$-biharmonic operator, *Bol. Soc. Paran. Mat.* **31** (2013), 179–192.

[112] P. W. Eloe and B. Ahmad, Positive solutions of a nonlinear nth order boundary value problem with nonlocal conditions, *Appl. Math. Lett.* **18** (2005), 521–527.

[113] P. W. Eloe and J. Henderson, Uniqueness implies existence and uniqueness conditions for a class of $(k+j)$-point boundary value problems for nth order differential equations, *Math. Nachr.* **284** (2011), 229–239.

[114] M. Fabian, P. Habala, P. Hájek, V. Montesinos, and V. Zizler, *Banach Space Theory*, Springer, New York, 2011.

[115] X. Fan and S. Deng, Remarks on Ricceri's variational principle and applications to the $p(x)$-Laplacian equations *Nonlinear Anal.* **67** (2007), 3064–3075.

[116] X. Fan and X. Han, Existence and multiplicity of solutions for $p(x)$-Laplacian equations in R^N, *Nonlinear Anal.* **59** (2004), 173–188.

[117] X. Fan and D. Zhao, On the spaces $L^{p(x)}(\Omega)$ and $W^{m,p(x)}(\Omega)$, *J. Math. Anal. Appl.* **263** (2001), 424–446.

[118] F. Faraci, Multiple periodic solutions for second order systems with changing sign potential, *J. Math. Anal. Appl.* **319** (2006), 567–578.

[119] S. Federica, A biharmonic equation in R^4 involving nonlinearities with critical exponential growth, *Commun. Pure Appl. Anal.* **12** (2013), 405–428.

[120] H. Feng and W. Ge, Existence of three positive solutions for M-point boundary-value problem with one-dimensional, *Taiwanese J. Math.* **14** (2010), 647–665.

[121] W. Feng and J. R. L. Webb, Solvability of m-point boundary value problems with nonlinear growth, *J. Math. Anal. Appl.* **212** (1997), 467–480.

[122] M. Ferrara and S. Heidarkhani, Multiple solutions for perturbed p-Laplacian boundary-value problems with impulsive effects, *Electron. J. Differential Equations* **2014** (2014), No. 106, 14pp.

[123] D. Franco and J. J. Nieto, Maximum principle for periodic impulsive first order problems, *J. Comput. Appl. Math.* **88** (1998) 149–159.

[124] D. Franco and J. J. Nieto, First order impulsive ordinary differential equations with anti-periodic and nonlinear boundary conditions, *Nonlinear Anal.* **42** (2000), 163–173.

[125] D. Franco, J. J. Nieto, and D. O'Regan, Anti-periodic boundary value problem for nonlinear first order ordinary differential equations, *Math. Inequal. Appl.* **6** (2003), 477–485.

[126] D. Franco, J. J. Nieto, and D. O'Regan, Existence of solutions for first order ordinary differential equations with nonlinear boundary conditions, *Appl. Math. Comput.* **153** (2004), 793–802.

[127] M. Galewski and J. Smejda, On variational methods for nonlinear difference equations, *J. Comput. Appl. Math.* **233** (2010), 2985–2993.

[128] J. García-Melián and J. Sabina de Lis, Maximum and comparison principles for operators involving the p-Laplacian, *J. Math. Anal. Appl.* **218** (1998), 49–65.

[129] M. Ghergu, A biharmonic equation with singular nonlinearity, *Proc. Edinburgh Math. Soc.* **55** (2012), 155–166.

[130] J. R. Graef, S. Heidarkhani, and L. Kong, A critical points approach to multiplicity results for multi-point boundary value problems, *Appl. Anal.* **90** (2011), 1909–1925.

[131] J. R. Graef, S. Heidarkhani, and L. Kong, A critical points approach for the existence of multiple solutions of a Dirichlet quasilinear system, *J. Math. Anal. Appl.* **388** (2012), 1268–1278.

[132] J. R. Graef, S. Heidarkhani, and L. Kong, Infinitely many solutions for systems of multi-point boundary value problems, *Topol. Methods Nonlinear Anal.* **42** (2013), 105–118.

[133] J. R. Graef, S. Heidarkhani, and L. Kong, Multiple solutions for a class of $(p_1, \ldots, p_n)$-biharmonic systems, *Commun. Pure Appl. Anal.* **12** (2013), 1393–1406.

[134] J. R. Graef, S. Heidarkhani, and L. Kong, Existence of nontrivial solutions to systems of multi-point boundary value problems, *Discrete Cont. Dynam. Sys.*, **Suppl.** (2013), 273–281.

[135] J. R. Graef, S. Heidarkhani, and L. Kong, Multiple solutions for systems of multi-point boundary value problems, *Opuscula Math.* **33** (2013), 293–306.

[136] J. R. Graef, S. Heidarkhani, and L. Kong, A variational approach to a Kirchhoff-type problem involving two parameters, *Results Math.* **63** (2013), 877–889.

[137] J. R. Graef, S. Heidarkhani, and L. Kong, Infinitely many solutions for systems of Sturm-Liouville boundary value problems, *Results Math.* **66** (2014), 327–341.

[138] J. R. Graef, S. Heidarkhani, and L. Kong, Nontrivial solutions for systems of Sturm-Liouville boundary value problems, *Differ. Equ. Appl.* **6** (2014), 255–265.

[139] J. R. Graef, S. Heidarkhani, and L. Kong, Infinitely many solutions for a class of perturbed second-order impulsive Hamiltonian systems, to appear.

[140] J. R. Graef, S. Heidarkhani, and L. Kong, Multiple solutions for systems of Sturm-Liouville boundary value problems, to appear.

[141] J. R. Graef, S. Heidarkhani, and L. Kong, Nontrivial periodic solutions of a class of second-order impulsive Hamiltonian systems, to appear.

[142] J. R. Graef, S. Heidarkhani, and L. Kong, Infinitely many periodic solutions to a class of perturbed second-order impulsive Hamiltonian systems, to appear.

[143] J. R. Graef, J. Henderson, and A. Ouahab, *Impulsive Differential Inclusions, A Fixed Point Approach*, De Gruyter Series in Nonlinear Analysis and Applications, Vol. 20, De Gruyter, Berlin, 2013.

[144] J. R. Graef and L. Kong, Existence of solutions for nonlinear boundary value problems, *Comm. Appl. Nonlinear Anal.* **14** (2007), 39–60.

[145] J. R. Graef, L. Kong, and Q. Kong, Higher order multi-point boundary value problems, *Math. Nachr.* **284** (2011), 39–52.

[146] J. R. Graef, L. Kong, and Q. Kong, On a generalized discrete beam equation via variational methods, *Commun. Appl. Anal.* **16** (2012), 293–308.

[147] J. R. Graef, L. Kong, and Q. Kong, Infinitely many solutions for a discrete fourth order boundary value problem, *Nonlinear Dyn. Syst. Theory* **13** (2013), 400–411.

[148] J. R. Graef, L. Kong, Y. Tian, and M. Wang, On a discrete fourth order periodic boundary value problem, *Indian J. Math.* **55** (2013), 163–184.

[149] J. R. Graef, L. Kong, and M. Wang, Solutions of a nonlinear fourth order periodic boundary value problem for difference equations, *Dynam. Contin. Discrete Impuls. Syst. Series A* **20** (2013), 53–63.

[150] J. R. Graef, L. Kong, and M. Wang, Existence of multiple solutions to a discrete fourth order periodic boundary value problem, *Discrete Contin. Dyn. Syst.* **Suppl.** (2013), 291–299.

[151] J. R. Graef, L. Kong, and M. Wang, Multiple solutions to a periodic boundary value problem for a nonlinear discrete fourth order equation, *Adv. Dyn. Syst. Appl.* **8** (2013), 203–215.

[152] J. R. Graef, L. Kong, and M. Wang, Infinitely many Homoclinic solutions for second order difference equations with p-Laplacian, *Commun. Appl. Anal.* **19** (2015), 95–102.

[153] J. R. Graef, L. Kong, and M. Wang, Existence of homoclinic solutions for second order difference equations with p-Laplacian, to appear.

[154] J. R. Graef, L. Kong, M. Wang, and B. Yang, Uniqueness and parameter dependence of positive solutions of a discrete fourth order problem, *J. Difference Equ. Appl.* **19** (2013), 1133–1146.

[155] J. R. Graef and B. Yang, Multiple positive solutions to a three point third order boundary value problem, *Discrete Contin. Dyn. Syst.* **Suppl.** (2005), 337–344.

[156] M. R. Grossinho, L. Sanchez, and S. A. Tersian, On the solvability of a boundary value problem for a fourth-order ordinary differential equation, *Appl. Math. Lett.* **18** (2005), 439–444.

[157] H. Gu, T. An, Existence of infinitely many periodic solutions for second-order Hamiltonian systems, *Electron. J. Differential Equations* **2013** (2013), No. 251, 10pp.

[158] D. D. Hai, On singular Sturm-Liouville boundary-value problems, *Proc.*

Roy. Soc. Edinburgh Sect. A **140** (2010), 49–63.

[159] G. Han and Z. Xu, Multiple solutions of some nonlinear fourth-order beam equation, *Nonlinear Anal.* **68** (2008), 3646–3656.

[160] A. Haraux, Anti-periodic solutions of some nonlinear evolution equations, *Manuscripta Math.* **63** (1989), 479–505.

[161] T. He and Y. Su, On discrete fourth-order boundary value problems with three parameters, *J. Comput. Appl. Math.* **233** (2010), 2506–2520.

[162] X. He and X. Wu, Periodic solutions for a class of nonautonomous second-order Hamiltonian systems, *J. Math. Anal. Appl.* **341** (2008), 1354–1364.

[163] Z. He and J. Yu, On the existence of positive solutions of fourth-order difference equations, *Appl. Math. Comput.* **161** (2005), 139–148.

[164] S. Heidarkhani, Multiple solutions for a class of multipoint boundary value systems driven by a one dimensional $(p_1, \ldots, p_n)$-Laplacian operator, *Abstr. Appl. Anal.* **2012** (2012), Article ID 389530, 15 pp.

[165] S. Heidarkhani, Infinitely many solutions for systems of n two-point boundary value Kirchhoff-type problems, *Ann. Polon. Math.* **107** (2013), 133–152.

[166] S. Heidarkhani, G. A. Afrouzi, A. Hadjian, and J. Henderson, Existence of infinitely many anti-periodic solutions for second-order impulsive differential inclusions, *Electron. J. Differential Equations* **2013** (2013), No. 97, 13pp.

[167] S. Heidarkhani, M. Ferrara, and S. Khademloo, Nontrivial solutions for one-dimensional fourth-order Kirchhoff-type equations, *Mediterr. J. Math.*, to appear.

[168] S. Heidarkhani and J. Henderson, Infinitely many solutions for a class of nonlocal elliptic systems of $(p_1, \ldots, p_n)$-Kirchhoff type, *Electron. J. Differential Equations* **2012** (2012), No. 69, 15 pp.

[169] S. Heidarkhani and J. Henderson, Multiple solutions for a nonlocal perturbed elliptic problem of p-Kirchhoff type, *Comm. Appl. Nonlinear Anal.* **19** (2012), 25–39.

[170] S. Heidarkhani and J. Henderson, Multiple solutions for a Dirichlet quasilinear system containing a parameter, *Georgian Math. J.* **21** (2014), 187–197.

[171] S. Heidarkhani and J. Henderson, Critical point approaches to quasilinear second order differential equations depending on a parameter, *Topol. Methods Nonlinear Anal.*, **44** (2014), 177–197.

[172] S. Heidarkhani and J. Henderson, Infinitely many solutions for nonlocal elliptic systems of $(p_1, \ldots, p_n)$-Kirchhoff type, *Electron. J. Differential Equations* **2015** (2015), No. 30, 11pp.

[173] S. Heidarkhani and J. Henderson, Infinitely many solutions for a perturbed quasilinear two-point boundary value problem, to appear.

[174] S. Heidarkhani, and D. Motreanu, Multiplicity results for a two-point boundary value problem, *Panamer. Math. J.* **19** (2009), 69–78.

[175] S. Heidarkhani, Y. Tian, Multiplicity results for a class of gradient systems depending on two parameters, *Nonlinear Anal.* **73** (2010), 547–554.

[176] S. Heidarkhani, Y. Tian, and C. Tang, Existence of three solutions for a class of $(p_1, \ldots, p_n)$-biharmonic systems with Navier boundary conditions, *Ann. Polon. Math.* **104** (2012), 261–277.

[177] J. Henderson, Solutions of multipoint boundary value problems for second order equations, *Dynam. Systems Appl.* **15** (2006), 111–117.

[178] J. Henderson, Existence and uniqueness of solutions of $(k+2)$-point nonlocal boundary value problems for ordinary differential equations, *Nonlinear Anal.* **74** (2011), 2576–2584.

[179] J. Henderson, B. Karna, and C. C. Tisdell, Existence of solutions for three-point boundary value problems for second order equations, *Proc. Amer. Math. Soc.* **133** (2005), 1365–1369.

[180] J. Henderson and S. K. Ntouyas, Positive solutions for systems of nth order three-point nonlocal boundary value problems, *Electron. J. Qual. Theory Diff. Equ.* **2007** (2007), No. 18, 12 pp. (electronic).

[181] J. Henderson and S. K. Ntouyas, Positive solutions for systems of nonlinear boundary value problems, *Nonlinear Stud.* **15** (2008), 51–60.

[182] A. Iannizzotto, Three critical points for perturbed nonsmooth functionals and applications, *Nonlinear Anal.* **72** (2010), 1319–1338.

[183] A. Iannizzotto and S. A. Tersian, Multiple homoclinic solutions for the discrete p-Laplacian via critical point theory, *J. Math. Anal. Appl.* **403** (2013), 173–182.

[184] M. Izydorek and J. Janczewska, Homoclinic solutions for a class of second order Hamiltonian systems, *J. Differential Equations* **219** (2005), 375–389.

[185] Y. Jabri, *The Mountain Pass Theorem, Variants, Generalizations and Some Applications*, Encyclopedia of Mathematics and its Applications, Vol. 95, Cambridge University Press, New York, 2003.

[186] J. Ji and B. Yang, Eigenvalue comparisons for boundary value problems of the discrete beam equation, *Adv. Difference Equ.* **2006** (2006) Art. ID 81025, 9 pp.

[187] R. Kajikiya, A critical point theorem related to the symmetric mountain pass lemma and its applications to elliptic equations, *J. Funct. Anal.* **225** (2005), 352–370.

[188] K. Kefi, $p(x)$-Laplacian with indefinite weight, *Proc. Amer. Math. Soc.* **139** (2011), 4351–4360.

[189] W. G. Kelly and A. C. Peterson, *Difference Equations, an Introduction with Applications*, Second Edition, Academic Press, New York, 2001.

[190] G. Kirchhoff, *Mechanik*, Teubner, Leipzig, 1883.

[191] H. Kleinert and A. Chervyakov, Functional determinants from Wronski Green function, *J. Math. Phys.* **40** (1999), 6044–6051.

[192] V. L. Kocic and G. Ladas, *Global Behavior of Nonlinear Difference Equations of Higher Order with Applications*, Kluwer, Dordrecht, 1993.

[193] L. Kong, Homoclinic solutions for a second order difference equation with p-Laplacian, *Appl. Math. Comput.* **247** (2014), 1113–1121.

[194] L. Kong, On a fourth order elliptic problem with a $p(x)$-biharmonic operator, *Appl. Math. Lett.* **27** (2014), 21–25.

[195] L. Kong, Eigenvalues for a fourth order elliptic problem, *Proc. Amer. Math. Soc.* **143** (2015), 249–258.

[196] O. Kováčik and Rákosník, On spaces $L^{p(x)}$ and $W^{m,p(x)}$, *Czechoslovak Math. J.* **41** (1991), 592–618.

[197] A. Kristály, Existence of two non-trivial solutions for a class of quasilinear elliptic variational systems on strip-like domains, *Proc. Edinburgh Math. Soc.* **48** (2005), 465–477.

[198] A. Kristály, W. Marzantowicz, and C. Varga, A non-smooth three critical points theorem with applications in differential inclusions, *J. Glob. Optim.* **46** (2010), 49–62.

[199] J. Kuang, Existence of homoclinic solutions for higher-order periodic difference equations with p-Laplacian, *J. Math. Anal. Appl.* **417** (2014), 904–917.

[200] V. Lakshmikantham, D. D. Bainov, and P. S. Simeonov, *Theory of Impulsive Differential Equations,* World Scientific, Singapore, 1989.

[201] A. C. Lazer and P. J. McKenna, Large amplitude periodic oscillations in suspension bridges: Some new connections with nonlinear analysis, *SIAM Rev.* **32** (1990), 537–578.

[202] M. Lazzo and P. G. Schmidt, Oscillatory radial solutions for subcritical biharmonic equations, *J. Differential Equations* **247** (2009), 1479–1504.

[203] E. K. Lee and Y. H. Lee, Multiple positive solutions of singular two point boundary value problems for second order impulsive differential equation, *Appl. Math. Comput.* **158** (2004), 745–759.

[204] Y. Li, Positive solutions of fourth-order periodic boundary value problems, *Nonlinear Anal.* **54** (2003), 1069–1078.

[205] Y. Li and H. Fan, Existence of positive periodic solutions for higher-order ordinary differential equations, *Comput. Math. Appl.* **62** (2011), 1715–1722.

[206] C. Li and C. L. Tang, Three solutions for a class of quasilinear elliptic systems involving the (p, q)-Laplacian, *Nonlinear Anal.* **69** (2008), 3322–3329.

[207] C. Li and C. L. Tang, Three solutions for a Navier boundary value problem involving the p-biharmonic, *Nonlinear Anal.* **72** (2010), 1339–1347.

[208] L. Li and C. L. Tang, Existence of three solutions for (p,q)-biharmonic systems, *Nonlinear Anal.* **73** (2010), 796–805.

[209] F. Liao and J. Sun, Variational approach to impulsive problems: a survey of recent results, *Abstr. Appl. Anal.* **2014** (2014), Article ID 382970, 11 pp.

[210] X. N. Lin and D. Q. Jiang, Multiple positive solutions of Dirichlet boundary value problems for second order impulsive differential equations, *J. Math. Anal. Appl.* **321** (2006), 501–514.

[211] J. L. Lions, On some questions in boundary value problems of mathematical physics, in: *Contemporary Developments in Continuum Mechanics and Partial Differential Equations,* (Proc. Internat. Sympos. Inst. Mat. Univ. Fed. Rio de Janeiro, 1977), North-Holland Mathematics Studies, Vol. 30, North-Holland, 1978, pp. 284–346.

[212] J. Liu, S. Chen, and X. Wu, Existence and multiplicity of solutions for a class of fourth-order elliptic equations in R^N, *J. Math. Anal. Appl.* **395** (2012), 608–615.

[213] S. Liu and S. Li, Infinitely many solutions for a superlinear elliptic equation, *Acta Math. Sinica* **46** (2003), 625–630.

[214] J. Q. Liu and J. B. Su, Remarks on multiple nontrivial solutions for quasilinear resonant problems, *J. Math. Anal. Appl.* **258** (2001), 209–222.

[215] S. Liu and M. Squassina, On the existence of solutions to a fourth-order quasilinear resonant problem, *Abstr. Appl. Anal.* **7** (2002), 125–133.

[216] X. L. Liu and W. T. Li, Existence and multiplicity of solutions for fourth-order boundary value problems with parameters, *J. Math. Anal. Appl.* **327** (2007), 362–375.

[217] Y. Long, Nonlinear oscillations for classical Hamiltonian systems with bi-even subquadratic potentials, *Nonlinear Anal.* **25** (1995), 1665–1671.

[218] R. Ma, Existence of positive solutions for superlinear m-point boundary value problems, *Proc. Edinburgh Math. Soc.* **46** (2003), 279–292.

[219] D. Ma and X. Chen, Existence and iteration of positive solutions for a multi-point boundary value problem with a p-Laplacian operator, *Portugal. Math. (N. S.)* **65** (2008), 67–80.

[220] M. Ma and Z. Guo, Homoclinic orbits for second order self-adjoint difference equations, *J. Math. Anal. Appl.* **323** (2006), 513–521.

[221] M. Ma and Z. Guo, Homoclinic orbits and subharmonics for second order difference equations, *Nonlinear Anal.* **67** (2007), 1737–1745.

[222] R. Ma and D. O'Regan, Solvability of singular second order m-point boundary value problems, *J. Math. Anal. Appl.* **301** (2005), 124–134.

[223] R. Ma and Y. Xu, Existence of positive solution for nonlinear fourth-order difference equations, *Comput. Math. Appl.* **59** (2010), 3770–3777.

[224] S. A. Marano and D. Motreanu, Infinitely many critical points of non-differentiable functions and applications to a Neumann-type problem involving the p-Laplacian, *J. Differential Equations* **182** (2002), 108–120.

[225] S. A. Marano and D. Motreanu, On a three critical points theorem for non-differentiable functions and applications to nonlinear boundary value problems, *Nonlinear Anal.* **48** (2002), 37–52.

[226] J. Mawhin, *Problèmes de Dirichlet variationnels nonlinéaires*, Les Presses de l'Université de Montréal, 1987.

[227] J. Mawhin and M. Willem, *Critical Point Theory and Hamiltonian Systems*, Applied Mathematics Sciences, Vol. 74, Springer, New York, 1989.

[228] A. M. Micheletti and A. Pistoia, Multiplicity results for a fourth-order semilinear elliptic problem, *Nonlinear Anal.* **31** (1998), 895–908.

[229] M. Mihǎlescu and V. Rǎdulesu, A multiplicity result for a nonlinear degenerate problem arising in the theory of electrorheological fluids, *Proc. R. Soc. A* **462** (2006), 2625–2641.

[230] M. Mihǎlescu and V. Rǎdulesu, On a nonhomogeneous quasilinear eigenvalue problem in Sobolev spaces with variable exponent, *Proc. Amer. Math. Soc.* **135** (2007), 2929–2937.

[231] M. Mihǎilescu, V. Rǎdulescu, and S. Tersian, Homoclinic solutions of difference equations with variable exponents, *Topol. Methods Nonlinear Anal.* **38** (2011), 277-289.

[232] G. Molica Bisci and D. Repovs, Nonlinear algebraic systems with discontinuous terms, *J. Math. Anal. Appl.* **398**, (2013), 846–856.

[233] J. J. Nieto and D. O'Regan, Variational approach to impulsive differential equations, *Nonlinear Anal. Real World Appl.* **10** (2009), 680–690.

[234] H. Okochi, On the existence of periodic solutions to nonlinear abstract

parabolic equations, *J. Math. Soc. Japan* **40** (1988), 541–553.

[235] K. Perera and Z. Zhang, Nontrivial solutions of Kirchhoff-type problems via the Yang index, *J. Differential Equations* **221** (2006), 246–255.

[236] S. Pinsky and U. Tritman, Antiperiodic boundary conditions to supersymmetric discrete light cone quantization, *Phys. Rev. D* **62** (2000), 087701, 4 pp.

[237] S. I. Pohožaek, A certain class of quasilinear hyperbolic equations, *Mat. Sb. (N. S.)* **96** (1975), 152–168.

[238] S. I. Pokhozhaev, On a constructive method of the calculus of variations, *Dokl. Akad. Nauk SSSR* **298** (1988), 1330–1333 (in Russian); translation in *Soviet Math. Dokl.* **37** (1988), 274–277.

[239] P. Pucci and J. Serrin, A mountain pass theorem, *J. Differential Equations* **60** (1985), 142–149.

[240] P. H. Rabinowitz, Homoclinic orbits for a class of Hamiltonian systems, *Proc. Roy. Soc. Edinburgh* **114** (1990), 33–38.

[241] P. H. Rabinowitz, *Minimax Methods in Critical Point Theory with Applications to Differential Equations*, CBMS Regional Conference Series in Mathematics, Vol. 65, American Mathematical Society, Providence, 1986.

[242] P. H. Rabinowitz, Variational methods for Hamiltonian systems, in: *Handbook of Dynamical Systems*, Vol. 1, North-Holland, 2002, Part 1, Chapter 14, pp. 1091–1127.

[243] M. Råužička, *Electrorheological Fluids: Modeling and Mathematical Theory*, Lecture Notes in Mathematics, Vol. 1748, Springer-Verlag, Berlin, 2000.

[244] B. Ricceri, A general variational principle and some of its applications, *J. Comput. Appl. Math.* **113** (2000), 401–410.

[245] B. Ricceri, On a three critical points theorem, *Arch. Math. (Basel)* **75** (2000), 220–226.

[246] B. Ricceri, Existence of three solutions for a class of elliptic eigenvalue problem, *Math. Comput. Modelling* **32** (2000), 1485–1494.

[247] B. Ricceri, Sublevel sets and global minima of coercive functionals and local minima of their perturbations, *J. Nonlinear Convex Anal.* **52** (2004), 157–168.

[248] B. Ricceri, A three critical points theorem revisited, *Nonlinear Anal.* **70** (2009), 3084–3089.

[249] B. Ricceri, On an elliptic Kirchhoff-type problem depending on two parameters, *J. Glob. Optim.* **46** (2010), 543–549.

[250] B. Ricceri, Nonlinear eigenvalue problems, in: *Handbook of Nonconvex Analysis and Applications*, D. Y. Gao and D. Motreanu eds., International Press, 2010, pp. 543–595.

[251] B. Ricceri, A further refinement of a three critical points theorem, *Nonlinear Anal.* **74** (2011), 7446–7454.

[252] A. M. Samoilenko and N. A. Perestyuk, *Impulsive Differential Equations*, World Scientific, Singapore, 1995.

[253] J. Simon, Regularitè de la solution d'une equation non lineaire dans R^N, in: Journés d'Analyse Non Linéaire (Proc. Conf., Besançon, 1977), (P. Bénilan,

J. Robert, eds.), Lecture Notes in Math., Vol. 665, pp. 205–227, Springer, Berlin-Heidelberg-New York, 1978.

[254] J. Su and Z. Liu, A bounded resonance problem for semilinear elliptic equations, *Discrete Contin. Dyn. Syst.* **19** (2007), 431–445.

[255] J. Sun, H. Chen, and J. J. Nieto, Infinitely many solutions for second-order Hamiltonian system with impulsive effects, *Math. Comput. Modelling* **54** (2011), 544–555.

[256] J. Sun, H. Chen, J. J. Nieto, and M. Otero-Novoa, The multiplicity of solutions for perturbed second-order Hamiltonian systems with impulsive effects, *Nonlinear Anal.* **72** (2010), 4575–4586.

[257] J. Sun, H. Chen, and L. Yang, The existence and multiplicity of solutions for an impulsive differential equation with two parameters via a variational method, *Nonlinear Anal.* **73** (2010), 440–449.

[258] J. Sun and W. Li, Multiple positive solutions to second-order Neumann boundary value problems, *Appl. Math. Comput.* **146** (2003), 187–194.

[259] C. Tang, Periodic solutions for nonautonomous second order systems with sublinear nonlinearity, *Proc. Amer. Math. Soc.* **126** (1998), 3263–3270.

[260] C. Tang and X. Wu, Periodic solutions for a class of nonautonomous subquadratic second order Hamiltonian systems, *J. Math. Anal. Appl.* **275** (2002), 870–882.

[261] X. Tang and X. Lin, Existence of infinitely many homoclinic orbits in discrete Hamiltonian systems, *J. Math. Anal. Appl.* **373** (2011), 59–72.

[262] T. Teramoto, On Positive radial entire solutions of second-order quasilinear elliptic systems, *J. Math. Anal. Appl.* **282** (2003), 531–552.

[263] Y. Tian, Applications of variational methods to anti-periodic boundary value problem for second-order differential system, to appear.

[264] Y. Tian, Z. Du, and W. Ge, Existence results for discrete Sturm-Liouville problem via variational methods, *J. Difference Equ. Appl.* **13** (2007), 467–478.

[265] Y. Tian and W. Ge, Multiple solutions for a second-order Sturm-Liouville boundary value problem, *Taiwanese J. Math.* **11** (2007), 975–988.

[266] Y. Tian and W. Ge, Periodic solutions of non-autonomous second-order systems with a *p*-Laplacian, *Nonlinear Anal.* **66** (2007), 192–203.

[267] Y. Tian and W. Ge, Applications of variational methods to boundary value problem for impulsive differential equations, *Proc. Edinburgh Math. Soc.* **51** (2008), 509–527.

[268] Y. Tian and W. Ge, Second-order Sturm-Liouville boundary value problem involving the one-dimensional *p*-Laplacian, *Rocky Mountain J. Math.* **38** (2008), 309–327.

[269] Y. Tian, W. Ge, and D. O'Regan, Sign-changing and multiple solutions of impulsive boundary value problems via critical point methods, *J. Dyn. Control Syst.*, to appear.

[270] Y. Tian, J. R. Graef, L. Kong, and M. Wang Existence of solutions to a multi-point boundary value problem for a second order differential system via the dual least action principle, *Discrete Contin. Dynam. Sys.* **Suppl.** (2013), 759–769.

[271] Y. Tian, J. R. Graef, L. Kong, and M. Wang, Three solutions for second-order impulsive differential inclusions with Sturm-Liouville boundary value conditions via nonsmooth critical point theory, to appear.

[272] Y. Tian and J. Henderson, Three anti-periodic solutions for second-order impulsive differential inclusions via nonsmooth critical point theory, *Nonlinear Anal.* **75** (2012), 6496–6505.

[273] Y. Tian and J. Henderson, Anti-periodic solutions for a gradient system with resonance via a variational approach, *Math. Nachr.* **286** (2013), 1537–1547.

[274] Y. Tian and J. Henderson, Anti-periodic solutions of higher order nonlinear difference equations: a variational approach *J. Difference Equ. Appl.* **19** (2013), 1380–1392.

[275] Y. Tian and D. Sun, The solutions of Sturm-Liouville boundary-value problem for fourth-order impulsive differential equation via variational methods, *Abstr. Appl. Anal.* **2014** (2014), Article ID 690381, 12 pp.

[276] Z. Wang, Nonradial positive solutions for a biharmonic critical growth problem, *Commun. Pure Appl. Anal.* **11** (2012), 517–545.

[277] Z. Wang and J. Zhang, Periodic solutions of a class of second order non-autonomous Hamiltonian systems, *Nonlinear Anal.* **72** (2010), 4480–4487.

[278] W. Wang and P. Zhao, Nonuniformly nonlinear elliptic equations of p-biharmonic type, *J. Math. Anal. Appl.* **348** (2008), 730–738.

[279] G. Warnault, Regularity of the extremal solution for a biharmonic problem with general nonlinearity, *Commun. Pure Appl. Anal.* **8** (2009), 1709–1723.

[280] J. R. L. Webb, Optimal constants in a nonlocal boundary value problem, *Nonlinear Anal.* **63** (2005), 672–685.

[281] J. R. L. Webb and G. Infante, Non-local boundary value problems of arbitrary order, *J. London Math. Soc.* (2) **79** (2009), 238–258.

[282] W. Wang and P. Zhao, Nonuniformly nonlinear elliptic equations of p-biharmonic type, *J. Math. Anal. Appl.* **348** (2008), 730–738.

[283] M. Willem, *Minimax Theorem*, Birkhäuser, 1996.

[284] L. Yang, H. Chen, and X. Yang, The multiplicity of solutions for fourth-order equations generated from a boundary condition, *Appl. Math. Lett.* **24** (2011), 1599–1603.

[285] J. Yu, Z. Guo, and X. Zou, Periodic solutions of second order self-adjoint difference equations, *J. London Math. Soc.* **71** (2005), 146–160.

[286] J. Yu, Y. Long, and Z. Guo, Subharmonic solutions with prescribed minimal period of a discrete forced pendulum equation, *J. Dynam. Differential Equations* **16** (2004), 575–586.

[287] A. Zang, $p(x)$-Laplacian equations satisfying Cerami conditions, *J. Math. Anal. Appl.* **337** (2008), 547–555.

[288] A. Zang and Y. Fu, Interpolation inequalities for derivatives in variable exponent Lebesgue-Sobolev spaces, *Nonlinear Anal.* **69** (2008), 3629–3636.

[289] E. Zeidler, *Nonlinear Functional Analysis and its Applications*, Vol. I, 1986; Vol. II A &, Vol. II B, 1990; Vol. III, 1985; Springer-Verlag, Berlin-Heidelberg-New York.

[290] B. Zhang, L. Kong, Y. Sun, and X. Deng, Existence of positive solutions

for BVPs of fourth-order difference equation, *Appl. Math. Comput.* **131** (2002), 583–591.

[291] D. Zhang, Multiple solutions of nonlinear impulsive differential equations with Dirichlet boundary conditions via variational method, *Result. Math.* **63** (2013), 611–628.

[292] D. Zhang and B. Dai, Existence of solutions for nonlinear impulsive differential equations with Dirichlet boundary conditions, *Math. Comput. Modelling* **53** (2011), 1154–1161.

[293] G. Q. Zhang, X. P. Liu, and S. Y. Liu, Remarks on a class of quasilinear elliptic systems involving the (p, q)-Laplacian, *Electron. J. Differential Equations* **2005** (2005), No. 20, 10pp.

[294] J. Zhang and S. Li, Multiple nontrivial solutions for some fourth-order semilinear elliptic problems, *Nonlinear Anal.* **60** (2005), 221–230.

[295] Q. Zhang and C. Liu, Infinitely many periodic solutions for second order Hamiltonian Systems, *J. Differential. Equations* **251** (2011), 816–833.

[296] Q. Zhang and X. Tang, New existence of periodic solutions for second order non-autonomous Hamiltonian systems, *J. Math. Anal. Appl.* **369** (2010), 357–367.

[297] X. Zhang and Y. Zhou, Periodic solutions of non-autonomous second order Hamiltonian systems, *J. Math. Anal. Appl.* **345** (2008), 929–933.

[298] Z. Zhang and R. Yuan, Homoclinic solutions for a class of non-autonomous subquadratic second order Hamiltonian systems, *Nonlinear Anal.* **71** (2009), 4125–4130.

[299] J. Zhao, *Structure Theory for Banach Space*, Wuhan University Press, Wuhan, 1991.

[300] Z. Zhao, Property of positive solutions and its applications for Sturm-Liouville singular boundary value problems, *Nonlinear Anal.* **69** (2008), 4514–4520.

[301] V. Zhikov, Averaging of functionals of the calculus of variations and elasticity theory, *Math. USSR Izv.* **29** (1987), 33–66.

[302] J. Zhou and Y. Li, Existence of solutions for a class of second order Hamiltonian systems with impulsive effects, *Nonlinear Anal.* **72** (2010), 1594–1603.

[303] W. Zou and S. Li, Infinitely many solutions for Hamiltonian systems, *J. Differential Equations* **186** (2002), 141–164.

Index

Printed in the United States
By Bookmasters